Imaging the Brain with Optical Methods

Anna Wang Roe
Editor

Imaging the Brain with Optical Methods

Springer

Editor
Anna Wang Roe
Department of Psychology
Vanderbilt University
Nashville, TN 37203
USA
anna.roe@vanderbilt.edu

ISBN 978-1-4419-0451-5 e-ISBN 978-1-4419-0452-2
DOI 10.1007/978-1-4419-0452-2
Springer New York Dordrecht Heidelberg London

Library of Congress Control Number: 2009930646

Cover illustration: Contributed by Barbara Martin. Back cover illustration: Left: Brain vasculature of visual cortex seen through optical window in macaque monkey. Right: Color activation map seen through the window: pattern of "blobs" in V1 and stripes in V2 while monkey is watching isoluminant color grating. Contributed by Haidong D. Lu and Anna W. Roe.

Printed on acid-free paper

Springer is part of Springer Science+Business Media (www.springer.com)

Preface

Monitoring brain function with light *in vivo* has become a reality. The technology of detecting and interpreting patterns of reflected light has reached a degree of maturity that now permits high spatial and temporal resolution visualization at both the systems and cellular levels. There now exist several optical imaging methodologies, based on either hemodynamic changes in nervous tissue or neurally induced light scattering changes, that can be used to measure ongoing activity in the brain. These include the techniques of intrinsic signal optical imaging, near-infrared optical imaging, fast optical imaging based on scattered light, optical imaging with voltage sensitive dyes, and two-photon imaging of hemodynamic signals. The purpose of this volume is to capture some of the latest applications of these methodologies to the study of cerebral cortical function.

This volume begins with an overview and history of optical imaging and its use in the study of brain function. Several chapters are devoted to the method of intrinsic signal optical imaging, a method used to record the minute changes in optical absorption due to hemodynamic changes that accompanies cortical activity. Since the detected hemodynamic changes are highly localized, this method has excellent spatial resolution (50–100 μm), a resolution sufficient for visualization of fundamental modules of cerebral cortical function. This methodology has led to huge advances in our understanding of cortical function and is now being used intraoperatively to study function and dysfunction (such as epilepsy and stroke) in human neocortex. At even higher spatial resolution, in vivo two-photon imaging permits incredible visualization of responses within cortical columns and within microvascular compartments. Other rapidly developing optical technologies and their applications are also introduced. Near-infrared imaging, also hemodynamically based, is particularly well suited for studying brain function in infants and children due to its noninvasiveness and ease of use. High temporal resolution optical imaging, capable of following neuronal activity on the millisecond timescale, is achieved with methods such as voltage-sensitive dye imaging and imaging based on scattered light signals.

Together, this volume provides a cross section of the current state of optical imaging methodologies and their contribution towards our understanding of the spatial and temporal organization of cerebral cortical function.

Nashville, TN

Anna Wang Roe

Contents

Contributors

W. Akemann
Laboratory for Neuronal Circuit Dynamics, Brain Science Institute, RIKEN, 2-1 Hirosawa, Wako-Shi, Saitama 351-0198, Japan
akemann@brain.riken.jp.

B.J. Baker
Department of Cellular and Molecular Physiology, Yale University School of Medicine, New Haven, CT 06520, USA
bradley.baker@yale.edu

P. Blinder
Department of Physics, UCSD, La Jolla, CA, USA
pb@physics.ucsd.edu

Christopher J. Cannistraci
Institute of Imaging Science, Vanderbilt University, Nashville, TN 37232, USA
c.j.cannistraci@Vanderbilt.Edu

Li Min Chen
Department of Radiology and Radiological Science & Institute of Imaging Science, Vanderbilt University, Nashville, TN 37232, USA
limin.chen@vanderbilt.edu

Lawrence B. Cohen
Department of Cellular and Molecular Physiology, Yale University School of Medicine, New Haven, CT 06520, USA
lawrence.cohen@yale.edu

Anna Devor
Departments of Neurosciences and Radiology, UCSD, La Jolla, CA, USA; Martinos Center for Biomedical Imaging, MGH, Harvard Medical School, Charlestown, MA, USA
adevor@nmr.mgh.harvard.edu

D. Dimitrov
Laboratory for Neuronal Circuit Dynamics, Brain Science Institute, RIKEN, 2-1 Hirosawa, Wako-Shi, Saitama 351-0198, Japan

Alissa L. Ferry
Psychology Department, Northwestern University, Evanston, IL 60208, USA
AlissaFerry2011@u.northwestern.edu

Robert Friedman
Department of Psychology, Vanderbilt University, Nashville, TN 37240, USA
robert.m.friedman@vanderbilt.edu

Andrew Geneslaw
Department of Neurological Surgery, Weill Medical College of Cornell University, New York, NY 10021, USA
andy.geneslaw@gmail.com

John S. George
Biophysics Group, Los Alamos National Laboratory, Mail Stop D454, Los Alamos, NM 87545, USA
jsg@lanl.gov

John Gore
Department of Radiology and Radiological Science, Vanderbilt University, Nashville, TN 37232, USA
john.c.gore@vanderbilt.edu

Susan J. Hespos
Psychology Department, Northwestern University, Evanston, IL 60208, USA
hespos@northwestern.edu

Xiaoying Huang
Department of Physiology and Biophysics, Georgetown University, Washington, DC 20057, USA
huangxiaoying@neuro.duke.edu

T. Hughes
Department of Cell Biology and Neuroscience, Montana State University, Bozeman, MT 59717, USA
thughesgfp@mac.com

E.Y. Isacoff
Department of Molecular and Cell Biology, University of California, Berkley, Berkley, CA 94720, USA
ehud@berkeley.edu

Y. Iwamoto
Laboratory for Neuronal Circuit Dynamics, Brain Science Institute, RIKEN, 2-1 Hirosawa, Wako-Shi, Saitama 351-0198, Japan

L. Jin
Department of Cellular and Molecular Physiology, Yale University
School of Medicine, New Haven, CT 06520, USA
lei.jin@yale.edu

D. Kleinfeld
Department of Physics and Graduate Program in Neurosciences, UCSD,
La Jolla, CA, USA
dk@physics.ucsd.edu

T. Knöpfel
Laboratory for Neuronal Circuit Dynamics, Brain Science Institute,
RIKEN, 2-1 Hirosawa, Wako-Shi, Saitama 351-0198, Japan
tknopfel@brain.riken.jp

Hongtao Ma
Department of Neurological Surgery, Weill Medical College of Cornell University,
New York, NY 10021, USA
hom2001@med.cornell.edu

H. Mutoh
Laboratory for Neuronal Circuit Dynamics, Brain Science Institute, RIKEN, 2-1
Hirosawa, Wako-Shi, Saitama 351-0198, Japan
mutoh@brain.riken.jp

Sohee Park
Department of Psychology, Vanderbilt University, Nashville, TN 37240, USA
sohee.park@vanderbilt.edu

A. Perron
Laboratory for Neuronal Circuit Dynamics, Brain Science Institute,
RIKEN, 2-1 Hirosawa, Wako-Shi, Saitama 351-0198, Japan

V.A. Pieribone
Department of Cellular and Molecular Physiology, Yale University
School of Medicine, New Haven, CT 06520, USA
John B. Pierce Laboratory, New Haven, CT 06520, USA
vincent.pieribone@cmp.yale.edu

David M. Rector
Department of Veterinary and Comparative Anatomy, Pharmacology, and
Physiology, Washington State University, Pullman, WA 99164, USA
drector@vetmed.wsu.edu

Anna W. Roe
Department of Psychology, Vanderbilt University, Nashville, TN 37240, USA
anna.roe@vanderbilt.edu

Per E. Roland
The Laboratory of Brain Research, Department of Neuroscience,
Karolinska Institutet, S171 77 Solna, Sweden
Per.Roland@ki.se

Takayuki Sato
Laboratory for Integrative Neural Systems, RIKEN Brain Science Institute,
Wako, Saitama 351-0198, Japan
takayuki@brain.riken.jp

Jennifer L. Schei
Department of Physics and Astronomy, Washington State University,
Pullman, WA 99164, USA
jlschei@gmail.com

James Schummers
Department of Brain and Cognitive Sciences, Picower Institute for Learning and
Memory, Massachusetts Institute of Technology, Cambridge, MA 02139, USA
schummej@mit.edu

Theodore H. Schwartz
Department of Neurological Surgery, Weill Medical College of Cornell
University, New York, NY 10021, USA
schwarh@med.cornell.edu

A.Y. Shih
Department of Physics, UCSD, La Jolla, CA, USA
a2shih@ucsd.edu

Minah Suh
Department of Neurological Surgery, Weill Medical College of Cornell University,
New York, NY 10021, USA
minah.suh@gmail.com

Mriganka Sur
Department of Brain and Cognitive Sciences, Picower Institute for Learning and
Memory, Massachusetts Institute of Technology, Cambridge, MA 02139, USA
msur@ai.mit.edu

Manabu Tanifuji
Laboratory for Integrative Neural Systems, RIKEN Brain Science Institute,
Wako, Saitama 351-0198, Japan
tanifuji@riken.jp

I.C. Teng
Department of Neurosciences, UCSD, La Jolla, CA, USA
iteng@ucsd.edu

P. Tian
Department of Neurosciences, UCSD, La Jolla, CA, USA
ptian@ucsd.edu

P.S. Tsai
Department of Physics, UCSD, La Jolla, CA, USA
ptsai@physics.ucsd.edu

Kazushige Tsunoda
Laboratory for Integrative Neural Systems, RIKEN Brain Science Institute, Wako, Saitama 351-0198, Japan; Laboratory of Visual Physiology, National Institute of Sensory Organs, Tokyo 152-8902, Japan
tsunodakazushige@kankakuki.go.jp

Go Uchida
Laboratory for Integrative Neural Systems, RIKEN Brain Science Institute, Wako, Saitama 351-0198, Japan
guchida@brain.riken.go.jp

Jian-Young Wu
Department of Physiology and Biophysics, Georgetown University, Research Bldg WP-263970 Ruervoir Rd NW
Washington DC 20007, USA
wuj@georgetown.edu

Yukako Yamane
Laboratory for Integrative Neural Systems, RIKEN Brain Science Institute, Wako, Saitama 351-0198, Japan
yukako@brain.riken.jp

Hongbo Yu
Department of Brain and Cognitive Sciences, Picower Institute for Learning and Memory, Massachusetts Institute of Technology, Cambridge, MA 02139, USA
hbyu@mit.edu

Mingrui Zhao
Department of Neurological Surgery, Weill Medical College of Cornell University, New York, NY 10021, USA
mrzhao2004@gmail.com

Chapter 1
Casting Light on Neural Function: A Subjective History

John S. George

Abstract Optical methods offer a number of advantages for the study of neural systems. Optical techniques are relatively noninvasive, and offer wide field of view, in addition to high resolution in time and in space. Improvements in optical sensor technologies and imaging techniques continually enhance imaging performance, and extend resolution into three dimensions. Digital signal processing strategies allow increasingly subtle signals to be extracted and visualized. Imaging methods allow large populations of cells to be examined simultaneously, while resolving individual cells. Differential absorption or fluorescence emission by endogenous biochemicals or exogenous reporters allows characterization of specific aspects of the chemical and physical environment of cells, and produces signals that are highly correlated with neural activation. Fast intrinsic optical signals, which appear to be tightly coupled to the biophysical processes of neural activation, hold great promise for dynamic imaging of function in large populations of neurons. Coupled with multi-channel electrophysiological and computational modeling techniques, optical imaging enables powerful new understanding of the function of the brain.

For over 400 years, since the development of the microscope by Leeuwenhoek, optical methods have been the principal tool for studying cellular anatomy and physiology. In principle and in practice, optical methods offer a number of advantages for the study of neural systems. Optical methods are typically non-contact and thus somewhat noninvasive, at least relative to the penetrating microelectrodes typically used for neurophysiology *in vivo* and *in vitro*. Optical imaging methods offer wide field of view, coupled with high resolution in time and in space. In typical applications, spatial resolution is limited by diffraction or scattering; temporal resolution is limited by noise, which can be addressed by improved electronics and by increasing illumination levels. Improved optical technologies and imaging techniques, ranging from large numerical aperture optics, confocal and multi-photon

J.S. George
Los Alamos National Laboratory, Biophysics Group, Mail Stop D454,
Los Alamos, NM, 87545, USA
e-mail: jsg@lanl.gov

A.W. Roe (ed.), *Imaging the Brain with Optical Methods*,
DOI 10.1007/978-1-4419-0452-2_1,

techniques, and emerging methods for optical tomography, increasingly extend resolution into three dimensions. Imaging methods allow large populations of cells to be examined simultaneously, while resolving individual cells. Electronic imagers are continually improving in speed, pixel count, dynamic range, noise performance, and contrast resolution, effectively supplanting conventional photographic techniques, especially for functional neuroimaging. Digital signal processing strategies allow increasingly subtle signals to be extracted and visualized. Differential absorption or fluorescence emission by endogenous biochemicals or exogenous reporters allows characterization of specific aspects of the chemical and physical environment of cells, even in complex mixtures of chromophores. Spectral sensitivity can be exploited for enhanced contrast in individual images or dynamic measurements, and micro-spectroscopic measures or spectral imaging can allow resolution of the distribution and dynamic state of specific chemical constituents of cells and of tissue.

Nevertheless, for over 50 years, since the seminal work of Hodgkin and Huxley (1952) electrophysiological methods have dominated the study of neural function. The essential role of "animal electricity" in the function of excitable tissues was suggested in the eighteenth century by Galvani, who noted that electrical currents applied to a frog's leg caused strong and immediate contraction. In the modern era, the study of neurophysiology has been driven by technology: first the explosion of electronics technology during and in the aftermath of World War II, then the development and continuing improvement of transistors which allowed the development of low-noise, ultrasensitive amplifiers. The emergence of digital computing technology has driven the move to digital signal acquisition and signal processing that enables much of the analysis presently used for sensitive physical measurements. Solid-state technologies for sensors and imagers have pushed back boundaries of speed, sensitivity, and dynamic range. This relentless push of technology has enabled the work described in this volume.

Studies of individual cells have dominated the experimental study of electrophysiology. This focus began with the study of the action potential, which in many respects defines the neuronal response. Model systems such as the squid giant axon allowed the measurement of transmembrane potential with relatively large electrodes and provided a remarkable degree of experimental freedom. The development of glass micropipettes that could penetrate a cell without destroying it allowed the measurement of electrical activity in individual neurons. The development of extracellular electrodes based on microwires has enabled more robust, longer-term recordings, and thus facilitated the study of information processing by neurons. Pioneers of neuroscience including Kuffler, Hubel and Wiesel, and Mountcastle employed such electrodes to study single- and multi-unit spiking activity from small collections of cells. Spike sorting techniques based on the characteristic dynamics and amplitude of the action potential allow us to identify the responses of individual cells. The ability to identify the activity of individual cells in single passes enables powerful paradigms for relating the response of a cell to its pattern of inputs, and this experimental capability has clearly influenced our concept of how neurons work.

A great deal of work has focused on the biophysical characterization of the molecular ionic conductances and second messenger systems that initiate and modulate the electrophysiological response. Much early work employed relatively heavy-handed techniques such as manipulation of the ionic composition of the bath, or the introduction of buffers or ionophores to perturb the intracellular ionic composition in predictable ways. The use of ion-selective electrodes has allowed the study of ionic fluxes associated with cells and tissue systems; however, the difficulties associated with producing practical intracellular ion-selective electrodes have motivated the development of optical indicators for ionic species. Pharmacological studies, employing a wide range of agonists and antagonists of cellular receptors have allowed us to explore in great detail the symphony of synaptic neurotransmitters and systemic neurohumoral agents that mediate the communications between neurons. Patch clamp techniques, originally demonstrated by Sakmann and Neher (1984), have revolutionized our ability to study the ionic channels that mediate electrophysiological response. This family of methods allows measurements of the integrated response of whole cells as well as the function of individual channels that exist with a small patch of neuronal plasma membrane, while allowing manipulation of the biochemical and ionic composition of either or both faces of the membrane.

Neural Population Activity: In spite of the stunning advances achieved through studies of individual neurons, a number of factors increasingly drive the study of neural populations. A principal motivation has been to explore the spatial architecture of neural function. Observations of the functional deficits produced by focal head injuries led to the assignment of broad classes of function (e.g., visual, vs. somatosensory, vs. motor) to specific anatomical regions of cortex. The application of electrophysiological techniques to experimental animals led to a further refinement of this picture. For example, in the primate's visual system, approximately three dozen cortical visual areas have been identified, each with distinct information processing functions. Such studies also disclosed organizational motifs within areas. Neighboring neurons often responded to similar features in the incoming sensory stream, and the sensory parameter space often was found to be arrayed systematically across its corresponding cortical projection. Primary visual cortical areas are organized *retinotopically* with a relatively smooth conformal projection from location in the retina (and thus the visual field) to location in cortex. Primary auditory cortex is arrayed *tonotopically* with a systematic cortical representation of the acoustic frequency spectrum, arising from the progressive distribution of mechanical resonance frequencies along the basilar membrane of the primary sensory organ, the cochlea.

The advent of *human functional neuroimaging* techniques such as Positron Emission Tomography (PET) and functional Magnetic Resonance Imaging (fMRI) over the last two-plus decades has allowed the system-level exploration of functional anatomy in normal human subjects. Such studies have confirmed and extended the functional atlas suggested by animal electrophysiology, identifying homologous functional and anatomical areas, and even disclosing gross organization themes. For example, the retinotopic organization of primary visual areas and the abrupt discontinuities in the map between discrete visual areas have allowed fMRI to rather precisely map the borders of adjacent visual areas. These imaging methods exploit changes in local tissue metabolism or blood flow associated with patterns of activity. Because such changes (and such measurements) are relatively slow, they present an essentially static picture of neural functional architecture and provide little information on the dynamics of neural systems on their characteristic timescales (Kleinschmidt et al. 1996). This has motivated considerable effort to integrate hemodynamic with macroscopic electrophysiological imaging modalities (George et al. 2001).

Electrophysiological studies suggest additional levels of spatial organization within cortical areas; however, the relatively sparse spatial sampling precluded an appreciation of the fine structure. Autoradiography techniques, exploiting differences in the baseline metabolism of different classes of cells, or differences imposed by manipulating patterns of stimulation, allowed direct visualization of the spatial/ functional architecture of neural tissue. Tootell used patterned visual stimulation to demonstrate the approximate polar/ log(radius) retinotopic projection of primary visual cortex. Other investigators demonstrated that ocular dominance stripes reflecting residual segregation of inputs from either one of the eyes. Livingstone showed that differences in patterns of physiological response between different types of neurons were reflected in distinct spatial constructs, blobs, interblobs, etc., observed in primary and secondary visual areas. A primary limitation of these techniques is that visualization of spatial structure required sacrifice of the animal, and was confined to a particular point in time, corresponding to a particular state and history of activation. The development of optical imaging techniques based on intrinsic optical correlates of cellular metabolism has allowed multiple classes of functional organization to be explored in detail within the same animal, and for spatial imaging maps to be correlated with patterns of electrophysiological response by investigators such as Roe, and Fitzpatrick (Roe and Ts'o 1999; Bosking et al. 2002). The ability to visualize optical maps during the course of an experiment enables a powerful interplay between imaging and electrophysiological techniques. Imaging studies have also disclosed striking patterns of spatial organization that were not suspected based on earlier electrophysiological studies.

Although the functional architecture of neural tissue has implications for possible mechanisms of information processing, such techniques are fundamentally an extension of the observational and descriptive tradition of biological science. Spatial maps probably tell us more about how cells choose to organize themselves than how they implement the functionality of neural systems. Decades of study of individual cells provide a picture of neurons as independent processors, which encode in their rate of firing, features detected in their pattern of input, or desired states to be achieved though motor activity or modulation of autonomous function. However, a number of recent studies underscore the possibility that information is encoded by dynamic, coordinated spatiotemporal patterns of activity across large populations of cells (Fries et al. 1997). The first suggestions of this idea came from studies of local field potentials (LFPs) recorded with microelectrodes. LFPs are relatively low frequency responses that reflect the integrated activity of neural populations across hundreds of microns to millimeters of tissue. Such measures are dominated by excitatory and inhibitory postsynaptic potentials and other subthreshold responses. These signals were often filtered out and eliminated by investigators interested in the spiking activity of individual neurons. Although field potentials reflect the activity of neurons over a substantial region of space, they are sampled from selected spatial locations with individual electrodes. As with microelectrode studies of single neurons, the use of arrays of electrodes can greatly enhance the efficiency of sampling, allow localization of neural sources, and enable studies of similarities and differences in correlated LFP activity across populations of cells. These correlations are dynamic and may vary as a function of the state of the cell and its dynamic pattern of input.

Electro- and Magnetoencephalography (EEG and MEG) are noninvasive techniques that are sensitive to the same physiological currents that mediate function at the level of individual neurons. Under some conditions, these techniques resolve the dynamic responses of neural systems at their characteristic (millisecond) resolution. However, because the sensors are placed at the head surface, typically centimeters away from the active neural sources, the measurements are most sensitive to the coordinated activity of large neural ensembles. Measurements vary as a function of location over the head surface and thus provide the possibility of localization of sources of observed activity. However, the inverse problem that must be solved to reconstruct source locations is ill-posed: ambiguous and typically underdetermined. Solutions are necessarily model-based, embodying assumptions about the nature and distributions of sources, and in present practice, often integrating information from multiple imaging modalities, including anatomical and functional MRI. EEG and MEG can monitor ongoing spontaneous activity, often apparent as spatially distributed patterns of oscillation synchronized over large regions of cortex (Lindsey et al. 1997). The most prominent is a low frequency oscillation (8–12 Hz) called the alpha rhythm, which is suppressed by opening the eyes. Subsequent work has identified an extended family of characteristic rhythms: beta, gamma, delta, theta, mu, and others that range from very low frequencies (0.1 Hz) to rather high frequencies (>600 Hz). The gamma band (roughly 25–80 Hz) has attracted particular interest as a correlate of neural population activity that appears to mediate certain information coding and processing activities within extended neuronal networks.

Event-related Responses: The development of electronic signal acquisition and processing capabilities enabled new forms of study of neural population activity. Evoked responses and event-related techniques typically involve the collection of data epochs pre- or post-triggered by a particular stimulus or response event (or alternatively the collection of continuous data with timestamps to allow *post hoc* processing). Signals are averaged across trials, time-locked to the event of interest. This process suppresses the signatures of physiological processes or other forms of noise (signal-not-of-interest) and often discloses response transients with characteristic patterns in time and space. In a few cases, this activity is spike-like and appears to reflect the arrival of a synchronous volley of action potentials at some relatively early sensory waystation. More commonly, cortical responses are tens of milliseconds wide, corresponding roughly to the timescale of postsynaptic potentials. These responses often have a characteristic temporal waveform of positive and negative extrema, which are often named as response components. This apparent structure leads to the notion that an evoked response is a characteristic, unitary signal elicited by the sensory event or control imperative from (presumably quiescent) neurons, but increasing evidence suggests that such responses may instead represent a reorganization of ongoing activity (George 2002).

Neural Stimulation: Although descriptive studies of neural responses to physiological stimuli are of particular value for the study of neural systems, for many purposes, it is useful to study responses of selected portions of a system by stimulating certain tracts or even specific cells. In addition to providing a useful tool for physiological study such methods are increasingly used for prosthesis to substitute for sensory or motor neural input lost through injury or disease. Such stimulation is most commonly achieved through direct application of current by intracellular, extracellular, or patch electrodes. Magnetic stimulation due to the eddy currents elicited by time-varying (typically pulse-like) patterns of current in coils or arrays or coils, is increasingly used for noninvasive stimulation and may have particular value for localized excitation in prosthetic devices. Yet another useful approach for neuronal stimulation is based on injection of neural transmitters or modulators by mechanical pressure or iontophoresis (of charged species) from micropipettes. It is also becoming increasingly feasible to deliver such agents from microfluidics systems.

Optical methods also offer an array of techniques for neural stimulation or modulation. Neurotransmitter agonists or antagonists can be encapsulated or modified by the addition of chemical constituents and subsequently released by a light pulse of suitable intensity and wavelength. In addition to biochemicals, such methods can be used to produce a localized concentration pulse of physiologically active ions, such as calcium, or chelators to produce a local deficit in ion concentration. Other investigators have produced chemicals that confer neurophysiological sensitivity that can be modulated by light. Some of the early organics (e.g., described by Grinvald and colleagues) could be used to elicit action potential, but continued light exposure produced cellular toxicity. More recently, several investigators have described the use of protein based expressible biochemical ionophores derived from the light regulated "channel rhodopsin." Transfection with such a species can confer light sensitivity on many excitable cells, though the sensitivity is typically limited by the extent of transfection and the efficiency of transcription. Recently, investigators have demonstrated the feasibility of direct stimulation of neurons by pulsed IR stimulation. The biophysical mechanism of this effect is not clear though it is possible that the localized thermal deposition might transiently disrupt the cell membrane, admitting ions to perturb the membrane potential and produce an action potential. Although peripheral nerves have been shown to tolerate repetitive stimulation without apparent tissue damage, it is not clear whether such methods would support the stimulation rates and duty cycle required for prosthetic applications.

1.1 Imaging of Neural Function

Given the numerous technical and conceptual advantages of optical measurement and imaging techniques for the study of biological tissue, it is not surprising that scientists have long attempted to apply these methods for the study of neural function. Indeed, such studies have been described for over a century. However, since neuronal activation is fundamentally an electrochemical process, it is not immediately obvious how light might be used to directly probe the biophysical mechanism or consequences of neuronal firing.

But, neural tissue is a living biological mass like every other constituent tissue of the body. Neurons require energy to go about their business: the essential generic business of cells including protein synthesis, cellular remodeling, ion pumping and the myriad physiological processes that sustain life, as well as the more specialized electrophysiological processes that support the neural role in information encoding, transmission, and processing. These neuronal processes are energetically demanding. The brain requires a disproportionate share of the body's metabolic resources, and these demands are met by the same cardiovascular transport systems that provide nutrients and oxygen and remove metabolic wastes in all of the body's organs and functional tissues. Because the energetic demands of neural tissue are so high, the associated transport systems have developed exquisite capabilities for local and global regulation to attempt to match metabolic supply and demand.

1.1.1 Endogenous Chromophores

Physicians and physiologists had long noted the spectral differences (i.e., the differences in perceived color) between arterial and venous blood. Blood is a strong and ubiquitous biological chromophore with substantial oxygen-dependent spectral changes in both the visible and near-infrared regions of the electromagnetic spectrum. Sherrington noted that experimentally-induced changes in the activity of neural tissue were associated with changes in the color of tissue. These observations were eventually explained in terms of hemodynamic changes in the volume and flow of blood within neural tissue, and changes in blood oxygenation due to changes in the balance between delivery of oxygenated blood to tissue and the extraction and utilization of oxygen by active neural tissue. This change in blood oxygenation state is the basis of functional MRI techniques and is surprisingly local and specific, at least at the scale of cerebral gross anatomy. However, the underlying autoregulatory responses are relatively slow, operating on a timescale of seconds, two to three orders of magnitude off the 1–100 ms timescale characteristic of neuronal transactions.

Other metabolic changes in neurons are more tightly linked to the metabolism of individual cells. Neural tissue energy requirements are satisfied primarily though aerobic glycolysis – the oxygen dependent catabolism of glucose. This process involves the transient production of a range of intermediate compounds including high energy intermediate and energy storage compounds such as NAD/NADH, as well as catalytic enzymes and electron transport molecules such as the cytochromes. Many of these molecules have spectral absorbance or fluorescence changes dependent on their own oxygenation state, which is tightly linked to local changes in tissue oxygenation. Such tightly coupled local systems respond rapidly, on the timescale of hundreds of milliseconds, to local changes in metabolic demand created by neural activity.

1.1.2 Optical Reporters

Although such naturally occurring biological chromophores offer the possibility of dynamic measurements of neural activation, they still constitute a relatively slow and indirect measure of the electrophysiological processes that are the basis of neural computation. For this reason, over the last three decades, a number of investigators have worked to develop absorbance or fluorescent dyes to serve as reporters of neuronal membrane potential. Some of the earliest examples, such as, described by Alan Waggoner and colleagues (Cohen et al. 1974; Waggoner 1979) responded relatively slowly, probably through changes in their partition between aqueous and hydrophobic membrane phases of excitable tissue as a function of membrane potential. Subsequent work by Grinvald and Hildesheim, Loew and others developed fast dyes, which operate via other mechanisms to closely track the membrane potential. Continuing work has optimized the optical response properties of such dyes and enhanced the sensitivity of molecular optical properties to membrane potential changes. However, the limited volume of the cellular membranes and the various barriers to labeling of neurons embedded in neural tissue typically limit the extent of labeling and the magnitude of signal that can be achieved. To address the resulting problems with sensitivity, as well as issues associated with the fast transient nature of the response, most investigators have used photodiodes to record the optical signals. Cohen and colleagues as well as others have pioneered the use of arrays of photodiodes for studies of populations of neurons. While these systems produce clear maps with crisp dynamic resolution, they typically require extensive parallel amplifier arrays, which limit the number of channels to levels roughly comparable to state of the art multi-electrode arrays.

1.1.3 Functional Imaging

In spite of the advantages of photodiodes for sensitivity and dynamic response, one of the compelling potential advantages of optical techniques is the prospect of high-resolution imaging. In the mid 1980s, Blasdel and colleagues attempted a different strategy, employing a video camera as sensor and designing stimulus protocols to produce steady state activation (Blasdel and Salama 1986; Blasdel 1989). By constructing difference images between conditions, they were able to detect the small optical changes (on the order of 0.1%) anticipated for tissue stained with membrane potential sensitive dyes. This approach allowed visualization of spatial patterns of activation involving large numbers of neurons, for example the ocular dominance columns in visual cortex in which cells are predominantly driven from either one of the eyes. Using pharmacological manipulations, these investigators showed that the observed functional organization, though reflecting segregation of "hardwired" projections in subcortical pathways, is somewhat plastic, depending on biochemical competition to establish and maintain boundaries between the territories associated with either one of the eyes.

Although this work established (for me and many others) the power of high resolution imaging of neural system function, subsequent work casts doubt on the mechanism of the observed activation-dependent optical changes. Motivated by surprising results on some control experiments, Grinvald et al. (1986) quickly showed that similar patterns of apparent functional activation could be observed without the addition of voltage sensitive dyes. These intrinsic optical signals were robust and surprisingly high contrast, providing sensitivity at least comparable to that expected for voltage sensitive dyes available at the time. The phenomenon was rapidly and widely exploited to explore meso-scale functional organization with cortical networks (Ts'o et al. 1990; Frostig et al. 1990). One of the most striking examples was the demonstration of a pinwheel-like arrangement of orientation selective clusters of neurons within visual cortex. In some cases, these signals can be visualized without exposing the tissue, facilitating routine or even longitudinal studies (Masino et al. 1993). These optical changes appear to have contributions from multiple sources, depending on the wavelengths of light used for imaging and the duration of exposure. Under some circumstances there is a strong hemodynamic component reflecting changes in blood flow and oxygenation (Boas et al. 2008). However, the earlier "mapping signal" described by Malonek and Grinvald (1996) based on time and spatially resolved spectroscopy appears to reflect changes in local tissue oxygenation and metabolism. This intrinsic signal imaging method has proven spectacularly successful and useful for neuroscience investigation. Indeed, it constitutes the principal methodology for much of the work reported in this volume.

1.1.4 Calcium Imaging

My own active involvement in optical imaging of neural function began about this time. For a number of years, I had been involved in studies of the role of cytoplasmic calcium in visual transduction by retinal photoreceptor cells (Yoshikami et al. 1980; George and Hagins 1983). Our most successful approach to these studies involved the use of miniature and micro-ion-selective electrodes, but these were effectively limited to studies in the extracellular space of tissue or in broken cells. Optical methods would have clear advantages, especially if effective techniques could be developed to get the dyes into cells without disrupting function. For our application, it would also be desirable to make measurements in the near infrared to avoid directly stimulating photoreceptors while making our measurements. However, only limited options were available. The calcium - dependent bioluminescent protein aequorin was useful for studies of perturbations of calcium ionic activity near typical cytoplasm resting levels, and signals could be detected with good contrast on a low background. However most applications required cellular injection techniques and the material was in very limited supply since it was isolated from jellyfish. Metallochromic indicators such as Arsenazo III had been employed by investigators such as Ross for studies of cellular calcium. Such dyes were readily synthesized, but offered no easy solution to the problem of introduction into the cytoplasm. Arsenazo and similar dyes responded to calcium with a change in absorbance spectrum. In principle, this implied that measurements were not light-limited, although our planned application was limited by the sensitivity of the tissue to visible light.

Tsien and colleagues subsequently addressed a number of the technical limitations of optical techniques for studies of cytoplasmic calcium, and the process enabled powerful new strategies for study of neural function (Grynkiewicz, et al. 1985). The first of these indicators to find wide application, Fura II, was a fluorescent dye excited in the mid UV with visible emission. Fluorescence measurements provided most of the contrast advantages of bioluminescence with the additional advantages of control of emission (through temporally structured excitation) and avoiding the "one shot" nature of many biological phosphoproteins whose emission depends on ATP. Calcium dependent emission changes in Fura II were modest, but the spectral shift permitted ratio measurements between fluorescence excited at different wavelengths for precise quantitative determination of calcium concentration levels and changes. Like many cation-binding molecules, Fura contained multiple carboxylic acid groups. Tsien and colleagues showed that by modifying these moieties with acetomethoxy (AM) esters, the normally hydrophilic molecules could cross cell membranes. Inside the cells, endogenous esterase enzymes cleaved the blocking groups, activating calcium sensitive fluorescence and trapping the dye inside the cell. This approach opened up the possibility of labeling large numbers of cells within tissue without adopting heroic experimental strategies.

Conner was one of the first to exploit the advantages of Fura II for imaging of functionally relevant changes in cytoplasmic calcium activity. In order to obtain sufficient sensitivity for electronic imaging, he employed a cooled CCD camera that allowed long integration times and provided high dynamic range. Excited by the possibilities present by Fura II, inspired by the images of Connor and others (Connor et al. 1988) and with a hunch that cytoplasmic calcium levels would probably change in response to neural activation, we decided to explore the feasibility of calcium imaging as a method to explore the spatiotemporal dynamics of activation in extended neural networks. Our initial studies employed a photomultiplier tube and demonstrated that there was indeed a fast calcium transient associated with individual electrophysiological responses in the rat hippocampal slice (George and Fowler 1988). The response began on the same timescale as the postsynaptic potential and peaked in a few tens of milliseconds, suggesting that it would be possible to capture response dynamics with standard video technology. However, the fluorescence signal we observed was very weak, approximately the same magnitude as tissue autofluorescence. Because we were interested in response dynamics, We employed an image intensifier fiber-optically coupled to a scientific CCD camera. and acquired image sequences time-locked to stimulation, so that we could employ signal averaging techniques to pull out weak signals in the noisy data (George and Fowler 1990; George et al. 1990). Some results from these experiments are illustrated in Fig. 1.1.

Although our experiments demonstrated the feasibility of video imaging of calcium fluorescence as a probe of neurophysiological activity, we were not able to achieve adequate signals for routine neuroscience investigation. We believed that the problem was due to limited penetration by the AM esters, but a strong response to manganese in a control experiment intended to visualize cellular structure suggested the possibility that the dye was in the cell but incompletely hydrolyzed. In any case, other investigators enjoyed more success with the method. O'Donovan achieved effective labeling of fetal spinal cord neurons, and by retrograde labeling strategies (O'Donovan et al. 1993), and others found similar success in preparations of juvenile tissue. Tank and colleagues achieved impressive results using intensified video imaging of individual neurons injected with Fura II (Tank et al. 1988).

The continuing advances in the development of indicators and or methodology have greatly improved the utility of calcium imaging. If the objective is to produce a signature of activation rather than to quantify cytoplasmic ionic activity, ratiometric imaging is not required. The Fluo dyes and, subsequently, other indicators provide a much stronger change in fluorescence emission for a given change in calcium. Changes in typical protocols for labeling have produced better and more consistent results. Some of the most recent calcium indicators are based on biological chromophores, the family of green fluorescent proteins. Through rational molecular design coupled with conventional mutation and screening techniques, Tsien and colleagues conferred calcium sensitivity on the molecule and optimized the fluorescence response. Such indicators can be expressed by transfected calls, raising the possibility that some strains of research animals may eventually be engineered and raised specifically to facilitate optical studies of neural tissue.

These advances have made calcium imaging an important tool for neuroscience investigation. Fine and colleagues have demonstrated that synaptic events consistently produce local changes in calcium that may be detected at the level of the individual dendritic spine. By expressing calcium sensitive GFP in retina and using multiphoton fluorescence imaging techniques, Reid has demonstrated the feasibility of imaging patterns of activation throughout the 3D matrix of retinal neurons.

1.1.5 Fast Intrinsic Signals

Although the intrinsic optical signals that are routinely exploited for neurophysiological investigations are relatively slow, other investigators have reported changes in light scattering and neural birefringence that are tightly coupled to the electrophysiological response. These signals were initially studied in preparations of isolated

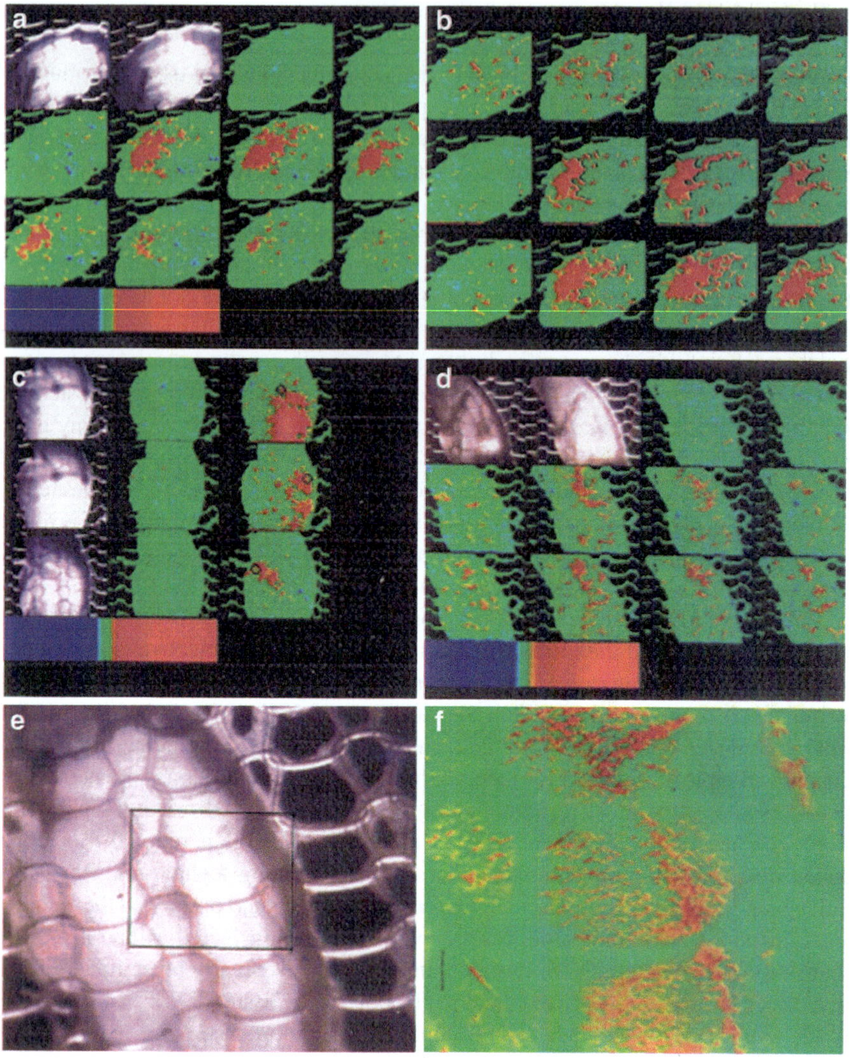

Fig. 1.1 Dynamic calcium imaging of electrophysiological responses in hippocampal slice. (**a**) *A sequence illustrating the response of CA1 to a single electrical stimulus*. Each image is subsampled from a video frame (33 ms). The first *two panels* in the first row are baseline fluorescence images excited at 350 and 380 nm. The electrical stimulus was delivered at the beginning of the first frame of the second row. Because of the continuous readout of the video imager and the relatively slow rise of the fluorescence transient (20–50 ms to peak), the response was typically not apparent until the following frame. The shadow of the wire electrode in the baseline images identifies the location of stimulation. The colorbar illustrates the range 0–2 in the ratio difference images. (**b**) *A family of responses to varying levels of electrical stimulation*. The figure illustrates a sequence of frames corresponding to the second row of figure **a** for stimuli of varying intensities. An electrical stimulus was delivered at the beginning of the first frame of each row. In the first row, stimulus intensity was adjusted to produce a population spike with an amplitude ~1/4 of the maximal response. In the second row, the stimulus was adjusted to produce ~1/2 maximal response. The maximal response is reproduced in the bottom row. (**c**) *The calcium response could be obtained with orthodromic (synaptic) or antidromic stimulation* and in several regions. *Upper panels* illustrate orthodromic stimulation of area CA1. *Middle panels* illustrate

nerves by Cohen and collaborators in Cambridge and environs (Cohen et al. 1968, 1972; Cohen 1973; also Lipton 1973). In the squid giant axon (which produces a single action potential in response to electrical stimulation), the optical response waveform for scattering or birefringence closely resembles the spike waveform. A number of investigators have studied these responses over the years. Tasaki and colleagues characterized fast optical signals as well as micromechanical responses that appear to be correlated with them. Several investigators with links to Cohen's lab have studied fast optical signals. Salzberg has focused on scattering changes that are associated with neurotransmitter exocytosis in synapse-rich regions of neural tissue (Salzberg et al. 1985). Landowne studied the fast birefringence responses and concluded that the dynamics were consistent with molecular rearrangement of the sodium channel itself (Landowne 1985, 1992). However, most investigators have concluded that given the small size of the signals (typically, changes ion the order of one part in 10^5–10^4 of the background light level), it would be difficult or impossible to image the responses. A notable exception came from the work of Keinfeld and colleagues, who observed signals on the order of 0.1% in localized photodiode measurements of isolated neurons in culture (Stepnoski et al. 1991).

However, in the 1990s, discouraged by the problems in labeling intact tissue and interested in signals that were more directly linked to the dynamic neuronal response, David Rector and I began to explore the possibility of imaging fast intrinsic signals. Rector, in the laboratory of Harper at UCLA, had previously detected fast optical transients associated with trans-collosal stimulation, using photodiodes (Rector et al. 1993, 1995). He also observed suggestive features in steady-state difference images. We set out to capture dynamic, cinematic image sequences of evoked responses. Initial studies examined the in vivo activation of the rat brainstem, elicited by electrical stimulation of the vagal nerve bundle. We employed a novel contact image probe based on a Gradient Index (GRIN) lens that allowed us to stabilize the tissue surface while focusing beyond the surface of the probe. LED illumination was delivered around the perimeter of the probe. We constructed a custom CCD camera, video amplifier, and data acquisition system in order to achieve the required sensitivity, dynamic range, and temporal resolution (Rector and George 2001). Using a normalized difference imaging procedure with signal averaging, we detected clear and

Fig. 1.1 (continued) antidromic stimulation of the corresponding region of the same slice. *Lower panels* illustrate antidromic stimulation of the Dentate Gyrus. On each row, *left panels* illustrate raw fluorescence images, Middle frames are averages of a pair of prestim ratio images. *Right panels* are averages of a pair of maximal response frames. (**d**) *Images capture the temporal dynamics of the electrical response*. The figure is a single time sequence illustrating the response to two electrical stimulation pulses, delivered at the beginning of the second and third rows (133 ms apart). Note that the optical response typically required 100–300 ms to return to baseline. (**e, f**) *Calibration and visualization of the distribution of the fluorescent indicator*. In **e** a control image was acquired as above using broadband fluorescence excitation (~330–370 nm). Increased fluorescence intensity allow operation of the intensifier at lower gain, with lower noise. Magnification was adjusted, and another image was collected corresponding to the region indicted by the rectangle in **e**. Ionophore and Mn2+ were added to the perfusion fluid, and another image was collected. A difference image **f** disclosed a periodic structure consistent with the distribution of neurons in the cell body layer reproducible dynamic signals that appeared to reflect several resolvable processes. The fastest response components appeared to track approximately the integral of the electrophysiological response, while the slowest probably reflected tissue hemodynamic changes (Rector et al. 1999, 2002).

Subsequent work examined fast optical responses elicited by physiological stimulation (twitches delivered to single or multiple vibrissae) of somatosensory (whisker barrel) cortex in rat (Rector et al. 2005) (Fig. 1.2). These experiments employed photodiodes as well as a simpler contact probe and camera incorporating a fiber optic image conduit, which provided much greater light collection efficiency that the GRIN lens but was limited to the resolution provided by "proximity focusing" at the tissue/probe interface. The imaging experiments disclosed the expected spatial arrangement of cortical projections of individual vibrissae. The photodiode measurements exhibited a fast oscillation that closely tracked a high frequency (400–600 Hz) oscillation in the field potential (Jones and Barth 1999) recorded with simultaneous electrophysiological and physiological measurements.

More recently, working with Yao, we have demonstrated the feasibility of imaging physiological activation of retina by light and electrical stimulation (Yao and George 2006a, b). These experiments employed isolated retina explanted onto a multi-electrode array for simultaneous electrophysiology. Illumination for functional imaging was in the near infrared at wavelengths that produced little or no detectable physiological response. Imaging employed a conventional upright microscope and a high resolution 14 bit camera. We typically read-out a limited sub-region of the array to keep frame rates in the range of 80–100 Hz. When the experimental system was optimized, we were surprised to routinely observe signals of a few percent, and in some cases, tens of percent. Such signals allowed routine, dynamic observation of responses of single cells in single trials (Fig. 1.3).

We were stunned by the size and contrast of these responses. The signals were readily observed in transmitted light and with greater contrast (defined as the ratio of the change in light intensity to background intensity: d*I*/*I*) in oblique darkfield imaging, which reduces background in favor of scattered light. Cross-polarized light measurements (operationally defined as birefringence) produced higher contrast still, but overall light levels were substantially reduced so that image quality often suffered. Although we made a number of technical and procedural changes to optimize these measurements, to our surprise, one of the most important changes was to go to higher resolution image acquisition, capturing features on the size

Fig. 1.2 (continued) *optical scattering response measured with a photodiode* (lower trace). (**c**) *Digitally filtered versions of the electrical and optical responses.* The wideband electrical response is shown in the upper trace. When filtered between 300 and 1,000 Hz, the ERP signal reveals a high frequency oscillation at 450–600 Hz that has maximal amplitude near the initial positive peak of the wideband response. The filtered optical signal (OP) contains an oscillation in the same frequency range that peaks later than the electrical response. Optical and electrical signals are highly correlated where the envelopes of oscillatory responses overlap. Similar results were observed in all three animals studied with the photodiode probe

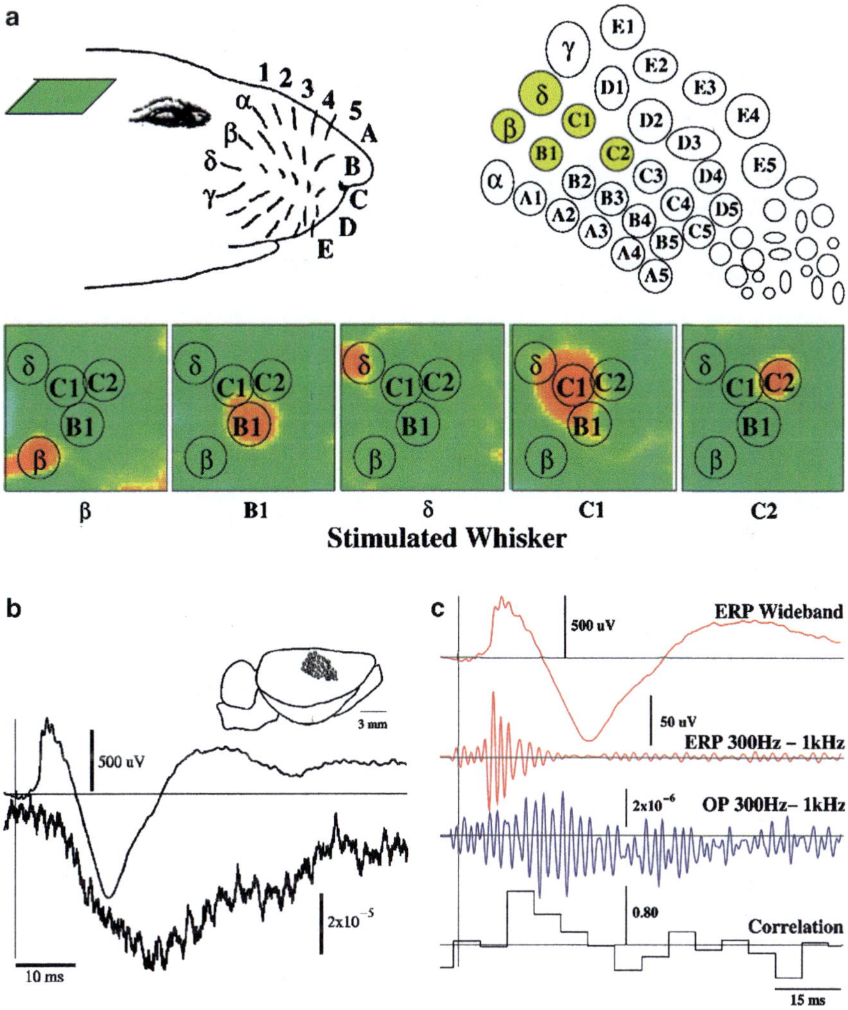

Fig. 1.2 Event Related Potential (ERP) and optical images in response to single whisker stimulation. Functional optical images acquired using fiber-optic coupled imager placed on the whisker barrel region of somatosensory cortex of a sedated rat. An average of 25 prestimulus baseline images were subtracted from each frame in an average image sequence based on 1,000 trials. (**a**) *The spatial map constructed during the early peak response* shows a discrete region corresponding to the cortical column associated with the stimulated whisker. A later epoch reveals a spreading of the optical response to cover a large portion of the imaged area. In pseudocolor images, warm colors represent decreases in scattered light, and cool colors represent scattered light increases. Stimulating particular whiskers produced early response images with discrete regions that localize to the expected map of cortical columns. Each of the five images was collected after stimulating the whisker indicated below the image. Each map was constructed from an average of 25 frames surrounding the first peak in the optical response from the baseline corrected average image sequences previously described. For spatial comparison, *circles* were drawn on all images, showing the activated regions. Qualitatively similar results were obtained in all animals, although other experiments compared 2–4 whiskers. (**b**) *Electrical evoked response potential (upper trace) and*

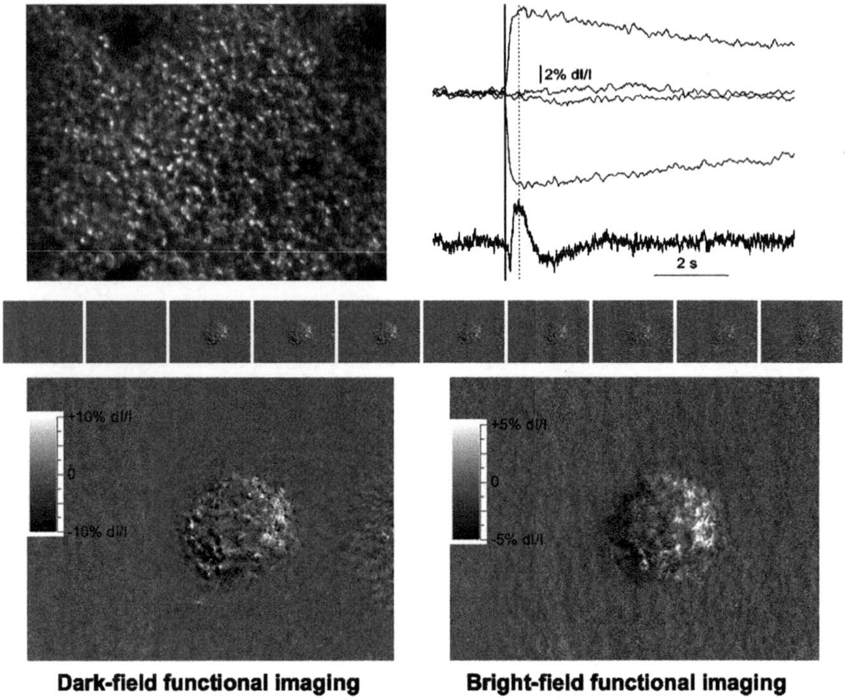

Fig. 1.3 *High resolution imaging of retinal responses to visible light stimulation.* The isolated retina was explanted onto a multi-electrode array placed on the stage of an upright microscope (Olympus BX61 WI). Raw frames were acquired at 320×240 pixels, 80 frames/s. The exposure time of CCD camera was 1.5 ms. The stimulus was a 100 ms, pentagonal, visible light flash delivered with 2 s pre-stimulus baseline recording. *Upper left:* Structure of retinal photoreceptor array disclosed by darkfield imaging. Portions of four electrodes *(dark regions)* can be seen in the image. *Middle panels:* Each illustrated frame is a darkfield image averaged over a one second interval with the average pre-stimulus baseline image subtracted. *Upper right:* Fast optical responses in darkfield for selected regions of interest consisting of 5×5 pixels (approximately square microns) *Lower left:* Enlarged image of the fourth frame of the difference image sequence illustrated above. These traces show different polarities and timescales of optical responses. *Lower right:* Fast function image collected in transmitted light (brightfield). Image corresponds exactly to the preceding darkfield image, and was acquired from the same retinal location

scale of individual neurons. Initially, we were concerned that these results might be a consequence of the transmitted light measurements coupled with the special geometrical configuration (predominantly axial) of retinal neurons relative to the light path. However, recent work by Yao has demonstrated high contrast responses in retina observed with a simple confocal microscope using reflected light. This suggests that the method may be widely applicable for studies of neural tissue.

We anticipate that there is room for continued improvement in the acquisition of fast intrinsic functional signals. Techniques such as Hoffman Modulation contrast, Differential Interference Contrast (DIC), and multi-path oblique illumination will enhance the contrast available in functional signals while improving our ability to resolve cellular structure in unstained samples. Some investigators have suggested the possible utility of terahertz radiation (very long wavelength IR) for functional neuroimaging, though the required illumination and imaging technologies are relatively immature. On the far edge of technology, the use of nanostructured materials may eventually enable practical optics exhibiting a negative index of refraction. Such optics would remove the diffraction limits that constrain the spatial resolution of imaging with conventional refractive optics.

The *mechanisms of intrinsic optical signals* associated with metabolic and hemodynamic responses of neural tissue can be identified with confidence based on their characteristic spectra. Our understanding of the mechanisms of fast changes in light scattering and birefringence is much less certain. The signals associated with neural secretion are well established, but this mechanism does not account for signals in isolated axons or nerve fiber bundles. The suggestion that birefringence signals might be due to configuration changes in channel proteins is appealing, but it is difficult to see how such a hypothesis might be tested with certainty.

One of the oldest and still most interesting hypotheses is that the fast optical signal changes might reflect the swelling and shrinking of neurons associated with the movement of ions and concomitant flux of water across cell membranes (Hill 1950; Van Harreveld 1958). The swelling of glial cells associated with potassium uptake has previously been implicated by MacVicar and Hochman (1991) in relatively large and slow optical signals observed in response to intense neural activation (Roper et al. 1992). Micromechanical changes – twitches – associated with neuronal firing have been reported by Tasaki and a number of other investigators (Tasaki et al. 1968, 1980; Tasaki and Byrne 1992). Using an optical lever, Yao et al. (2003) observed a transient swelling associated with excitation that propagated along an isolated nerve bundle, with a time course that approximated that observed in scattering and birefringence changes (Fig. 1.4). Similar responses have recently been observed using interferometric techniques. Although the transient was small, on the order of nanometers, because the movement of water would predominantly reflect a shift between intracellular and extracellular spaces (Holtoff and white, 1996), the fractional changes in the diameter of individual axons might be considerably larger. Recent work has visualized the propagation of the optical response along an isolated nerve fiber (Schei et al. 2008). Theoretical analysis suggests that even changes in apparent optical rotation might be produced by swelling of neuronal structures (Yao et al. 2005b), although subtle differences in the dynamics of scattering and birefringence changes suggests that different or additional processes may be involved these two classes of optical response (Foust and Rector 2007).

Of course, water movements are not the only possible explanation for the micromechanical responses of nerves. In artificial membrane systems,Tasaki has demonstrated the possibility of propagating phase changes, and such changes might be more likely in the complex cytoskeletal and membrane matrix consisting of many components with known sensitivity to calcium and other cytoplasmic ions (Oldenbourg et al. 1998). Indeed, other investigators have proposed major revisions to the theory of neuronal firing, suggesting that spikes reflect a soliton-like physical

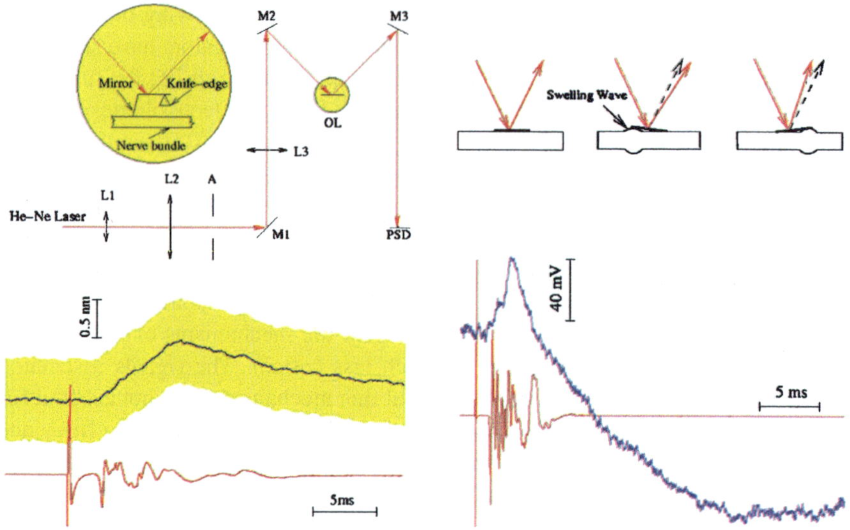

Fig. 1.4 Swelling of an isolated nerve demonstrated with an optical lever. *Upper left: Diagram of the optical lever* (OL) shows the light path of the experimental setup. A beam expander consists of two lenses, L1 and L2 and a 8-mm-diameter aperture A. M1, M2, and M3, are mirrors that are used to redirect the laser. One edge of the OL mirror rests on a knife-edge, and the opposite edge rests on the surface of the nerve bundle. The PSD is a quadrant photodiode. The nerve is located at the focal point of lens L3; the distance between the PSD and the nerve is approximately 200 mm, so the light spot at the photosensitive area of the PSD is approximately 8 mm in diameter. *Lower left: Time-triggered average plots show electrical and displacement signals*. The thin lines are electrical potential; the thick lines show the swelling displacements; and the *gray* backgrounds are *SEM,* i.e., the Standard Error of the Mean. This measure is dominated by very low frequency noise. When the low-frequency component is removed by means of high-pass filtering or ac coupling, or by numerical resetting of the baseline dc level, the shaded band is approximately equal to the noise within the measurement passband. Thus, the signal is detectable, but not statistically significant in single trials. The nerve bundles were stimulated with 0.1-ms long current pulses. The records were taken from stimulus occurred at the arrow. *Upper right: Schematic depiction of the path of reflected light perturbed by the propagation of the swelling wave* when the mirror is placed entirely on the nerve. (*Left panel)* no nerve swelling. (*Middle panel)* swelling wave is at the left of the mirror center. (*Right panel)* swelling wave is at the right of the mirror center. The *solid arrows* indicate the observed path of light propagation. The *dashed line* indicates the path of light propagation with no nerve swelling. *Lower right: Displacement signals associated with the propagation of the micromechanical response*. When the mirror was placed entirely on the nerve, we saw a fast upward deflection (corresponding to propagation of swelling) and a slower downward displacement (recovery). The *black solid curves* are electrical potential, and the *solid gray curves* show the change of output signal *S*. The nerve bundle was stimulated with 0.1-ms-long, 3-mA 1 Hz current pulses. The record was taken from has average of 100 signals. Stimulus occurred at the arrow. We estimate the signal returns to baseline after 30 ms

wave that propagates along the cell membrane (Heimburg and Jackson 2005). But to date, the most persuasive data in support of such an interpretations might alternatively be explained in terms of the predictable movement of water in response to the ionic fluxes described by Hodgkin and Huxley.

1.1.6 Neural Investigation

Although characterization and investigation of mechanisms of the optical responses of neural tissue during activation are intriguing topics in their own right, the more compelling motivation is to employ the methods for studies of neurons and of neural systems. The methods can provide unique insight into the biochemical and ionic mechanisms operating within individual neurons. As in other types of tissue, second messenger systems are critical to neuronal function, and optical techniques offer the specificity, sensitivity, and resolution required to localize responses and track the spatiotemporal dynamics of such systems, including waves of biochemical diffusion within tortuous spaces, or the localized ionic fluxes of dendritic spines.

Given the increased interest in the role of distributed representations and neural population dynamics in biological computation, there is little doubt that optical applications will continue to grow. Intrinsic signal imaging has already become the tool of choice for studies of functional architecture at the micro and meso scales. The increasing utility of dynamic optical imaging methods coupled with emerging technologies will enable new sophistication in studies of the relationship between tissue functional architecture, cellular microanatomy, and electrophysiology. Coupled with multi-electrode physiological techniques (and perhaps eventually on their own), high resolution dynamic imaging strategies offer our best prospect for exploring the coordinated function of large, extended systems of neurons.

1.1.7 Technical Progress

Over the last quarter century, optical imaging techniques have undergone a renaissance unprecedented since the introduction of photography for scientific investigation. Beginning with the widespread availability of electronic imaging and digital image acquisition and processing technology, the revolution now encompasses sweeping changes in optics, organic synthesis and bioengineering, and increasingly sophisticated understanding of the biochemical and biophysical machinery of life. There is no question that technological advances large and small have driven and continue to sustain the development of functional neuroimaging by optical techniques. Although CCD camera technology is relatively mature, the sensors and systems continue to improve, offering larger dynamic range, greater sensitivity, lower noise, and higher speed. The consumer market for digital photography and increasing utility for surveillance and physical measurements has driven the development of highly capable imagers. Back-thinned sensors offer reduced light losses due to absorbance or scattering in the silicon substrate; on-chip electron amplification offers the advantages of image intensifiers without the noise and compromises in quantum efficiency associated with conventional photocathodes. The use of multiple amplifiers for parallel readout in CCD and CMOS cameras, and on-board image storage allows much faster dynamic imaging.

The use of contact image probes or recording chambers that serve as windows on the brain facilitate *in vivo* imaging by enabling more routine access and by stabilizing optical interfaces that constitute a major source of noise. Endoscopes, and borescopes incorporating GRIN or conventional relay optics offer the prospect of imaging deep brain structures using minimally invasive surgical procedures. Confocal and multi-photon imaging techniques provide unprecedented capabilities for 3D fluorescence imaging, and in principle, can be employed for functional imaging of intrinsic signals by reflected light. These methods are also adaptable for endoscopic imaging. As is the case with optical sectioning in a conventional microscope, 3D image resolution is enhanced by the use of objectives with a large acceptance angle/numerical aperture. Although most confocal systems depend on mechanical scanning of laser illumination, we have described systems that can employ video projection or other solid-state spatial light modulator technologies to implement spatially and temporally structured illumination with no moving parts. The image aperture, which rejects out of focus light in a conventional confocal system, is implemented in digital signal processing to provide a fast, flexible, and inexpensive confocal image, particularly suited for functional imaging based on intrinsic signals (Rector et al. 2003).

Optical coherence tomography, an interferometric technique employing relatively broadband sources, is the technique of choice for structural imaging of the retina *in situ*. Recently, the technique has been used for functional imaging of retina based on both fast hemodynamic signals as well as the fast intrinsic optical responses (Yao et al. 2005a; Jonnal et al. 2007). Advances in the technology allow much faster acquisition of depth resolved imagery, employing a spectral acquisition technique to collect data from many depths in parallel rather than sequential scanning through physical movement of a reference mirror. Optical coherence microscopy, employing a video camera as detector instead of the usual photodiode, allows direct *en face* imaging, which is of more routine interest for mapping function than the series of slices through depth that are typically acquired in time domain OCT. Parallel OCT typically employs a liner detector in a spectrograph but in principle can employ an area imager to collect an additional simultaneous image dimension. Because OCT is fundamentally sensitive to scattered light, it should be applicable for functional imaging in a variety of tissues.

In addition to confocal, multi-photon and OCT techniques, digital deconvolution strategies employing stacks of conventional axial images can be used to improve the resolution and contrast of the 3D imagery. The general strategy is to predict and remove the contribution of out of focus light based on analysis of adjacent images within the stack and using an explicit model of the optical point spread function.

For retinal imaging, adaptive optics techniques originally developed for astronomy applications allow correction of the aberrations and other optical imperfections introduced by the physiological optics of the eye. These systems improve the resolution and contrast of acquired images, providing clear images of the photoreceptors on the back side of the retinal neural matrix. The system characterizes the optical wavefront using a sensor based on a microlens array coupled to a video camera and uses the information on phase distortion to correct the wavefront using a flexible mirror distorted by multiple actuators. Although the original adaptive optics mirror systems were large and expensive, a number of efforts are targeting development of MEMS-based systems that will be relatively inexpensive to manufacture and easy to apply. Although physical adaptive optics are essential for some purposes, if structured illumination is used, e.g., for confocal imaging, it may not be necessary to physically correct the wavefront. By characterizing the optical aberrations, it should be possible to correct them computationally to produce the desired improvements in image quality. We anticipate that adaptive optics will prove useful for functional imaging in systems other than retina, for example, correcting for the consequences of overlying fluid, macroscopic membranes or tissue matrix that perturb the bulk refractive properties of the system.

Near IR photons penetrate deeply into biological tissue, though they are highly scattered (Villringer and Chance 1997). Absorbance changes due to hemodynamic processes as well as the consequences of the fast scattering processes can be detected from the surface of the head (Franceschini et al. 2008; Gratton et al. 1997; Steinbrink et al. 2000). By employing structured illumination, sensitive detection (in some cases allowing capture of "time of flight information") and employing numerical models of the diffusion-like propagation of photons in a highly scattering medium, it is possible to reconstruct low resolution tomographic images of the optical properties of tissue. Such systems deployed for near IR spectroscopy allow noninvasive monitoring of cerebrovascular oxygenation and intracranial bleeding as well as neural function, and there are promising strategies to increase the spatial resolution of diffuse optical tomography. Though it is unlikely to ever provide the isotropic 3D resolution provided by MRI, optical methods offer a less expensive and more accessible alternative, with considerable biochemical specificity. Direct optical imaging techniques have been applied interoperatively by Toga and colleagues and Ojemann et al. as an alternative to the electrophysiological techniques used to map eloquent cortex during neurosurgical procedures (Cannestra et al. 2000; Haglund et al. 1992). Optical imaging procedures could certainly provide much higher resolution than existing procedures which employ electrical stimulation to assess disruption of language or motor function using behavioral techniques.

1.1.8 Future Directions

Given the confluence of need and opportunity, the future is bright for optical imaging of neural function. Improved instrumentation, signal processing, and computational reconstruction techniques will enhance the spatial resolution of the methods. Dynamic imaging will gain increasing importance, given the accessibility of dynamic signals (both intrinsic and conferred by molecular reporters of membrane potential and cytoplasmic ionic composition) and the availability of cameras with adequate sensitivity and speed to record them. Most significantly, these techniques enable new areas of investigation of fundamental importance to our evolving understanding of the function of large neural systems.

The ability to continuously monitor the responses of individual neurons will allow us to apply to optical data many of the sophisticated tools presently employed for analysis of multi-channel electrophysiology. These include auto and cross correlation, and tests of connection and causality. We will be able to assign neurons to functional classes based on response characteristics and reconstruct receptive field properties by reverse correlation. In addition to functional classification, a great deal of existing knowledge regarding neurons is based on taxonomies tied to neuronal microanatomy and biochemical and pharmacological criteria. Optical tagging strategies will allow us to fluorescently label selected neurons and to characterize their morphology. Labeling strategies presently used for fixed tissue allow us to classify cells according biochemical criteria such as neurotransmitter identity. New reporters incorporating suitable ligands should make such analysis possible in living tissue. There is little question that labeling technology will continue to advance, especially materials based on biologically derived macromolecules. Fluorescent nanoparticles such as quantum dots or dye-loaded silicon microspheres can be conjugated to recognition molecules to greatly enhance visibility. Proteins incorporate molecular recognition sites for most small molecules that are biologically significant, while antibodies provide a powerful and generic mechanism for specific binding of targeted macromolecules. By coupling these molecules or suitable molecular fragments to appropriate organic or biologically derived chromophores, it will be possible to harness the power of genetic design specification and biosynthesis.

Ultimately, in order to understand the function of the complex systems and systems of systems that constitute the brain, it will be necessary to develop detailed computational models. Although a century of neuroscience has provided a wealth of information on the structure and function of individual neurons, the computational function of microcircuits and of extended systems within the brain depends on the activity of extensive array of neuronal types, their patterns of interconnection, and the dynamic interplay that emerges from their interactions. Large scale computational models allow us to move beyond conjecture on the general motifs of neural computation to quantitative studies of the mechanisms and characteristic modes of information encoding and processing by neural populations (Kenyon et al. 2003, 2004; Stephens et al. 2006). Indeed, such models may eventually enable us to capture the unique functionality of biological intelligence in engineered hardware and software systems. But such models also require a quantity and quality of information not generally assessable, especially for the large and complex systems that constitute the human brain. In addition to identifying the types and functional characteristic of component neurons and characterizing their connectivity, we must eventually reconcile the function of brain models *in silico* with the observed responses of biological wetware, both to optimize the implicit parameters that we cannot easily measure and to test and validate the performance of our models. There is little question that optical techniques can be used to address many of the most difficult remaining issues, and thereby enable a powerful new understanding of the function of the brain.

References

Blasdel GG (1989) Visualization of neuronal activity in monkey striate cortex. Ann Rev Physiol 51:561–581

Blasdel GG, Salama G (1986) Voltage Sensitive dyes reveal a modular organization in monkey striate cortex. Nature 321(6070):579–585

Boas DA, Jones SR, Devor A, Huppert TJ, Dale AM (2008) A vascular anatomical network model of the spatio-temporal response to brain activation. NeuroImage 40(3):1116–1129

Bosking WH, Crowley JC, Fitzpatrick D (2002) Spatial coding of position and orientation in primary visual cortex. Nat Neurosci 5(9):874–888

Cannestra AF, Bookheimer SY, Pouratian N, O'Farrell A, Sicotte N, Martin NA, Becker D, Rubino G, Toga AW (2000) Temporal and topographical characterization of language cortices using intraoperative optical intrinsic signals. NeuroImage 12(1):41–54

Cohen LB (1973) Changes in neuron structure during action potential propagation and synaptic transmission. Physiol Rev 53(2):373–413

Cohen LB, Keynes RD, Hille B (1968) Light scattering and birefringence changes during nerve activation. Nature 218:438–441

Cohen LB, Keynes RD, Landowne D (1972) Changes in light scattering that accompany the action potential in squid giant axons: potential-dependent components. J Physiol 224:701–725

Cohen LB, Salzberg BM, Davila HV, Ross WN, Landowne D, Waggoner AS, Wang CH (1974) Changes in axon fluorescence during activity: molecular probes of membrane potential. J Membr Biol 19:1–36

Connor JA, Wadman WJ, Hockberger PE, Wong RKS (1988) Sustained dendritic gradients of calcium induced by excitatory amino acids in Ca1 hippocampal neurons. Science 240(4852):649–653

Foust AJ, Rector DM (2007) Optically teasing apart neural swelling and depolarization. Neuroscience 145(3):887–899

Franceschini MA, Nissilä I, Wu W, Diamond SG, Bonmassar G, Boas DA (2008) Coupling between somatosensory evoked potentials and hemodynamic response in the rat. NeuroImage 41(2):189–203

Fries P, Roelfsema PR, Engel AK, Konig P, Singer W (1997) Synchronization of oscillatory responses in visual cortex correlates with perception in interocular rivalry. Proc Nat Acad Sci USA 94(23):12699–12704

Frostig RD, Lieke E, Ts'o DY, Grinvald A (1990) Cortical functional architecture and local coupling between neuronal activity and the microcirculation revealed by in vivo high-resolution optical imaging of intrinsic signals. Proc Natl Acad Sci USA 87:6082–6086

George JS (2002) Event-related neural responses. In: Ramachandran, ed. Encyclopedia of the Human Brain. Academic, New York

George JS, Fowler JC (1988) A calcium transient is associated with hippocampal slice excitation. Biophys J 53(2):A560

George JS, Fowler JC (1990) Intensified video imaging of neural population activity. Soc Neurosci Abstr 16(1):55

George JS, Hagins WA (1983) Control of Ca2+ in rod outer segment disk membranes by light and cGMP. Nature 303:344–348

George JS, Fowler JC, Ranken DM (1990) Imaging the calcium transient associated with hippocampal slice excitation. Biophys J 53(2):A131

George JS, Schmidt DM, Rector DM, Wood CC (2001) Dynamic functional neuroimaging integrating multiple modalities. In: Jezzard, Matthew, Smith (eds) Functional MRI: an introduction to methods, Oxford, pp 353-382

Gratton G, Fabiani M, Corballis PM, Gratton E (1997) Noninvasive detection of fast signals from the cortex using frequency-domain optical methods. Ann NY Acad Sci 820:286–299

Grinvald, A, Lieke, E, Frostig, RD, Gilbert, CD, and Wiesel (1986), TN. Functional architecture of the cortex revealed by optical imaging of intrinsic signals. Nature, 324:361–364

Grynkiewicz G, Poenie M, Tsien RY (1985) A new generation of calcium indicators with greatly improved fluorescence properties. J Biol Chem 260:3440-3450

Haglund MM, Ojemann GA, Hochman DW (1992) Optical imaging of epileptiform and functional activity in human cerebral cortex. Nature 358:668–671

Heimburg T, Jackson AD (2005) On soliton propagation in biomembranes and nerves. Proc Nat Acad Sci USA 102(28):9790–9795

Hill DK (1950) The volume change resulting from stimulation of a giant nerve fibre. J Physiol 111:304–327

Hodgkin AL, Huxley AF (1952) Propagation of electrical signals along giant nerve fibers. Proc R Soc Lond B Biol Sci 140(899):177–183

Holthoff K, Witte OW (1996) Intrinsic optical signals in rat neocortical slices measured with near-infrared dark-field microscopy reveal changes in extracellular space. J Neurosci 16(8):2740–2749

Jones MS, Barth DS (1999) Spatiotemporal organization of fast (>200 Hz) electrical oscillations in rat Vibrissa/Barrel cortex. J Neurophysiol 82(3):1599–1609

Jonnal RS, Rha J, Zhang Y, Canse B, Gao W, Miller DT (2007) In vivo functional imaging of human cone photoreceptors. Opt Express 15(24):16142–16160

Kenyon GT, Moore B, Jeffs J, Denning KS, Stephens GJ, Travis BJ, George JS, Theiler J, Marshak DW (2003) A model of high-frequency oscillatory potentials in retinal ganglion cells. Vis Neurosci 20(5):465–480

Kenyon GT, Travis BJ, Theiler J, George JS, Stephens GJ, Marshak DW (2004) Stimulus-specific oscillations in a retinal model. IEEE Trans Neural Netw 15:1083–1091

Kleinschmidt A, Obrig H, Requardt M, Merboldt KD, Dirnagl U, Villringer A, Frahm J (1996) Simultaneous recording of cerebral blood oxygenation changes during human brain activation by magnetic resonance imaging and near-infrared spectroscopy. J Cereb Blood Flow Metab 16(5):817–826

Landowne D (1985) Molecular motion underlying activation and inactivation of soduim channels in squid giant axons. J Membr Biol 88:173–185

Landowne D (1992) Optical activity change with nerve impulses. Biophys J 61:A109

Lindsey BG, Morris KF, Shannon R, Gerstein GL (1997) Repeated patterns of distributed synchrony in neuronal assemblies. J Neurophysiol 78(3):1714–1719

Lipton P (1973) Effects of membrane depolarization on light scattering by cerebral cortical slices. J Physiol 231:365–383

MacVicar BA, Hochman D (1991) Imaging of synaptically evoked intrinsic optical signals in hippocampal slices. J Neurosci 11(5):1458–1469

Malonek D, Grinvald A (1996) Interactions between electrical activity and cortical microcirculation revealed by imaging spectroscopy: implications for functional brain mapping. Science 272(5261):551–554

Masino SA, Kwon MC, Dory Y, Frostig RD (1993) Characterization of functional organization within rat barrel cortex using intrinsic signal optical imaging through a thinned skull. Proc Natl Acad Sci USA 90(21):9998–10002

O'Donovan MJ, Ho S, Sholomenko G, Yee W (1993) Real-time imaging of neurons retrogradely and anterogradely labelled with calcium-sensitive dyes. J Neurosci Methods 46(2):91–106

Oldenbourg R, Salmon ED, Tran PT (1998) Birefringence of single and bundled microtubules. Biophys J 7:645–654

Rector DM, George JS (2001) Continuous image and electrophysiological recording with real time processing and control. Methods 25:151–163

Rector DM, Poe GR, Harper RM (1993) Fiber optic imaging of subcortical neural tissue in freely behaving animals. In: Dirnagl U et al (eds) Optical imaging of brain function and metabolism. Plenum, New York, pp 81–86

Rector DM, Poe GR, Kristensen MP, Harper RM (1995) Imaging the dorsal hippocampus: light reflectance relationships to electroencephalographic patterns during sleep. Brain Res 696:151–160

Rector DM, Rogers RF, George JS (1999) A focusing image probe for assessing neural activity in-vivo. J Neurosci Methods 91:135–145

Rector DM, Harper RM, George JS (2002) In-vivo observations of rapid scattered light changes associated with electrical events. In: Frostig RD (ed) Methods for in vivo optical imaging of the central nervous system. CRC, Boca Raton, pp 113–171

Rector DM, Ranken DM, George JS (2003) High-performance confocal system for microscopic applications. Methods 30(1):16–27

Rector DM, Carter KM, Volegov P, George JS (2005) Spatio-temporal mapping of rat whisker barrels with fast scattered light signals. Neuroimage 26(6):619–627

Roe AW, Ts'o DY (1999) Specificity of color connectivity between primate V1 and V2. J Neurophysiol 82(5):2719–2730

Roper SN, Obenaus A, Dudek FE (1992) Osmolality and nonsynaptic epileptiform bursts in rat CA1 and dentate gyrus. Ann Neurol 31(1):81–85

Sakmann B, Neher E (1984) Patch clamp techniques for studying ionic channels in excitable membranes. Ann Rev Physiol 46:455–472

Salzberg BM, Obaid AL, Gainer H (1985) Large and rapid changes in light scattering accompany secretion by nerve terminals in the mammalian neurohypophysis. J Gen Physiol 86:395–411

Schei JL, McCluskey MD, Foust AJ, Yao XC, Rector DM (2008) Action potential propagation imaged with high temporal resolution near-infrared video microscopy and polarized light. NeuroImage 40(3):1034–1043

Steinbrink J, Kohl M, Obrig H, Curio G, Syre F, Thomas F, Wabnitz H, Rinneberg H, Villringer A (2000) Somatosensory evoked fast optical intensity changes detected non-invasively in the adult human head. Neurosci Lett 291(2):105–108

Stephens GJ, George JS, Theiler J, Marshak D, Neuenschwander S, Singer W, Kenyon G (2006) See globally, spike locally: oscillations in a retinal model allow fast, local encoding of visual features. Biol Cybern 95(4):327–348

Stepnoski RA, LaPorta A, Raccuia-Behling F, Blonder GE, Slusher RE, Kleinfeld D (1991) Noninvasive detection of changes in membrane potential in cultured neurons by light scattering. Proc Natl Acad Sci USA 88:9382–9386

Tank DW, Sugimori M, Connor JA, Llinas RR (1988) Spatially resolved calcium dynamics of mammalian Purkinje cells in cerebellar slice. Science 242(4879):773–777

Tasaki I, Byrne PM (1992) Rapid structural changes in nerve fibers evoked by electrical current pulses. Biochem Biophys Res Commun 188(2):559–564

Tasaki I, Watanabe A, Sandlin R, Carnay L (1968) Changes in fluorescence, turbidity, and birefringence associated with nerve excitation. Proc Natl Acad Sci USA 61:883–888

Tasaki I, Iwasa K, Gibbons RC (1980) Mechanical changes in crab nerve fibers during action potentials. Jpn J Physiol 30:897–905

Ts'o DY, Frostig RD, Lieke EE, Grinvald A (1990) Functional organization in the primate visual cortex as revealed by optical imaging of intrinsic signals. Science 249:417–420

Van Harreveld A (1958) Changes in the diameter of apical dendrites during spreading depression. Am J Physiol 192:457–463

Villringer A, Chance B (1997) Non-invasive optical spectroscopy and imaging of human brain function. Trends Neurosci 20(10):435–442

Waggoner AS (1979) Dye indicators of membrane potential. Ann Rev Biophys Bioeng 8:47–68

Yao XC, George JS (2006a) Optical imaging of fast light-evoked fast neural activation in amphibian retina. Proc SPIE 6078(1):457–464

Yao XC, George JS (2006b) Dynamic neuroimaging of retinal light responses using fast intrinsic optical signals. Neuroimage 33(3):898–906

Yao XC, Rector DM, George JS (2003) Optical lever recording of displacements from activated lobster nerve bundles and *nitella* internodes. Appl Opt 42(16):2972–2978

Yao XC, Yamauchi A, Perry B, George JS (2005a) Rapid optical coherence tomography and recording scattering changes from activated frog retina. Appl Opt 44:2019–2023

Yao XC, Foust A, Rector DM, Barrowes B, George JS (2005b) Cross-polarized reflected light measurement of fast optical responses associated with neural activation. Biophys J 88:4170–4177

Yoshikami S, George JS, Hagins WA (1980) Light-induced calcium fluxes from outer segment layer of vertebrate retinas. Nature 286:395

Chapter 2
Fluorescent Sensors of Membrane Potential that Are Genetically Encoded

L. Jin, B.J. Baker, Lawrence B. Cohen, H. Mutoh, D. Dimitrov, A. Perron, Y. Iwamoto, E.Y. Isacoff, V.A. Pieribone, T. Hughes, T. Knöpfel, and W. Akemann

Abstract Imaging activity of neurons in intact brain tissue was conceived several decades ago, and, after many years of development, voltage-sensitive dyes now offer the highest spatial and temporal resolution for imaging neuronal functions in the living brain. Further progress in this field is expected from the emergent development of genetically encoded fluorescent sensors of membrane potential. These fluorescent protein voltage sensors overcome some drawbacks of organic voltage-sensitive dyes, such as nonspecificity of cell staining and the low accessibility of the dye to some cell types. In a transgenic animal, a genetically encoded sensor could, in principle, be expressed specifically in any cell type and would have the advantage of staining only the cell population determined by the specificity of the promoter used to drive expression. Here we, critically review the current status of these developments.

2.1 Introduction

Optical imaging is a remarkably flexible method for studying various cellular activities. Organic dyes have been developed that can faithfully report many biological variables including calcium concentration, pH, or membrane potential (Brown et al. 1975; MacDonald and Jobsis 1976; Davila et al. 1973). These optical probes enable simultaneous measurements from many locations and have been used to study the physiology of single neurons, as well as large populations of cells (for recent reviews see Zochowski et al. (2000); Grinvald and Hildesheim (2004); and Baker et al. (2005)).

The classical result defining a voltage-sensitive dye is illustrated in Fig. 2.1. The dots represent the measurement of transmitted light through a squid giant axon stained with the merocyanine-rhodanine dye shown in the insert. The thin line (which runs right through the dots) is the simultaneous microelectrode recording of

L. Jin (✉)
Physiology, 333 Cedar St, New haven, CT, 06510, USA
e-mail: lei.jin@yale.edu

A.W. Roe (ed.), *Imaging the Brain with Optical Methods*,
DOI 10.1007/978-1-4419-0452-2_2,

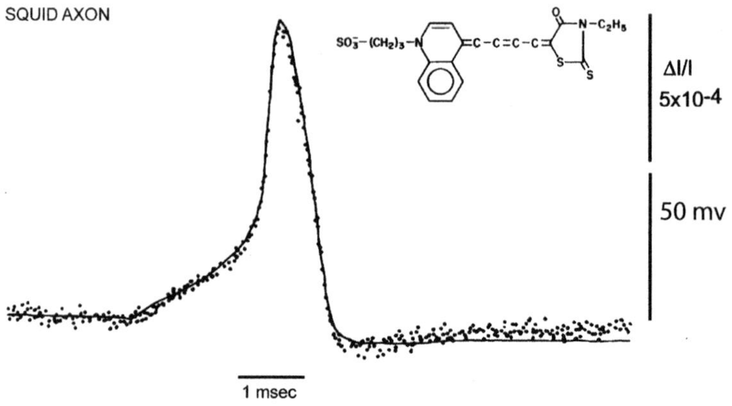

Fig. 2.1 Absorption change during the action potential of a squid giant axon. The dots are the change in transmitted light and the smooth line the simultaneous microelectrode recording of the action potential modified from Ross et al. (1977)

the action potential. This result illustrates some of the properties of organic voltage-sensitive dyes. First, they are outstandingly fast. Their response time constants are less than 2 μs (Loew et al. 1985), faster than all known neurobiological membrane potential changes. Second, in almost every case where it has been measured, the optical signals were linearly related to the change in membrane potential. This has greatly facilitated the interpretation of these signals in many circumstances. One example where this is true is the interpretation of signals measured from soma and processes of individual cells where the dye was injected into the inside of the cell and allowed to diffuse throughout the cell and its processes (Zecevic 1996; Canepari et al. 2007). This linearity, however, can be a drawback in other circumstances discussed below.

Two additional drawbacks of these organic dyes are the nonspecificity of cell staining and the low accessibility of the dye to some cell types. They either stain all cells in a tissue to which they are exposed, or if delivered via a patch pipette, they label one or a small number of cells. In the absence of targeting of the organic dyes to specific cell types, optical signals of interest are often drowned out either by background fluorescence from inactive cells, or by signals in cells that are not the focus of interest. In addition, with the exception of preparations with sparse cell staining, diffuse labeling limits the spatial resolution so that single cell responses cannot be readily resolved. One general solution to this problem is to make optical reporters from proteins. Since protein-based reporters are encoded in DNA, they can be placed under the control of cell-specific promoters and introduced in vivo using gene transfer techniques. Moreover, protein-based reporters can, in principle, be rationally "tuned" by the modification of their functional domains with mutations that adjust their dynamic range of operation. In a transgenic animal, a genetically encoded sensor could, in principle, be expressed in any cell type and would have the advantage of staining only the cell population determined by the specificity of the promoter used to drive expression.

Protein-based reporters are generally constructed from two parts: a "sensor" protein that undergoes a conformational rearrangement that depends on the parameter measured (e.g., a voltage-gated ion channel that senses membrane potential), and a fluorescent protein (FP) reporter fluorophore whose optical output is modulated by changes in the sensor protein. The first FP-voltage sensor, denoted FlaSh, was obtained by inserting green fluorescent protein (GFP) downstream of the pore region in the *Drosophila* voltage-gated potassium channel, Shaker (Siegel and Isacoff 1997). When FlaSh was expressed in *Xenopus* oocytes, changes in membrane potential were reported by changes of the fluorescence. However, this initial success was not followed by published reports of signals from mammalian brains using FlaSh or the two subsequent first generation voltage sensors, VFSP-1 (Sakai et al. 2001; Knöpfel et al. 2003), and SPARC (Ataka and Pieribone 2002).

A group of laboratories at Yale University, the University of California, Berkeley, the John B. Pierce Laboratory, Montana State University, and RIKEN has begun a collaborative effort to improve genetically encoded FP-voltage sensors for the use in mammalian preparations. Inefficient targeting to mammalian plasma membrane was identified as a major limitation of the first generation FP-voltage sensors. Therefore, the group started a concerted effort to find a more suitable signaling protein and FP reporter-combination, while also fine tuning emerging sensor proteins. Here we review the historical development of the first generation FP voltage sensors, describe the state of the art second generation FP voltage sensors, and outline efforts for the next generation FP voltage sensors . We then compare these traditional fully genetically encodable sensors with novel partially genetically encoded probes. Finally, we critically discuss the practical and theoretical limitations of current FP voltage sensor-based imaging techniques.

2.2 First Generation FP Voltage Sensors

The first generation FP voltage sensors were developed by molecular fusion of a GFP-based fluorescent reporter to voltage-gated ion channels or components thereof. The first prototype, FlaSh, was generated in the Isacoff laboratory and obtained by inserting wtGFP into the C-terminus of the *Drosophila* Shaker potassium channel (Siegel and Isacoff 1997). When expressed in oocytes, an 90 mV depolarization of the plasma membrane resulted in a 5% decrease in fluorescence (Fig. 2.2, right panel) this was designated FlaSh, from fluorescent Shaker.

To reduce unwanted effects on the cell's physiology, FlaSh was rendered non-conducting by introducing a W434F mutation, preventing ions from moving through the pore while maintaining voltage-dependent rearrangements (Perozo et al. 1993). In order to resolve action potentials, an ideal sensor would generate a robust signal on a millisecond timescale. The signal strength from FlaSh is comparable to that of the voltage-sensitive organic dye, di4-ANEPPS; however, the on and off rates are rather slow (τ-on ~ 100 ms; τ-off ~ 60 ms). In an effort to improve the kinetics of FlaSh, Guerrero et al. (2002) replaced the wtGFP with several different

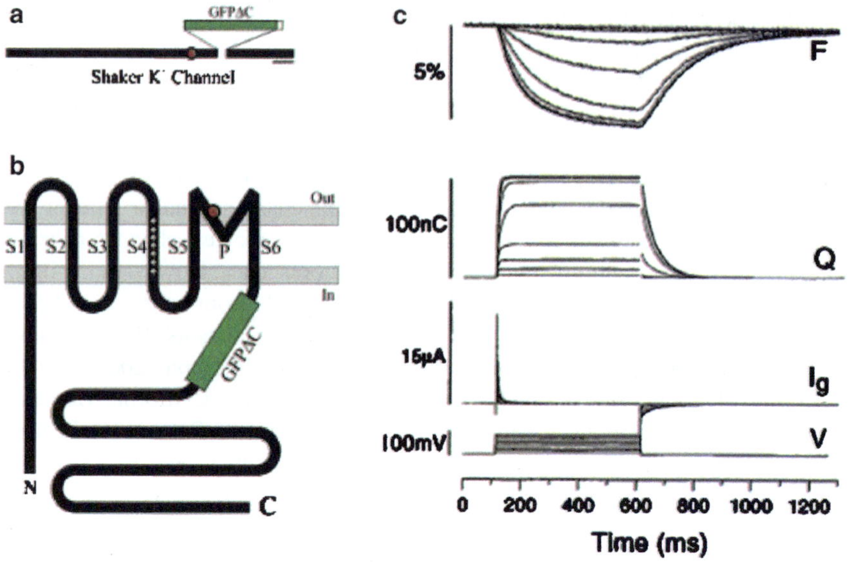

Fig. 2.2 (**a**) *GFPΔC* (*green*) was inserted in-frame into the Shaker K$^+$ channel (*black*). A point mutation *W434F* (*red circle*) was made in the pore of the channel to eliminate ionic current through the sensor. (**b**) Putative orientation of the FlaSh protein in the cell membrane (*gray*). The fourth transmembrane segment S4 is positively charged. Note that *GFPΔC* is intracellular and FlaSh is targeted to the cell membrane. (**c**) Simultaneous two-electrode voltage-clamp recording and photometry show current and fluorescence changes in response to voltage steps (*V*) between –60 mV and 10 mV, in 10 mV increments. Holding potential was –80 mV. FlaSh exhibits on and off gating currents (*Ig*) but no ionic current. Integrating the gating current gives the total gating charge (*Q*) moved during the pulse. FlaSh fluorescence (*F*) decreases reversibly in response to membrane depolarizations. Traces are the average of 20 sweeps. Fluorescence scale, 5% ΔF/F. Modified from Siegel and Isacoff (1997)

fluorescent proteins (FPs). The change in optical characteristics mediated by the different FPs was substantial. The signal intensity, the direction of the fluorescent change, and the speed of FlaSh were all altered by the chromophore in an unsystematic way (Fig. 2.3). Variants, involving wtGFP and uvGFP both have decreased fluorescence in response to depolarization steps, while Ecliptic variants of GFP, YFP, and CFP exhibit increased fluorescence in response to depolarization. Remarkably, upon depolarization, an increase in fluorescence is seen when eGFP is excited at 450 nm, but there is a decrease in fluorescence when excited at 480 nm. The speed of the response is also governed by the chromophore, with the Ecliptic variant of GFP generating the fastest response (τ-on ~ 5 ms). While the mechanism underlying the fluorescence change that results from an alteration in membrane potential remains poorly understood, it is clear that the fluorescent reporter contributes significantly to the kinetics of the optical response and is an important parameter to vary in any attempt to improve genetically encoded voltage sensors.

The second prototypic design, realized in the Knopfel laboratory, and termed VSFP1, exploits the voltage-dependent conformational changes around the fourth transmembrane segment (S4) of the voltage-gated potassium channel Kv2.1 and

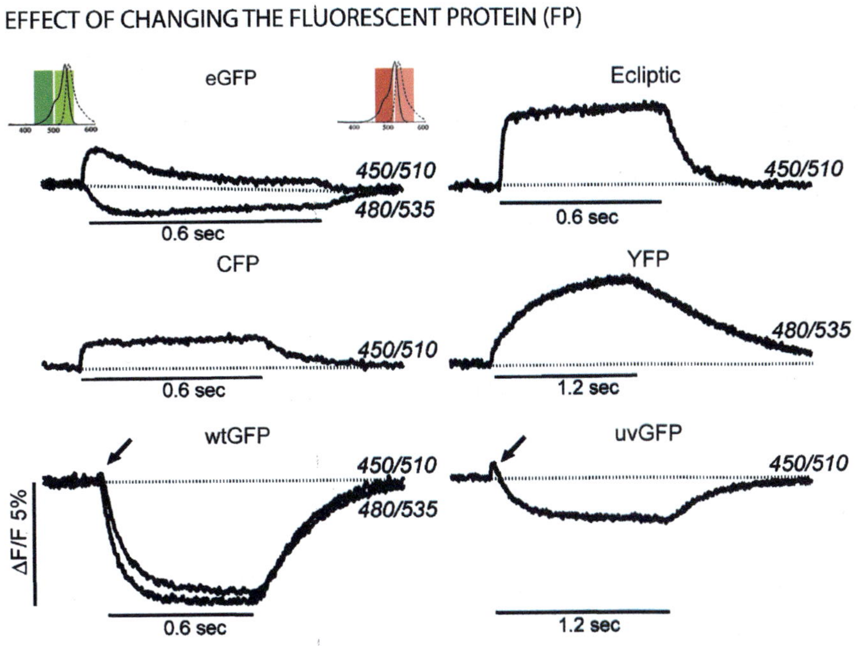

Fig. 2.3 Distinct polarity and kinetics of fluorescence response of six variant GFPs in the FlaSh construct. Variant GFPs in FlaSh have various degrees of prominence of early, intermediate, and late components of ΔF. Response to a depolarizing step (−80–0 mV for duration of time bar) is shown for optimal excitation and emission wavelengths. Where other excitation and emission wavelengths gave different responses (rather than simply smaller ones), these are shown (eGFP and wtGFP). Filters: *450/510* _ 425–475ex, 480D, 485–535em; *480/535* _ 460–500ex, 505D, 510–560em. *Arrow* indicates early upward ΔF in wtGFP and GFPuv, which we attribute to a decrease in self-quenching. Redrawn from Guerrero et al. (2002), Fig. 1

uses either fluorescence resonance energy transfer (FRET) (Sakai et al. 2001) or a permuted FP (Knöpfel et al. 2003). The third prototype, SPARC, from the Pieribone laboratory, was generated by inserting a FP between domains I and II of the rat skeletal muscle Na^+ channel and generated fast (1 ms) signals in frog oocytes (Ataka and Pieribone 2002).

These first generation voltage sensors are capable of optically reporting changes in membrane potential, but they have not been useful in mammalian systems because of their very poor plasma membrane expression in mammalian cells. When expressed in HEK 293 cells, the expression of these constructs is primarily intracellular and little if any of these first generation FP voltage sensors are colocalized at the cell surface with di8-ANEPPS (Fig. 2.4; top three sections). Figure 2.4b shows the profiles of the FP in green and that of di8-ANEPPS in red along the red lines in the right column of Fig. 2.4a. A much better colocalization occurs with Kv 1.4 with GFP at the N-terminus and with the cation/chloride cotransporter, NKCC1, with YFP fused near the carboxyl terminus (Fig. 2.4; bottom two sections). There does seem to be some plasma membrane expression for Flare (a Kv1.4 variant of FlaSh), but the best case scenario still exhibits fluorescence of a predominantly intracellular origin. More importantly, no functional

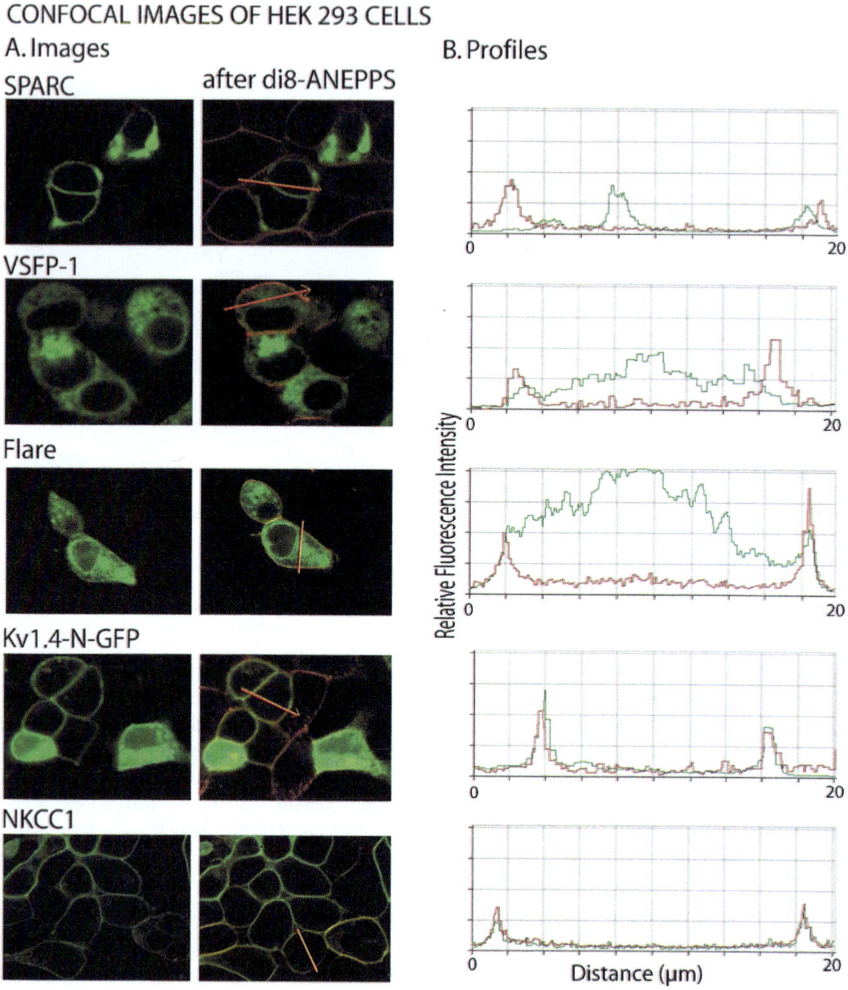

Fig. 2.4 Confocal images of HEK 293 cells. (**a**) SPARC, VSFP-1, Flare, Kv1.4-N-GFP, or NKCC1-YFP were expressed in HEK 293 cells and imaged via confocal microscopy. The images on the left show HEK 293 cells expressing the fluorescent construct. The images on the right are the same cells after the addition of di8-ANEPPS to the bathing medium. Di8-ANEPPS functions as a fluorescent plasma membrane marker. (**b**) The profiles show the *green* (FP-voltage sensor) and *red* (di8-ANEPPS) fluorescence along the *arrow* in the images. The *arrows* indicated the location of the external membrane. Modified from Baker et al. (2007)

optical signals could be detected in the average of 16 sweeos using either Flare, VSFP1, or SPARC in either HEK293 cells or in acutely dissociated hippocampal neurons (Baker et al. 2007). The absence of a signal from the first generation probes is, in part, due to their low membrane expression and, in part, due to a large, nonresponsive background fluorescence that would tend to mask voltage-dependent signals.

Because Kv1.4 with an N-terminal GFP exhibits excellent membrane expression (Fig. 2.4, fourth section), several strategies to release Flare or its Kv2.1-based homologues from the ER have been tried. Unfortunately, mutagenesis of potential ER retention signals, addition of ER release motifs, and expression in hippocampal neurons that might endogenously express trafficking partners all failed to significantly improve the plasma membrane expression (Ray, Tomita, Iwamoto, Dimitrov, Perron and Knöpfel, unpublished observations; Baker and Cohen unpublished observations). We hoped that the second generation of FP voltage sensors would overcome this poor targeting to neuronal membranes.

2.3 Second Generation FP Voltage Sensors

Recently, several lines of evidence supported the concept of self-contained voltage sensor domains in naturally occurring proteins such as Kv channels. A voltage sensor domain may function in the absence of the obligate formations seen in the complete Kv channel complex. The first body of evidence pointing toward this possibility came from crystal structures and structure-function modeling of KvAP and Kv1.2 that indicated that the secondary structure of the S1–S4 portion of Kv channels is relatively independent from neighboring protein (Lu et al. 2001; Jiang et al. 2003a, b; Long et al. 2005a, b). The second line of evidence was based on the identification of voltage sensor domains in non ion channel proteins. Ci-VSP (*Ciona intestinalis* Voltage-Sensor-containing Phosphatase), for example, is a voltage-controlled enzyme consisting of a transmembrane voltage sensor domain and a cytosolic phosphoinositide phosphatase domain (Murata et al. 2005). The voltage sensor domain from this protein was shown to be functional when the enzyme domain was removed (Murata et al. 2005). Very recently, it was also shown by the Isacoff laboratory that the Ci-VSP can exist as a monomer in the plasma membrane (Kohout et al. 2007). A third line of evidence for voltage sensor domains that can function in isolation emerged with the discovery of a new protein family termed Hv1 or voltage sensor domain only proteins, VSOPs. These membrane proteins mediate voltage-dependent proton transport (Sasaki et al. 2006; Ramsey et al. 2006) and have a domain homologous to the S1–S4 portion of Kv channels but lack the putative pore forming S5–S6 domain.

A self-contained voltage sensor domain that functions without additional protein components or without the need for subunit multimerization could be an interesting candidate for a FP voltage sensor. In addition, because the Ci-VSP is of heterologous origin, it might be less susceptible to posttranslational modifications and mistargeting in mammalian expression systems. With the improvement of plasma membrane targeting in mind, the use of Ci-VSP was explored by the Knopfel. Make laboratories singular (Dimitrov et al. 2007). Because the voltage sensor domain of Ci-VSP is homologous to that of canonical Kv channel subunits, the first step was to generate constructs similar to the first generation VSFPs (VSFP1s), but with the voltage sensor domain of Kv2.1 replaced by that of Ci-VSP. Since the exact role of the linkers coupling the voltage sensor domain to its operant is unclear, the CFP and YFP FRET pair was fused to the C-terminus of the voltage sensor domain conserving four different lengths of the intrinsic Ci-VSP S4-segment-downstream sequence. With this approach, the cytosolic phosphatase domain was replaced with the FRET pair. When expressed in PC12 cells or hippocampal neurons, all of the initial series of second generation VSFP (VSFP2) variants displayed bright fluorescence and clear apparent targeting to the plasma membrane (Dimitrov et al. 2007). More importantly, all of these first four constructs showed voltage-dependent modulation of cyan and yellow fluorescent output (Fig. 2.5). The FRET fluorescence change could result from a change in the distance separating the two chromophores or from a change in the angle between them or both. Furthermore, the Cyan signal is substantially larger than the yellow signal for VSFP2A (Fig. 2.3) suggesting that non-FRET mechanisms are also occurring. At present, we have very little understanding of the mechanism coupling the voltage-dependent structural changes in the voltage sensor domain to the changes in fluorescence of the chromophores.

Consistent with "gating" currents measured in Ci-VSP, the fluorescence voltage curves for VSFP2A-D (Fig. 2.5d) exhibited half maximal activation ($V_{1/2}$) at potentials far above the physiological range of mammalian neurons. To shift the voltage dependency to a more physiological range, a series of mutational alterations were explored in the putative S4 segment of Ci-VSP's voltage sensor domain. The most successful mutation was R217Q which resulted in the protein termed VSFP2.1 (Fig. 2.6a, b). Like the initial constructs, VSFP2.1 displayed bright fluorescence and clear targeting to the plasma membrane. VSFP2.1 exhibited a $V_{1/2}$ value

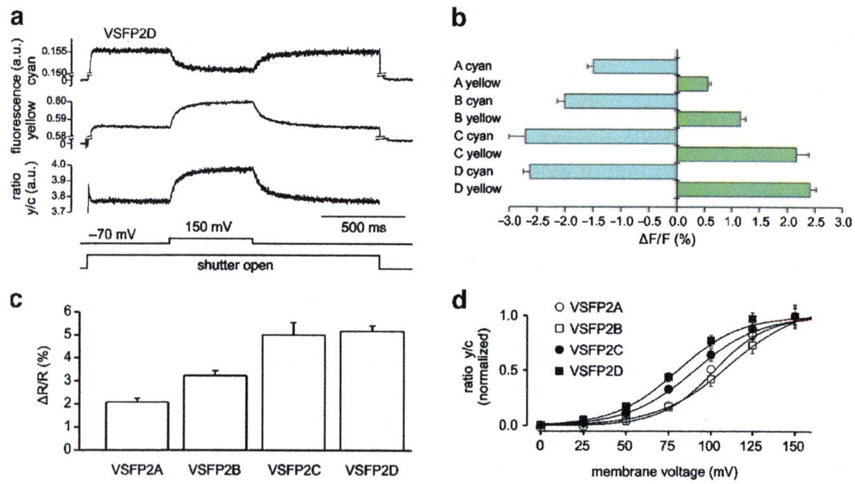

Fig. 2.5 Fluorescence signals induced by membrane depolarization in PC12 cells. (**a**) Sample traces of cyan fluorescence, yellow fluorescence, and the ratio of yellow/cyan fluorescence (average of 27 traces). Lower traces indicate times of shutter opening and membrane depolarization from −70 mV to 150 mV. (**b**) Average changes in cyan fluorescence and yellow fluorescence induced by depolarization to 150 mV. Labels **a–d** indicate VSFP2A (11 cells), VSFP2B (7 cells), VSFP2C (7 cells), VSFP2D (7 cells). (**c**) Average changes in the ratio of yellow and cyan fluorescence induced by depolarization to 150 mV. (**d**) Ratio of yellow/cyan fluorescence versus test membrane voltage. Lines are Boltzmann fits. Redrawn from Fig. 2 of Dimitrov et al. (2007)

of ~−70 mV both at 22°C and at 35°C (Fig. 2.6). At 22°C, activation and deactivation kinetics were similar to those of VSFP2A-D. At 35°C, activation was considerably faster (Fig. 2.6b).

The response-voltage relationship with a $V_{1/2}$ value in the physiological range of neuronal membrane fluctuations together with moderately fast kinetics makes VSFP2.1, a candidate for optical measurements of neuronal activity, such as large synaptic potentials, action potentials, and bistabilities in resting membrane potential. To demonstrate this prediction, PC12 cells were voltage clamped with membrane potential traces obtained from mouse mitral cells. The fluorescent output of VSFP2.1 could clearly resolve individual action potentials (arrows in Fig. 2.7), as well as the slower underlying membrane depolarization. As expected from the response kinetics of VSFP2.1, the optical readout of the fast action potentials was reduced relative to the slower components of membrane potential change.

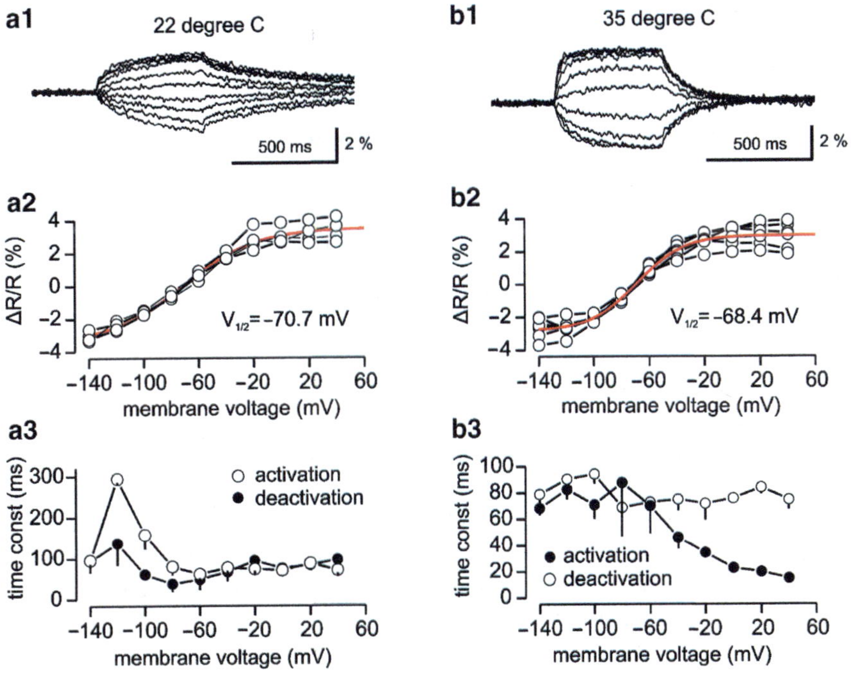

Fig. 2.6 Properties of VSFP2.1. Response–voltage relationship and kinetics of VSFP2.1 at 22°C (**a1**–**a3**) and at 35°C (**b1**–**b3**). (**a1**, **b1**) Ratio of yellow/cyan fluorescence during a family of 500 ms voltage steps from a holding potential of −70 mV to test the potentials of −140 mV to +40 mV (20 mV increments). Traces are grand averages over average responses from 4 cells (**a1**) and 6 cells (**b1**). (**a2**, **b2**) Ratio of yellow/cyan fluorescence versus test membrane voltage. Connected symbols are data from individual cells. Red lines are Boltzmann fits with $V_{1/2}$ values as indicated. (**a3**, **b3**) Activation and deactivation time constants. Figure 2.6b was taken from Dimitrov et al. (2007)

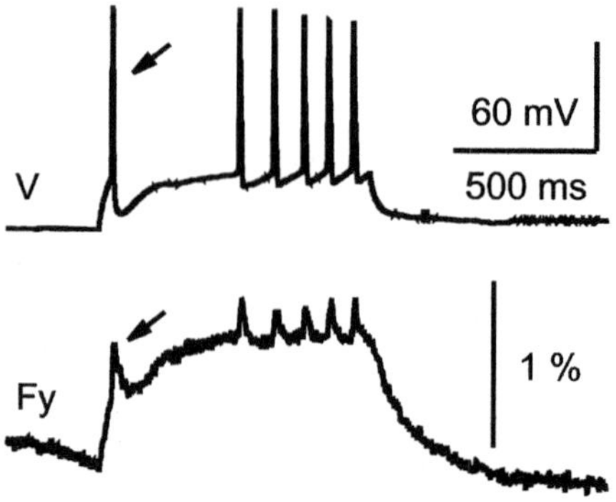

Fig. 2.7 VSFP2.1 can monitor physiological neuronal membrane voltage dynamics. PC12 cells expressing VSFP2.1 were voltage clamped with a voltage trace obtained from a current-clamped mouse olfactory bulb mitral cell. The mitral cell was stimulated to generate a series of action potentials by intracellular injection of a current pulse. Traces are the averages of 50 sweeps. Traces show membrane potential (*V*), yellow fluorescence (*Fy*). The fluorescence signals were digitally low pass filtered (0.2 kHz) and were not corrected for dye bleaching. Recordings were done at 35°C. Taken from Fig. 4 of Dimitrov et al. (2007)

2.4 Next Generation FP Voltage Sensors

Further development of VSFP2.1 is progressing on several lines: linker optimization, alternative FP variants, and the search for additional locations for fusing or inserting FPs into Ci-VSP or alternative sensor proteins.

2.4.1 *Linker Optimized Variants*

It is well established (Shimozono and Miyawaki 2008) for other types of FP sensors that sensitivity of the probe ($\Delta F/F$ or $\Delta R/R$ values) can be increased by optimizing the length and amino acid composition of the linker between the sensor protein and the FP. In such effect, the performance of VSFP2.1 was improved by removal of five amino acids that initially originated from engineered restriction sites in VSFP2.1 (Mutoh et al, 2009; Villalba-Galea et al. 2008).

2.4.2 *Alternative FP Colors*

Cyan- and yellow-emitting variants of GFP from *Aequorea victoria* (CFP and YFP, respectively) are the most widely used components for FRET-based biosensors. However, there are reasons why alternative colors, in particular, red shifted versions are more desirable: (1) red variants may be more suitable for deep tissue imaging, (2) reduced interference from tissue autofluorescence and hemoglobin absorption, and (3) spectral separation when used in combination with other fluorescent probes. Several efficient variants of these were very recently generated in the Knopfel laboratory, the most promising comprises combinations of either green FP with red FP or yellow FP with far-red FP (Dimitrov, Perròn, Mutoh and Knopfel, unpublished observations).

In addition, for some applications, non-FRET monochromatic probes are more favorable. Single FP sensors have been demonstrated previously including the FlaSh type of voltage sensors and several types of Ca^{2+} sensors (Hughes Baird et al. 1999; Nakai et al. 2001) Thus, we recently developed analogous single FP voltage sensors based on the *Ciona* protein (Lundby et al 2008, Perron et al 2009, Baker, Hughes Jin, and Cohen, unpublished observation). Finally, having available additional colors allows the possibility of recording simultaneously from two different neurons populations each with its own color.

2.4.3 Alternative Designs

In a canonical voltage sensor domain, it is not only the S4 segment that responds to voltage shifts, but rather the whole domain undergoes conformational changes in a complex manner. It follows, therefore, that finding alternative sites for insertions of FPs along the voltage sensor domain that would not disturb the function and folding of either the voltage sensor domain or the FP may lead to new designs of improved functional voltage sensors. To explore this approach, the Hughes laboratory used its transposon technology to generate a library of positions in the Ci-VSP that would allow the insertion of a FP without disturbing the folding of the Ci-VSP (critical for efficient membrane targeting) nor the folding of the FP itself (FP-specific fluorescence).

In addition to the above, there are many other, only partially explored, approaches for conformational tuning of the parts constituting the sensor. These include addition of prenylation signals or other signals for posttranslational modifications that may help improve the plasma membrane targeting and/or conformational states in a way that would be advantageous for sensing membrane potential.

2.5 Genetic Targeting of Neurons

Perhaps the most important feature of genetically encodable sensors is the possibility to achieve indicator loading in specific cell populations. This is the case because cell type-specific protein expression can be achieved by the use of appropriate regulatory gene sequences (promoters). This concept had been proposed many years ago in conjunction with the bioluminescent calcium-sensitive reporter protein aqueorin. However, the light output of bioluminescent probes was rather limited and consumed a cofactor that needed to be applied to the preparation. Recently, cell type-specific expression of FP variants, or fusion proteins containing them has been demonstrated in numerous transgenic and gene targeted mouse lines. For some experimental questions, it is desirable to label cells sparsely so that individual cells can easily be identified by fluorescence imaging in densely packed tissue. Sparse labeling can be achieved with "shotgun" promoters such as Thy1 (Feng et al. 2000), virus-based transfection (Dittgen et al. 2004), or transgenic strategies for combinatorial expression of FP components (Livet et al. 2007), but is associated with considerable uncertainty about the identity of the labeled cells and/or whether the labeled cells are representative of the given target population.

The specificity of virus-based transfections can be increased by cell-specific promoters (but this technique is limited by the size of recombinant sequence which can be packaged in to the virus). Perhaps more promising is the combination of viruses with mouse lines expressing recombinases, such as Cre and Flp, in specific subsets of neurons. This technology permits targeted expression of FP sensors in any subset neuron, for which, one of the now many recombinase-mice lines is available. For that purpose, the sensor gene in a lox-stop-lox cassette is inserted into a virus and then expressed under lox recombination as defined by the recombinase mouse.

A more expensive and time consuming approach is direct transgenic or genetic targeted expression of the FP sensor. The principal advantage being, reproducible preparation for experimental convenience and for a greater certainty in establishing what is stained (and – equally important – what is not). With such genetic models, the precise targeting allows optical signals to be reliably attributed to specific subsets of cells, while also allowing imaging to be performed without optically resolving the labeled cells (low noise bright-field imaging techniques). One can attribute the recorded signals to specific cellular elements even when the optical configuration does not resolve the elements themselves.

2.6 Genetically Encoded Sensors of Membrane Potential Compared to Alternative Targeting Approaches

In addition to entirely genetically encoded voltage sensors, there are two novel approaches that use genetics in combination with conventional organic chromophores. The first is a hybrid approach which combines the advantages of conventional voltage-sensitive dyes (relatively large ΔF/F) with genetic targeting (Chanda et al. 2005; DiFranco et al. 2007). The hybrid probe hVOS employs FRET between membrane-anchored GFP and a dye, dipicrylamine, that partitions into the plasma membrane in a membrane voltage-dependent manner. While this hybrid probe produces the largest yet reported ΔF/F values among all (at least partially) genetically encodable voltage-sensitive probes, the required high concentration of the lipophilic anion, dipicrylamine, has a significant effect on the activity of excitable membranes (Chanda et al. 2005; Zimmermann et al. 2008; Fernández et al. 1983). Thus, reduced dipicrylamine concentrations were required to obtain action potential responses (Chanda et al. 2005) resulting in much smaller fractional fluorescence changes (ΔF/F).

The second semi-genetic method uses the genetically targetable expression of an enzyme with an organic dye as a substrate. The idea is that only the enzymatically processed dye stains the plasma membrane where it functions as a conventional voltage-sensitive dye. In a proof of principle study (Hinner et al. 2006), a lipophilic dye was made water-soluble by the addition of phosphate groups to the hydrophobic tail. Cell-specific staining was achieved by genetically expressing an outward-facing membrane-bound phosphatase that cleaves the hydrophilic phosphate group from the engineered dye resulting in a closely localized population of lipophilic product dye that integrates and stains the membrane of the specified target cells.

In its current state, the applicability in neuronal tissues is unproven. The question arises whether the phosphatase would not catalyze intrinsic processes disturbing the physiological state of the tissue, or the dye itself, would not be a substrate to intrinsic proteins resulting in unwanted staining. Also, in a tissue where the individual cells are closely packed together, the "released" dye might diffuse and bind to plasma membranes of "not-selected-for" cells that are adjacent to the cell expressing the enzyme.

2.7 Signal-to-Noise Considerations

A second, important benefit of an FP voltage sensor is an increase in the signal-to-noise ratio because undesired staining contributes to noise but not to signal. Yet, the signal-to-noise ratio remains a critical parameter. For in-vivo experiments in mammals, a low signal-to-noise ratio often results from a contamination of the sensor signal by movement induced changes in the measured light intensity (blood vessel pulsations), changes in blood volume or hemoglobin oxygenation, light scattering, or endogenous tissue fluorescence (for a recent review see Baker et al. 2005). Synchronization with breathing and/or heartbeat along with careful selection of excitation and emission wavelengths can reduce these noise signals significantly (Baker et al. 2005)

There is, however, a physical limit for the signal-to-noise ratio in optical recordings due to the statistics of photon detection (Malmstadt et al. 1974; Knöpfel et al. 2006; Sjulson and Miesenböck 2007). In the best case, the signal-to-noise ratio increases with the square root of the number of photons detected per temporal and spatial bin. Accordingly, it can be predicted that the first successful (because least demanding) application of FP voltage sensors in intact tissue will be voltage imaging from populations of subsets of neurons, as previously demonstrated with FP calcium sensors (Díez-García et al. 2005, 2007) . Spatial averaging over many neurons that have a common voltage signal will provide a larger emission intensity for detecting small modulations. The signal-to-noise ratio also increases with slower sampling rates. For these reasons along with the fact that imaging sampling frequency is limited for some imaging techniques (like two-photon laser scanning), a FP voltage sensor with somewhat slow response characteristics may be more suitable for certain applications.

2.8 Capacitative Load and Other Possible Caveats

Genetically encoded optical sensors make it possible to record from multiple cells in undamaged tissue. However, as a recombinant gene product, the sensor itself can interfere with intrinsic biochemical and/or physical cell functions. Hence, as in the case of any artificially introduced probe, FP sensor-derived data require controls to demonstrate the absence of probe-induced undesired effects. Unlike synthetic potentiometric probes, for example, cyanine, merocyanine, and oxonal dyes, which exhibit phototoxicity in some situations (Grinvald et al. 1982; Kalyanaraman et al. 1987), FP-based sensors are not known and are not expected to form toxic photo byproducts. Similar to synthetic probes, however, they add to the nonlinear dielectric polarization of the membrane resulting in increased membrane "capacitive load" (Sjulson and Miesenböck 2007). The effect is based on a straight-forward physical mechanism: the voltage sensor domain moves charges across the membrane in response to changes in membrane potential. This charge movement, manifested as a transmembrane capacitive current (well known as gating currents for ion channels), redistributes charge between the inner and outer membrane surface as a function of voltage, resulting in a voltage- and time-dependent nonlinear capacitance. Expression of voltage sensor domain-containing proteins like VSFPs, therefore, adds an extra capacitance to the membrane. Depending on its magnitude and activation time course, this extra capacitance may influence the subthreshold and spiking properties of target membranes. The physiological effect of extra mobile membrane charge can be inferred from the behavior of excitable membranes after uptake of lipophilic anions, such as dipicrylamine, causing a dose-dependent increase of membrane capacitance and, at high doses, an increase in threshold, decrement of action potential amplitude, increased action potential width, and slowing of spike propagation (Fernández et al. 1983; Chanda et al. 2005). As the biophysical mechanism is fairly basic, simulation models can be used to predict the effects of extra mobile charges on the electrical response properties (Fernández et al. 1983; DiFranco et al. 2007).

2.9 Future Directions

In this review, we focused on the approach for FP voltage sensor engineering based on a fusion between a voltage sensor protein and a FP reporter domain. However, the theoretical "ideal" FP voltage sensor would consist of only one GFP-like protein acting as both sensor and reporter. This protein would be minimally invasive since it would have the smallest possible size and minimal interaction with other organism-intrinsic proteins. To achieve that, the GFP-like molecule needs to be localized at the water-lipid interface of the plasma membrane or even largely within the lipid phase. Sensitivity to membrane voltage could be based on either voltage-dependent changes in conformation and lipid-protein interactions; or be a direct effect of the transmembrane electric field on the optical properties of the chromophore, a molecular stark effect. The challenge, however, is to change the amino acid composition of the FP molecule, so it would intrinsically translocate to the hydrophobic phase without interrupting the maturation, constitution, and existence of a functional chromophore.

FP sensors will open paths to yet unexplored territories of functional neuroimaging. In addition to the FP voltage sensors on which this review focuses, there are now many different FP sensors for monitoring biochemical processes and transmitter dynamics (Tsien 2005) at the level of neuronal circuits. Reading neuronal circuit activity can also be complemented with emerging optical methods to write activity into neuronal circuits (Miesenbock and Kevrekidis 2005; Boyden et al. 2005) In future, these techniques may be combined to yield a bidirectional optical interface to brain function.

Clearly, both the signal size and the speed of FP voltage sensors need to be improved. Attempts to express them in vivo may reveal additional unanticipated requirements. Hopefully other laboratories and we will be able to contribute to meeting these goals.

Acknowledgments Supported by NIH grants U24-NS057631, DC05259, NS050833, and MH064214, and an intramural grant from RIKEN BSI.

References

Akemann W, Lundby A, Mutoh H, Knöpel T (2009) Effect of voltage sensitive fluorescent proteins on neuronal excitability. Biophys. J. 96:3959–76

Ataka K, Pieribone VA (2002) A genetically targetable fluorescent probe of channel gating with rapid kinetics. Biophys J 82:509–516

Baird GS, Zacharias DA, Tsien RY (1999) Circular permutation and receptor insertion within green fluorescent proteins. Proc Natl Acad Sci USA 96:11241–11246

Baker BJ, Kosmidis EK, Vucinic D, Falk C, Cohen L, Djurisic M, Zecevic D (2005) Imaging brain activity with voltage- and calcium-sensitive dyes. Cell Mol Neurobiol 25:245–282

Baker BJ, Lee H, Pieribone VA, Cohen LB, Isacoff EY, Knopfel T, Kosmidis EK (2007) Three fluorescent protein voltage sensors exhibit low plasma membrane expression in mammalian cells. J Neurosci Methods 161:32–38

Boyden ES, Zhang F, Feng J, Bamberg E, Nagel G, Deisseroth K (2005) Millisecond-timescale, genetically targeted optical control of neural activity. Nat Neurosci 8:1263–1268

Brown JE, Cohen LB, De Weer P, Pinto LH, Ross WN, Salzberg BM (1975) Rapid changes in intracellular free calcium concentration. Detection by metallochromic indicator dyes in squid giant axon. Biophys J 15:1155–1160

Canepari M, Djurisic M, Zecevic D (2007) Dendritic signals from rat hippocampal CA1 pyramidal neurons during coincident pre- and post-synaptic activity: a combined voltage- and calcium-imaging study. J Physiol 580:463–484

Chanda B, Blunck R, Faria LC, Schweizer FE, Mody I, Bezanilla F (2005) A hybrid approach to measuring electrical activity in genetically specified neurons. Nat Neurosci 8:1619–1626

Davila HV, Salzberg BM, Cohen LB, Waggoner AS (1973) A large change in axon fluorescence that provides a promising method for measuring membrane potential. Nat New Biol 241:159–160

Díez-García J, Matsushita S, Mutoh H, Nakai J, Ohkura M, Yokoyama J, Dimitrov D, Knöpfel T (2005) Activation of cerebellar parallel fibers monitored in transgenic mice expressing a fluorescent Ca^{2+} indicator protein. Eur J Neurosci 22:627–635

Díez-García J, Akemann W, Knöpfel T (2007) In vivo calcium imaging from genetically specified target cells in mouse cerebellum. Neuroimage 34:859–869

DiFranco M, Capote J, Quiñonez M, Vergara JL (2007) Voltage-dependent dynamic FRET signals from the transverse tubules in mammalian skeletal muscle fibers. J Gen Physiol 130:581–600

Dimitrov D, He Y, Mutoh H, Baker BJ, Cohen L, Akemann W, Knöpfel T (2007) Engineering and characterization of an enhanced fluorescent protein voltage sensor. PLoS ONE 2(5):e440. doi:10.1371/journal.pone.0000440

Dittgen T, Nimmerjahn A, Komai S, Licznerski P, Waters J, Margrie TW, Helmchen F, Denk W, Brecht M, Osten P (2004) Lentivirus-based genetic manipulations of cortical neurons and their optical and electrophysiological monitoring in vivo. Proc Natl Acad Sci USA 101(52): 18206–18211

Feng G, Mellor RH, Bernstein M, Keller-Peck C, Nguyen QT, Wallace M, Nerbonne MN, Lichtman JW, Sanes JR (2000) Imaging neuronal subsets in transgenic mice expressing multiple spectral variants of GFP. Neuron 28:41–51

Fernández JM, Taylor RE, Bezanilla F (1983) Induced capacitance in the squid giant axon. Lipophilic ion displacement currents. J Gen Physiol 82:331–46

Grinvald A, Hildesheim R (2004) VSDI: a new era in functional imaging of cortical dynamics. Nat Rev Neurosci 5:874–85

Grinvald A, Hildesheim R, Farber IC, Anglister L (1982) Improved fluorescent probes for the measurement of rapid changes in membrane potential. Biophys J 39:301–308

Guerrero G, Siegel MS, Roska B, Loots E, Isacoff EY (2002) Tuning FlaSh: redesign of the dynamics, voltage range and color of the genetically-encoded optical sensor of membrane potential. Biophys J 83:3607–18

Hinner MJ, Hübener G, Fromherz P (2006) Genetic targeting of individual cells with a voltage-sensitive dye through enzymatic activation of membrane binding. Chembiochem 7:495–505

Jiang Y, Lee A, Chen J, Ruta V, Cadene M, Chait BT, MacKinnon R (2003a) X-ray structure of a voltage-dependent K^+ channel. Nature 423:33–41

Jiang Y, Ruta V, Chen J, Lee A, MacKinnon R (2003b) The principle of gating charge movement in a voltage-dependent K+channel. Nature 423:42–48

Kalyanaraman B, Feix JB, Sieber F, Thomas JP, Girotti AW (1987) Photodynamic action of merocyanine 540 on artificial and natural cell membranes: involvement of singlet molecular oxygen. Proc Natl Acad Sci USA 84:2999–3003

Knöpfel T, Tomita K, Shimazaki R, Sakai R (2003) Optical recordings of membrane potential using genetically targeted voltage-sensitive fluorescent proteins. Methods 30:42–48

Knöpfel T, Díez-García J, Akemann W (2006) Optical probing of neuronal circuit dynamics: genetically encoded versus classical fluorescent sensors. Trends Neurosci 29:160–166

Kohout SC, Ulbrich MH, Bell SC, Isacoff EY (2007) Subunit organization and functional transitions in Ci-VSP. Nat Struct Mol Biol 15:106–108

Livet J, Weissman TA, Kang H, Draft RW, Lu J, Bennis RA, Sanes JR, Lichtman JW (2007) Transgenic strategies for combinatorial expression of fluorescent proteins in the nervous system. Nature 450(7166):56–62

Loew LM, Cohen LB, Salzberg BM, Obaid AL, Bezanilla F (1985) Charge-shift probes of membrane potential. Characterization of aminostyrylpyridinium dyes on the squid giant axon. Biophys J 47:71–77

Long SB, Campbell EB, MacKinnon R (2005a) Crystal structure of a mammalian voltage-dependent *Shaker* family K^+ channel. Science 309:897–903

Long SB, Campbell EB, MacKinnon R, (2005b) Voltage sensor of Kv1.2: structural basis of electromechanical coupling. Science 309:903–908

Lu Z, Klem AM, Ramu Y (2001) Ion conduction pore is conserved among potassium channels. Nature 413:809–813

Lundby A, Mutoh H, Dimitrov D, Akemann W, Knöpfel T (2008) Engineering of a genetically encodable fluorescent voltage sensor exploiting fast Ci-VSP voltage-sensing movements. PLoS ONE 3 e2514

MacDonald VW, Jöbsis FF (1976) Spectrophotometric Studies on the pH of Frog Skeletal Muscle. pH Change during and after Contractile Activity. J Gen Physiol 68:179–195
Malmstadt HV, Enke CG, Crouch SR, Harlick G (1974) Electronic measurements for scientists. Benjamin, Menlo Park, CA
Miesenbock G, Kevrekidis IG (2005) Optical imaging and control of genetically designated neurons in functioning circuits. Annu Rev Neurosci 28:533–563
Murata Y, Iwasaki H, Sasaki M, Inaba K, Okamura Y (2005) Phosphoinositide phosphatase activity coupled to an intrinsic voltage sensor. Nature 435:1239–1243
Mutoh A, Perron A, Dimitrov D, Iwamoto Y, Akemann W, Chudakov DM, Knöpfel T (2009) Spectrally-resolved response properties of the three most advanced FRET based fluorescent voltage probes. PLoS ONE 4 e4555
Nakai J, Ohkura M, Imoto K (2001) A high signal-to-noise Ca^{2+} probe composed of a single green fluorescent protein. Nat Biotechnol 19:137–141
Perozo E, MacKinnon R, Bezanilla F, Stefani E (1993) Gating currents from a nonconducting mutant reveal open-closed conformations in Shaker K^+ channels. Neuron 11:353–8
Ramsey IS, Moran MM, Chong JA, Clapham DE (2006) A voltage-gated proton-selective channel lacking the pore domain. Nature 440:1213–1216
Ross WN, Salzberg BM, Cohen LB, Grinvald A, Davila HV, Waggoner AS, Wang CH (1977) Changes in absorption, fluorescence, dichroism, and birefringence in stained giant axons : optical measurement of membrane potential. J Membr Biol 33:141–183
Sakai R, Repunte-Canonigo V, Raj CD, Knöpfel T (2001) Design and characterization of a DNA-encoded, voltage-sensitive fluorescent protein. Eur J Neurosci 13:2314–8
Sasaki M, Takagi M, Okamura Y (2006) A voltage sensor-domain protein is a voltage-gated proton channel. Science 312(5773):589–592
Shimozono S, Miyawaki A (2008) Engineering FRET constructs using CFP and YFP. Methods Cell Biol. 85:381–393
Siegel MS, Isacoff EY (1997) A genetically encoded optical probe of membrane voltage. Neuron 19:735–41
Sjulson L, Miesenböck G (2007) Optical recording of action potentials and other discrete physiological events: a perspective from signal detection theory. Physiology 22:47–55
Tsien RY (2005) Building and breeding molecules to spy on cells and tumors. FEBS Lett 579:927–932
Villalba-Galea CA, Dimitrov D, Mutoh H, Lundby A, Sadntner W, Bezanilla F, Knöpfel T (2009) Charge movement of the voltage sensitive fluorescent protein. Biophys J 96(2):L19–21
Yaksi E, Friedrich RW (2006) Reconstruction of firing rate changes across neuronal populations by temporally deconvolved Ca^{2+} imaging. Nat Methods 3(5):377–83
Zecevic D (1996) Multiple spike-initiation zones in single neurons revealed by voltage-sensitive dyes. Nature 381:322–325
Zimmermann D, Kiesel M, Terpitz U, Zhou A, Reuss R, Kraus J, Schenk WA, Bamberg E, Sukhorukov VL (2008) A combined patch-clamp and electrorotation study of the voltage- and frequency-dependent membrane capacitance caused by structurally dissimilar lipophilic anions. J Membr Biol 221:107–121
Zochowski M, Wachowiak DM, Falk CX, Cohen LB, Lam Y-W, Antic S, Zecevic D (2000) Imaging membrane potential with voltage-sensitive dyes. Biol Bull 198:1–21

Chapter 3
The Influence of Astrocyte Activation on Hemodynamic Signals for Functional Brain Imaging

Hongbo Yu, James Schummers, and Mriganka Sur

Abstract Hemodynamic signals enable functional brain imaging, yet their origin and the mechanism by which they report neural activity are unresolved. Astrocytes are a major class of nonneuronal cell in the brain that receive inputs at excitatory synapses and link to the vasculature via endfeet on capillaries. Recent work utilizing in vivo high resolution cellular imaging of calcium signals in astrocytes and neurons with two-photon microscopy has revealed that astrocytes in visual cortex have sharply tuned response features that match the features of adjacent neurons. The spatially restricted, stimulus-specific, blood volume component of hemodynamic signals is exquisitely sensitive to astrocyte but not neuronal activation, demonstrating that astrocytes are responsible for a critical component of neurovascular coupling and hemodynamic signaling.

3.1 Brief Review of Hemodynamic Signals

Functional brain imaging relies heavily on evoked hemodynamic signals. In 1878, the Italian physiologist Mosso observed that brain pulsations over the right prefrontal cortex of a patient increased during the performance of a calculation task. Since then, multiple imaging techniques based on region-specific blood flow changes have been established and revealed various aspects of functional localization in the brain. Alongside, numerous studies have increasingly revealed details of the mechanisms that enable these forms of functional brain imaging. It has now become clear that when neuronal populations are active in any region of the brain, the region consumes more oxygen and energy immediately following activation. This oxygen consumption leads to an initial increase of local deoxyhemoglobin (dHb) (Frostig et al. 1990; Grinvald et al. 1999), followed by an increase in blood flow, carrying fresh blood with abundant supply of oxyhemoglobin (Hb). Often the late

M. Sur (✉)
Department of Brain and Cognitive Sciences, Picower Institute for Learning and Memory, Massachusetts Institute of Technology, Cambridge, MA, 02139, USA
e-mail: msur@mit.edu

A.W. Roe (ed.), *Imaging the Brain with Optical Methods*,
DOI 10.1007/978-1-4419-0452-2_3, © Springer Science+Business Media, LLC 2010

phase blood flow overcompensates the need for oxygen and causes the local dHb to be lower than baseline (Fox et al. 1988; Fox and Raichle 1986), followed by a slower return to baseline. This sequence of events, referred to as neurovascular coupling, forms the basis of modern functional brain imaging techniques, such as functional magnetic resonance imaging (fMRI) based on blood oxygen level-dependent (BOLD) signals, and intrinsic signal optical imaging.

3.1.1 The BOLD Signal and Its Components

Ogawa and Lee first took advantage of the fact that dHb is paramagnetic, and thus in the presence of dHb, the magnetic resonance signal decreases quadratically owing to the diffusion of field gradients with a concomitant variation in blood vessel contrast (Ogawa and Lee 1990; Ogawa et al. 1990a; Ogawa et al. 1990b). During activity-induced increases in local blood flow, called functional hyperemia, dHb concentration decreases seconds after brain activation and is reflected as an increase of BOLD signal in fMRI (Kwong et al. 1992; Turner et al. 1991). In detail, the primary physiological means by which neural activity causes changes in blood oxygenation are oxygen consumption (extraction) due to increased metabolic demand and increased blood flow which brings oxygen-saturated hemoglobin and also leads to increased local blood volume. The balance of these opposing processes in space and in time is the primary determinant of local hemoglobin oxygen concentration, and thus BOLD signal magnitude. Activity-driven oxygen extraction precedes activity-driven increases in blood flow, but the magnitude of the blood flow effect on dHb typically exceeds the oxygen extraction effect, so that the predominant signal measured is related to the increased blood flow. However, the extraction component (the so-called "initial dip") can be detected under some circumstances (Buxton 2001). Increasingly refined models of these processes have been developed (Stephan et al. 2007; Friston et al. 2000; Buxton et al. 1998) and have proved generally successful in describing these phenomena quantitatively.

Since these pioneering studies, BOLD fMRI has become one of the most prominent modalities for noninvasive imaging of neural activity in human and nonhuman subjects alike. Particularly for human studies, where more invasive measurements are typically not feasible, BOLD fMRI has played a major role in our understanding of the localization of brain activity related to specific sensory, motor, and cognitive functions.

3.1.2 Intrinsic Signal Imaging Relies on Similar Signals as BOLD

Grinvald et al. (1986) found that the small evoked reflectance changes of the exposed cortex can be employed to functionally map the barrel cortex in rats and visual cortex in cats. Brain tissue exhibits activity-dependent changes in the reflectance of specific wavelengths, which faithfully indicate neuronal activity levels. These reflectance changes depend on natural physiological processes in brain tissue and are, therefore, referred to as "intrinsic signals." Compared to BOLD fMRI, the signals of intrinsic optical imaging are more complicated, and they include the absorption of both dHb and Hb as well as light scattering (Grinvald et al. 1999). Different components of the intrinsic signal become quantitatively more prominent when imaging at different wavelengths. This wavelength dependence can be exploited to tease apart different components of the hemodynamic signal, which are presumed to arise from distinct underlying physiological processes. For example, at certain wavelengths of green light (such as 546 nm), the absorption rates of Hb and dHb are identical, and the reflectance change at this isobestic point is closely related to the overall total hemoglobin concentration (including dHb and Hb) and thus cerebral blood volume; the activity related blood volume signal results in a decrease in reflectance. At orange to red wavelengths (such as ~600–650 nm), the absorption of dHb dominates over Hb, so that the measured signals predominantly reflect the amount of dHb. Thus, the time course of the signals measured in orange light shows a biphasic curve: the initial dip from the local deoxygenation and the subsequent upward deflection from the hyperemia seconds after stimulation (Vanzetta et al. 2004). In far-red light, where hemoglobin absorption is minimal, light scattering-induced changes in reflectance dominate the overall signal (Frostig et al. 1990), whereas in green or orange light, light scattering signals contribute less than 10% of the overall reflectance change.

3.1.3 Origin and Complexity of Hemodynamic Signal Components

The broad application of functional brain imaging techniques has facilitated neuroscience research over the last several decades; however, after many years of intensive study, the origin of the imaging signal remains largely unsolved and basic questions remain. In particular, how does local neural activity influence blood flow, and what are the mechanisms that couple neural activity to hemodynamic signals? Which aspects of neural activity (presynaptic, postsynaptic, spiking, inhibitory, excitatory, etc) are most closely related to BOLD magnitude? On what spatial scale does the neurovascular coupling occur?

It is widely accepted that vascular modulation of blood flow involves the dilatation of arterioles (Cox et al. 1993; Ngai et al. 1988), but it is not clear how the signal is delivered from neurons to neighboring blood vessels. One hypothesis is that the synaptically triggered increase of postsynaptic calcium is central to the initiation of the production of vasoactive agents (Iadecola 2004). The potential vasoactive agents include extracellular diffusible hydrogen and potassium ions (Kuschinsky and Wahl 1978; Paulson and Newman 1987), nitric oxide (Dreier et al. 1995; Niwa et al. 1993), adenosine (Rubio and Berne 1975), and arachidonic acid metabolites (Niwa et al. 2000; Takano et al. 2006). In neocortex, some interneurons directly contact vascular processes, and the activation of these interneurons can evoke dilation or constriction of adjacent vessels in vitro (Cauli et al. 2004; Hamel 2006; Hirase et al. 2004a; Vaucher et al. 2000). On the other hand, recent studies are not in favor of a direct link between postsynaptic neurons and local blood flow: when the spiking activity of adjacent neurons is abolished in cerebellum (Mathiesen et al. 1998) or olfactory cortex (Petzold et al. 2008), the local blood flow does not decrease. Similarly, blood flow changes in visual cortex do not reflect alterations of neuronal activity but rather closely follow changes in astrocyte activation (Schummers et al. 2008; see also below). Furthermore, blood flow signals seem to be most closely related to local field potentials (Logothetis et al. 2001), and thus a large proportion of the hemodynamic signal appears to be linked to presynaptic potentials (Logothetis et al. 2001). It has been argued that this is consistent with the greater metabolic consumption involved in synaptic transmission compared with that in spiking (Iadecola 2004). It has been shown that over a narrow range, there is a linear relationship between the local field potentials and BOLD contrast signals (Hewson-Stoate et al. 2005). However, a predominant nonlinearity exists over a wider range, especially when using low stimulus intensities (Sheth et al. 2004). This complicated coupling between neural activity and hemodynamic signals makes the interpretation of functional brain imaging difficult and indicates an indirect pathway from neural activity to local blood flow control. A clear mechanistic explanation of the coupling will go a long way toward advancing our interpretation of hemodynamic imaging data in terms of the underlying neural activity.

3.2 Astrocytes and Their Link with Neurons and the Vasculature

Astrocytes are known to be closely linked to blood vessels. Golgi first noticed that glial cells are connected to blood vessels either directly at the soma or at the end of long processes termed endfeet. Astrocytes send processes that extend to cover nearby synapses, as well as endfeet which contact vessel walls. Thus, anatomically, astrocytes are well positioned to link neural activity to hemodynamic activity. Furthermore, each astrocyte has its own nonoverlapping territory (Bushong et al. 2002; Bushong et al. 2003), suggesting the possibility that individual astrocytes might constitute the functional unit of neurovascular coupling . However, owing to the fact that astrocytes are electrically nonspiking (i.e., they do not generate action potentials - Volterra and Meldolesi 2005), they have long been thought to be inactive cells in the brain, whose only role is to provide metabolic support to neurons. Several recent pieces of evidence have now emerged to challenge this stereotyped role of astrocytes.

3.2.1 *Synaptic Inputs to Astrocytes*

Recent advances in staining methods have demonstrated that the morphology of astrocytes is closely related to neurons. Astrocytes send processes that conjoin most excitatory cortical synapses: as many as 90% of spines in somatosensory cortex are contacted by astrocytic processes (Genoud et al. 2006). Astrocyte processes can typically span ca 200 μm and are arranged in tessellated, largely nonoverlapping domains (Bushong et al. 2003), so that a single astrocyte can make contact with more than 100,000 synapses (Bushong et al. 2002). These processes are motile on the timescale of minutes to hours. Spontaneous motility of astrocytic processes is common, and is coupled to dynamics of abutting dendritic spines (Haber et al. 2006; Hirrlinger et al. 2004). Changes in neuronal activity, in vitro or in vivo, result in remodeling of the fine structure of astrocytic processes surrounding synapses. For example, preferential stimulation of a single whisker leads to an increase in the coverage of synaptic contacts by astrocyte processes in rodent somatosensory cortex (Genoud et al. 2006). Furthermore, during development, astrocyte maturation in visual cortex correlates with the critical period for neuronal plasticity, and disruption of visual activity during this time can influence the number, morphology, and receptor expression of cortical astrocytes (Hawrylak and Greenough 1995; Muller 1990; Muller 1992; Nakadate et al. 2001). The fine structure of astrocyte morphology at synapses is likely to have important significance for synaptic transmission and plasticity. Astrocytes sense synaptic glutamate through a number of means – their processes have high concentrations of glutamate receptors and transporters, enabling them to control the kinetics of synaptic transmission by regulating the amount of glutamate available in the synaptic cleft (Anderson and Nedergaard 2003; Haydon and Carmignoto 2006).

3.2.2 *Activation of Calcium Signaling in Astrocytes*

In contrast to previous views of their role, it is now recognized that astrocytes are responsive to activity in nearby neurons. The major signature of astrocytic activation is mobilization of intracellular calcium. Astrocytes in vitro exhibit spontaneous calcium activity, which can be in the form of individual spontaneous events, oscillations, or waves. Calcium signaling can be initiated by a number of stimuli, including synaptic glutamate (Volterra and Meldolesi 2005). Brief exposure to glutamate leads to sustained calcium increases lasting several seconds following a delay or ramp time of a few seconds (Cornell-Bell et al. 1990; Porter and McCarthy 1996). In cell culture and in vitro slice preparations, calcium signals can propagate through the astrocytic network over distances of hundreds of microns under certain conditions (Cornell-Bell et al. 1990), though the long-range propagation may be pathological (Volterra and Meldolesi 2005). Furthermore, astrocyte calcium signaling can be differentially triggered by synaptic inputs from different sources, suggesting that they may be involved in processing the information content of neural activity (Perea and Araque 2005).

With few exceptions, the activity of astrocytes has been characterized in vitro. With the recent combination of a specific in vivo astrocyte marker (Nimmerjahn et al. 2004) and in vivo two-photon imaging (e.g., Schummers et al. 2008), it is now possible to monitor the activity of astrocytes in vivo. The early studies using particular astrocyte labels and cellular imaging provided support for the notion that astrocytes participate in neuronal representations and processing. Astrocytes show correlated calcium waves (Hirase et al. 2004b; Nimmerjahn et al. 2004), though the propensity of astrocytes to exhibit spontaneous calcium waves in vivo under healthy conditions is open to debate (Wang et al. 2006; Takata and Hirase 2008). It has recently been demonstrated that cortical activation by sensory stimulation (whisker stimulation) can evoke calcium responses in astrocytes mediated by mGluRs (Wang et al. 2006), though an early study demonstrated astrocyte glycogen utilization following whisker stimulation (Swanson et al. 1992). Together with the in vitro evidence above, these findings suggest that astrocytes may actively sample local synaptic inputs and interact with neuronal network activity on a fine scale.

3.3 Role of Astrocytes in Hemodynamic Signaling

3.3.1 Astrocytes and Hemodynamic Responses

It is well recognized that neural activation in the brain is closely coupled with vascular activity, and local hemodynamics provide the key mapping signals used for functional imaging methods such as intrinsic signal optical imaging and functional MRI (Vanzetta et al. 2005; Thompson et al. 2003; Sheth et al. 2004; Duong et al. 2001). However, until recently it has been unclear how the fast electrical activity of neurons is linked to the relatively slow vascular changes and hemodynamic signals. As described above, several kinds of evidence support the idea that astrocytes form a key link between neuronal activity and hemodynamic responses. The intrinsic optical mapping signal, for instance, has three major components. Among these, the flush-in of blood flow is the slowest signal component; it appears with a delay of a few seconds after stimulation, which is very similar to the response delay of astrocytes (Wang et al. 2006). Because astrocytes send processes to neighboring synapses and also endfeet to the local microvasculature (Simard et al. 2003), activation of astrocytes can directly modulate the dilatory state of local arterioles (Takano et al. 2006) by release of vasoactive substances such as cyclooxygenase, nitric oxide and ATP, and triggering of prostaglandin synthesis and arachidonic acid metabolites (reviewed in Haydon and Carmignoto 2006). Furthermore, in vivo work in mice reveals that the intracellular calcium concentration of astrocytes increases after whisker stimulation (Wang et al. 2006). Astrocytes may therefore regulate local blood delivery in an exquisite way, and thus directly modulate the late component of functional imaging signals.

However, this proposal is not without complexity since astrocytes are also thought to be linked together by strong gap junctional connections, and large scale calcium waves have been described in astrocytes in vitro, which in turn do not favor spatially restricted activation of astrocytes and their suggested highly localized control of blood flow. Thus, fundamental questions about the relationship between neuronal networks, astrocytes, and hemodynamic responses need to be answered: How closely matched are astrocyte responses to adjacent neuronal responses? Specifically, are the calcium responses of astrocyte narrowly or broadly tuned relative to those of adjacent neurons? Are the maps of astrocyte tuning properties (e.g., in visual cortex) sharply divided into subregions in space, as are neuronal maps? Are hemodynamic signals separable from astrocyte responses, and from neuronal responses?

3.3.2 Response Specificity of Astrocytes

The precise orderly mapping of orientation preference in visual cortex of higher mammals provides a model system to study the interactions among neurons, astrocytes, and hemodynamic responses. Neurons with similar preferred orientations cluster and form a columnar structure in primary visual cortex; preferred orientation of neuronal columns generally varies systematically and smoothly across cortical space (Fig. 3.1a), with several sparsely distributed focal regions, such as pinwheel centers, where preferred orientation changes rapidly on a scale of less than 50 microns (examples in Fig. 3.1d, also see Bonhoeffer and Grinvald 1991; Yu et al. 2005; Grinvald et al. 1999). The detailed structure of this mapping was first revealed by intrinsic signal imaging, which is based on indirect measurement of cortical electrophysiology (Grinvald et al. 1999; Frostig et al. 1990). It is remarkable that methods such as intrinsic signal optical imaging, based on putatively coarse hemodynamic signals, nonetheless provide reports of neuronal maps at high spatial resolution. More recently, in vivo two-photon calcium imaging has been applied in cortex (Stosiek et al. 2003; Helmchen and Denk 2005), by injecting a small amount of the calcium indicator dye OGB1 into the cortex and labeling all cells in a small volume (Fig. 3.1c). By inferring changes in neural firing rates from fluorescent readings of changes in calcium concentration, the visual response properties of large populations of individual neurons can be described. Thus, single cell resolution orientation maps can be obtained, and they prove to be highly precise and organized at the level of individual neurons, even at pinwheel centers (Fig. 3.1e, g–i, also see Ohki et al. 2006). Importantly, when orientation maps from neuronal calcium signals and hemodynamic signals from the same cortex are aligned carefully by local vascular pattern (Fig. 3.1b, c), they match very well spatially, even at pinwheel centers (Fig. 3.1d, e), suggesting a very precise neurovascular coupling. As a potential mediator of this coupling, understanding the response specificity of astrocytes with respect to neighboring neurons and local blood volume control is crucial .

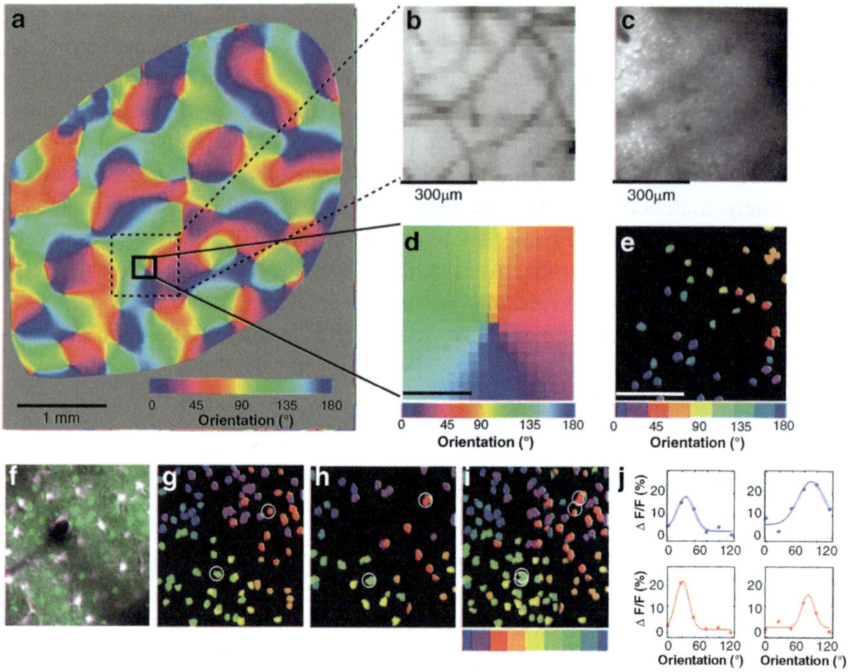

Fig. 3.1 Matched orientation preference maps revealed by intrinsic signal imaging, and two-photon imaging of astrocytes and neurons in the ferret visual cortex. (**a**) Orientation preference map generated by intrinsic signal optical imaging. (**b**) Surface blood vessel pattern captured by the CCD camera during optical imaging, covering a region 750 μm square (indicated by the *dashed box* in **a**). (**c**) Fluorescence image captured with the two-photon microscope after injection of OGB1 in the region indicated in (**b**). Note the similarity in the vascular pattern between panels **b** and **c**. (**d**) Expanded view of the orientation preference map from the small boxed area indicated in **a**. Scale bar, 100 μm. (**e**) Single cell orientation preference map of a group of neurons in the same cortical area shown in **d**. Note that the preferred orientation of the neurons closely matches that of the optical imaging signal in **d**. (**f**) Merged image of SR101 and OGB1 label in a 250 μm × 250 μm patch of cortex from a single plane120 μm below the pial surface. Astrocytes appear *white*; neurons appear *green*. Scale bar, 100 μm. (**g**) Single cell-based orientation preference map for the population of neurons labeled in (**f**). Neurons from multiple planes are included in this image. Orientation preference was determined by Gaussian fits to the data and is coded according to the scale at (**i**). (**h**) Single cell-based orientation preference map for the population of astrocytes labeled in (**f**). (**i**) Overlaid orientation preference map for neurons and astrocytes. (**j**) Example tuning curves from two neurons (*blue traces*; indicated by *circles* in **g**) and two astrocytes (*red traces*; indicated by *circles* in **h**)

For this purpose, parallel two-photon calcium imaging of neurons and astrocytes and optical imaging of blood volume changes was performed in primary visual cortex of ferrets (Schummers et al. 2008). Astrocytes were labeled by infiltration with the specific astrocyte marker SR101 (Nimmerjahn et al. 2004), while neurons and astrocytes were loaded with the calcium indicator OGB1. In this double labeled preparation, astrocytes and neurons are interleaved with each other in visual cortex

(Fig. 3.1f). Astrocytes do respond to visual stimuli, and the calcium elevation of the cell body is sharply tuned to the orientation of the drifting gratings, even at pinwheel centers (Fig. 3.1j). Importantly, the single cell resolution astrocyte orientation map is also highly organized with distinct pinwheel centers (Fig. 3.1h) that are as precise as that of neurons (Fig. 3.1g). The overlay of the two maps (Fig. 3.1i) shows that the alignment of the pinwheel center is matched perfectly between neurons and astrocytes. In summary, orientation maps of neurons, astrocytes, and hemodynamic signals coexist in primary visual cortex of ferret: astrocyte orientation preference maps resemble both neuronal and hemodynamic orientation maps with extremely high spatially resolution, consistent with the potential role of astrocytes in mediating the neurovascular coupling.

3.3.3 Role of Astrocytes in Hemodynamic Signaling

This similarity of the astrocyte map and the neuronal and hemodynamic maps is suggestive of a potential role for astrocytes in mediating the coupling between the latter. It is also noteworthy that the calcium response of astrocytes are delayed 2–4 s after visual stimulation (Fig. 3.2b; see Schummers et al. 2008), as also reported in barrel cortex (Wang et al. 2006). The timing of the onset astrocyte responses thus coincides with the onset of the hyperemic response, providing further correlative evidence for a role for astrocytes in this coupling. In order to dissect the influence of astrocytes more directly, hemodynamic orientation maps were measured while the activation of astrocytes was manipulated independently from neuronal activity by blocking astrocytic responses without interfering with neuronal synaptic transmission. One mechanism to trigger astrocyte responses is the activation of glutamate transporters (De Saint Jan and Westbrook 2005). Astrocyte glutamate transporters provide the major mechanism for glutamate clearance from the synaptic cleft, and their activity tightly regulates the amplitude and kinetics of synaptic transmission in vitro (Anderson and Swanson 2000). When the glutamate transporter antagonist DL-threo-β-benzyloxyaspartate (TBOA) was applied via a visualized pipette, the responses of astrocytes were clearly and significantly reduced (Fig. 3.2a, b). The responses of neurons were unchanged or increased to a lesser extent, and some neurons which were unresponsive in the control condition become measurably responsive during TBOA application (Fig. 3.2a). Furthermore, neuronal responses were prolonged during TBOA application (Fig. 3.2b), consistent with an increase in glutamate availability at synapses, because it is not cleared by astrocyte transporters. These data demonstrate a key role for astrocytes in regulating the strength and time course of neuronal responses to incoming synaptic inputs.

Having demonstrated that TBOA is an effective means to silence astrocytes without any potential confound from reducing neuronal responses, the effects of TBOA on stimulus-specific blood volume responses was examined by intrinsic signal imaging. It is known that intrinsic signals measured at the near-isobestic green wavelength of 546 nm are closely related to the overall hemoglobin concentration

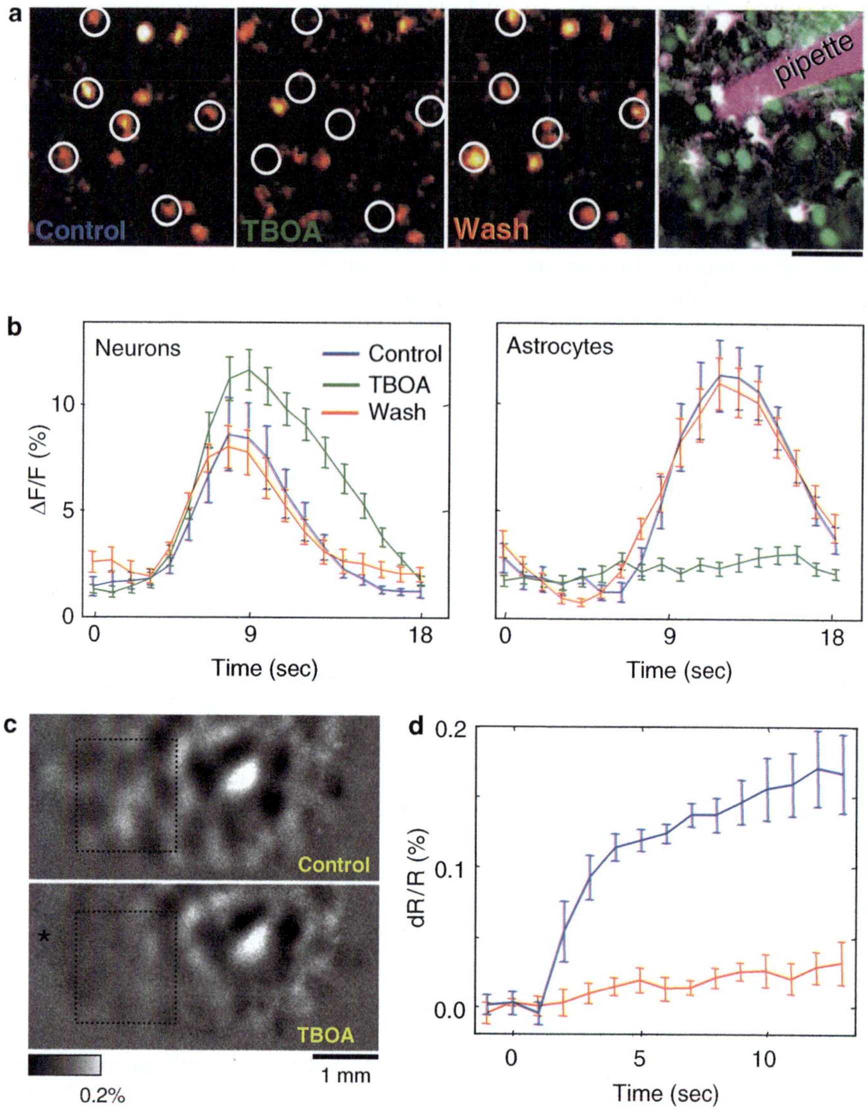

Fig. 3.2 Astrocyte calcium responses and the intrinsic optical signal are selectively affected by the glutamate transporter antagonist TBOA. (**a**) Magnitude map for visually driven responses in a field of cells, before, during and after application of TBOA. Astrocytes are circled in *white*. Position of TBOA pipette, and dual labeling of astrocytes (*white*) and neurons (*green*) are shown in rightmost panel. (**b**) Mean (± SEM) responses of a population of 13 astrocytes and 25 neurons from the same experiment as in (**a**) to a continuously changing orientation stimulus, before, during, and after TBOA application. Note that the response duration of neurons is prolonged and response magnitude slightly increased during TBOA application, while the response of astrocytes is abolished. (**c**) Differential intrinsic signal maps (derived from the response to a grating at 0 degrees minus that at 90 degrees) obtained with light of 546 nm, before and during TBOA application. TBOA was applied from a cannula positioned at the *. (**d**) Time course of the mapping signal magnitude (mean ± SEM), calculated from the portion of the map indicated by the rectangle in (**c**). TBOA nearly abolishes the mapping signal, as evident by the severe reduction of differential response (*contrast*) within the rectangle in (**c**) after TBOA application

and thus local blood volume (Grinvald et al. 1999; Frostig et al. 1990). Under green light, the differential map of two orthogonal stimulus orientations thus reflects the orientation-specific control of local blood volume. TBOA reduced these signals almost to baseline (Fig. 3.2c, d) – a striking effect, given that neuronal responses are actually increased following TBOA application (Fig. 3.2b). On an average, the mapping signal was reduced to a similar extent as the astrocyte calcium response. This demonstrates that blocking astrocyte calcium responses greatly weakens orientation-specific local blood volume regulation. It is of interest to know whether this weakening was due to a nonspecific, general, effect on the cortex caused by TBOA injection or a specific effect on blood volume control. To clarify this issue, the global intrinsic signals were evaluated simultaneously at two unique wavelengths. The global signals are derived from the "cocktail" signals of two orthogonal stimulus orientations, and they reflect the overall visually evoked intrinsic signals. At 546 nm, in the center of TBOA application site (Fig. 3.3a, star), the green light reflectance changed little compared to the control (Fig. 3.3a, left box; Fig. 3.3b, left), suggesting little blood volume increase evoked by the visual stimulus. Imaging under red light illumination (630 nm) emphasizes oximetric components of intrinsic signals. A typical visually driven reflectance change at this wavelength includes an initial decrease due to increased oxygen consumption, followed 2–4 s later by an upward deflection in reflectance, due to increased blood flow, which brings additional, oxygenated blood to offset the oxygen consumption (examples in Fig. 3.3c, blue curves, also see Grinvald et al. 1999; Frostig et al. 1990). Following TBOA application, close to the application site, the reflectance curve decreased monotonically (Fig. 3.3c, left, red curve), suggesting strong visually evoked consumption of oxygen in a well functioning cortex. However, the upward deflection which reflects blood flow increase disappeared (Fig. 3.3c, left), suggesting that the late increase in blood flow was dependent on astrocyte activity. Furthermore, the inflection point at 630 nm (Fig. 3.3c), which indicates the balance between oxygen consumption and additional oxygen brought by increased blood flow, occurred 2–4 s after visual stimulation, consistent with the delay of peak calcium responses in astrocytes (Figs. 3.2b, 3.4c, 3.5b). With increasing distance from the injection site, the signal at 546 nm became progressively stronger (Fig. 3.3b, middle and right panels), and the upward deflection in the signal at 630 nm gradually increased in magnitude (Fig. 3.3c). The amplitudes of the downward signal at 546 nm and upward component at 630 nm were well matched at each distance from the TBOA application site (Fig. 3.3b, c). Taken together, the effect of astrocyte glutamate transporter blockade by TBOA correlated well with the reduction of evoked blood volume increase, as revealed by the spatial and temporal resolution of activation of hemodynamic signals when investigated by both green and red light.

Instead of physical intervention by TBOA injection, we found that manipulation of the level of inhalation anesthesia of the animal can also specifically block the astrocyte calcium response. Under slightly higher isoflurane concentration, the neuronal responses were nearly unchanged whereas the astrocyte responses were sharply reduced (Fig. 3.4a, b). The time course and amplitude of the neuron response were nearly identical under both high and low concentrations (Fig. 3.4c),

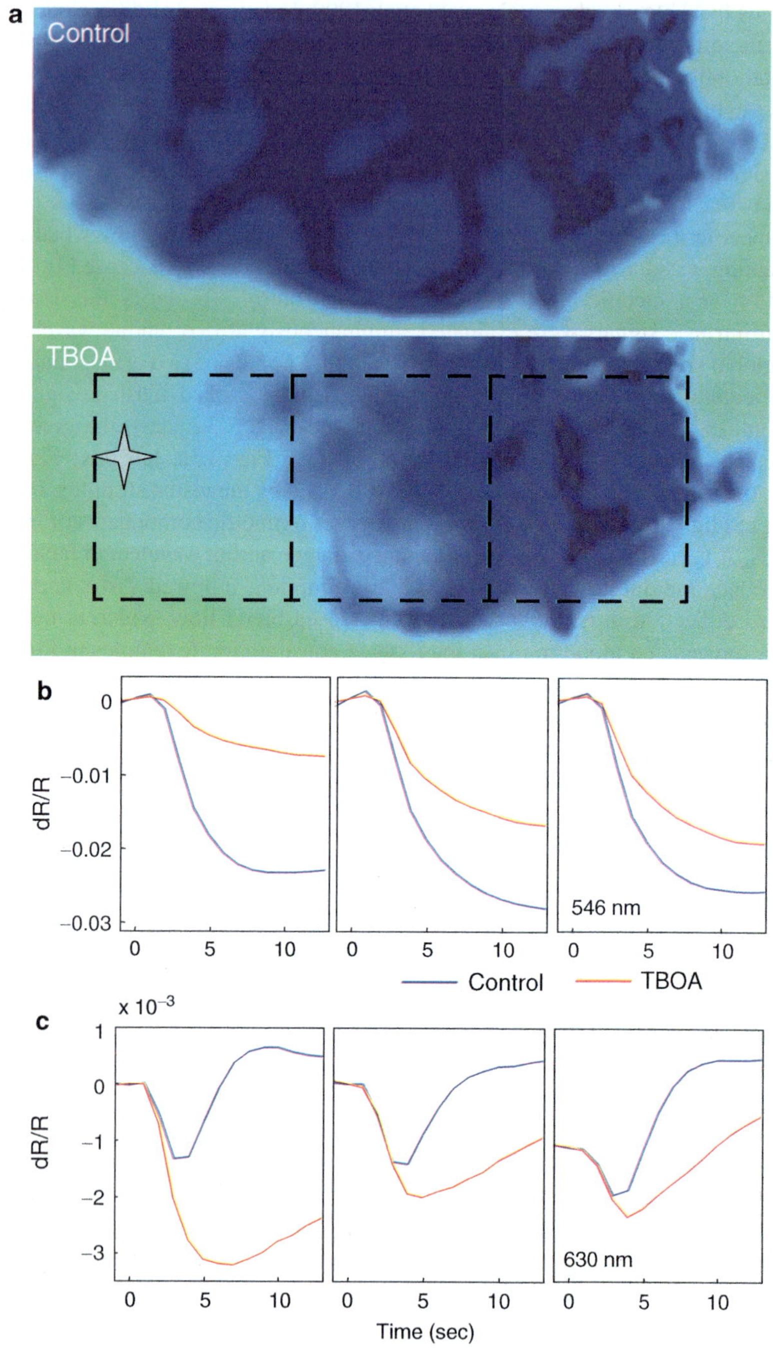

Fig. 3.3 Spatial localization and wavelength dependence of TBOA effect on global signal in the visual cortex. (**a**) Global signal map (derived from the sum of the response to a grating at 0 degrees and that at 90 degrees) in the control condition and after TBOA application. TBOA was applied

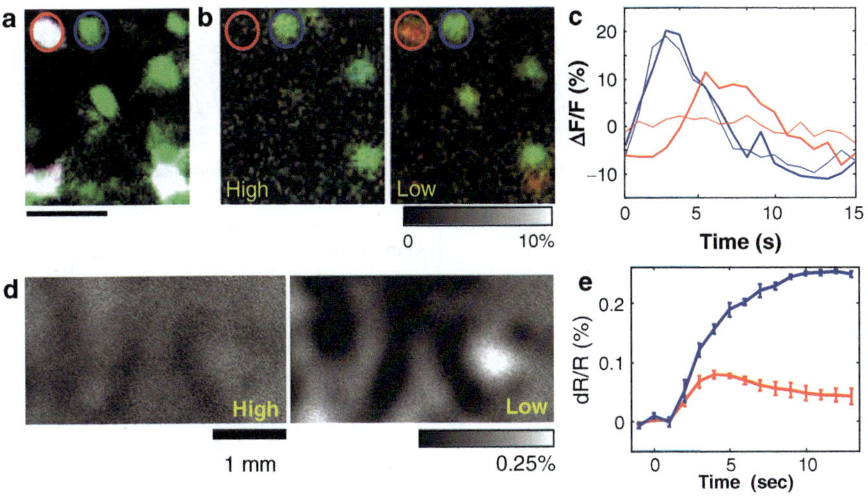

Fig. 3.4 Astrocyte calcium responses and the intrinsic optical signal are selectively affected by increased isoflurane . (**a**) Merged image of SR101 and OGB1 label in a small patch of cortex. *Blue* and *red circles* mark a neuron (*labeled green*) and an astrocyte (*labeled white*), respectively. Scale bar: 50 μm. (**b**) Cycle averaged visually evoked responses to a periodically rotating grating for the two cells in (**a**), under high (1.2%; left image) and low (0.8%; right image) concentrations of isoflurane. The images are color coded such that brightness indicates response amplitude; the amplitude bar applies to both images. (**c**) Orientation tuned responses of the two cells circled in (**a**) to a rotating grating. The neuron is plotted in *blue*, and the astrocyte in *red*. *Thick* and *thin lines* indicate low and high isoflurane, respectively. (**d**) Example of intrinsic signal optical imaging differential (0 minus 90 degrees) maps at 546 nm, computed from the response at 4–13 s after stimulation, during low and high isoflurane conditions. (**e**) Plots of the time course of reflectance change (dR/R; mean ± SEM) in the example shown in (**d**). Stimulus was turned on at time 0 s. *Blue line* depicts low isoflurane, *red line* depicts high isoflurane. Each line shows the average ± SEM of five traces under each condition

whereas the astrocyte response was nearly abolished under high concentrations (Fig. 3.4c). Consistent with the TBOA experiments, at 546 nm, concentrations of isoflurane that preferentially reduced astrocyte responses led to a large reduction in the differential orientation maps and in the mapping signal (Fig. 3.4d, e).

The dose-dependent effects of isoflurane on the responses of neurons, astrocytes, and intrinsic signals provide further evidence for the role of astrocytes in neurovascular coupling. Small changes in isoflurane concentration, over a narrow range of concentrations around ~1%, produced a modest reduction of neuronal responses but a dramatic reduction of astrocyte responses (Fig. 3.5a, b). The responses

Fig. 3.3 (continued) from a cannula positioned at the *. (**b**) Time course of global signal, measured at 546 nm, from the three regions depicted in (**a**), demonstrating a graded effect of TBOA application. The signal after TBOA (*red curve*) is reduced compared to the control signal (*blue curve*), in particular close to the application site (*left*). (**c**) Time course of global signal, measured at 630 nm. The early decrease in reflectance, indicative of oxygen consumption, is unaltered. The later overshoot, caused by increased perfusion of oxygenated blood, is reduced in a graded manner

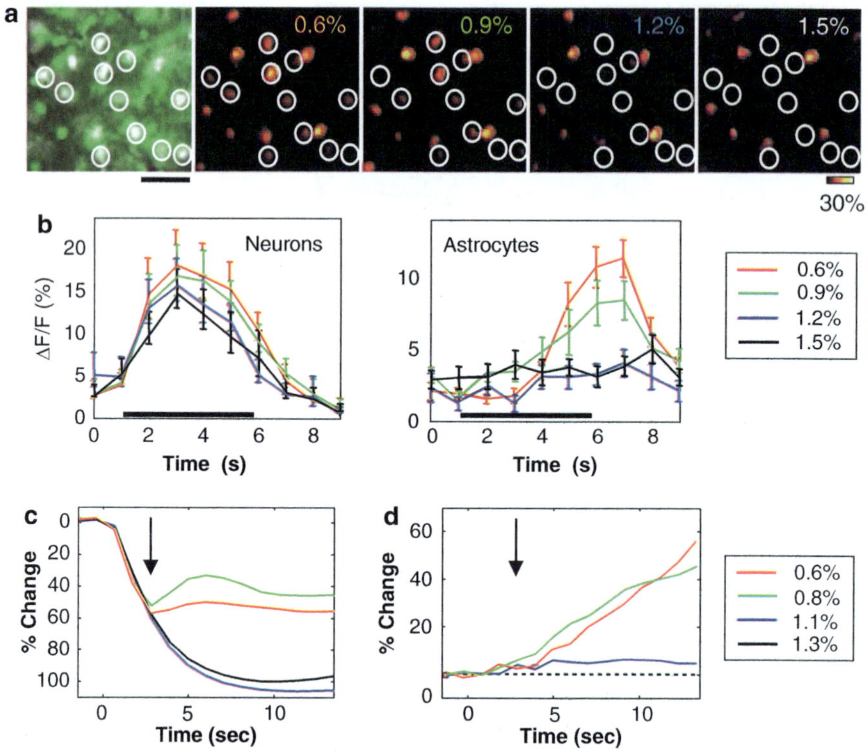

Fig. 3.5 Astrocyte but not neuronal calcium responses are nonlinearly influenced by isoflurane in a dose-dependent manner, as is the intrinsic signal. (**a**) Dose-dependent effect of isoflurane on the responses of neurons and astrocytes. The response amplitude of a field of neurons and astrocytes at different isoflurane levels (0.6%, 0.9%, 1.2%, and 1.5%). *White circles* indicate astrocytes. Scale bar: 50 μm. (**b**) The mean response time courses of the neurons ($n=8$) and astrocytes ($n=11$) from the field of view in (**a**). Stimulus time is indicated by the *black bar*. (**c**) Plot of global intrinsic signal strength at 630 nm (shown as %, normalized by the maximal reflectance change in 1.3% isoflurane condition) as a function of time after stimulus onset under four isoflurane concentrations (red, 0.6%; green, 0.8%; blue, 1.1%; black, 1.3%). (**d**) Plot of the difference in the 630 nm mapping signal strength (%, normalized by the maximal reflectance change in 1.3% isoflurane condition) between different isoflurane concentrations (black, baseline, 1.3–1.3%; blue, 1.1–1.3%; green, 0.8–1.3%; red, 0.6–1.3%). The *arrow* in (**c**) and (**d**) shows the divergent points of intrinsic signals ~3 s after stimulation, consistent with the astrocyte calcium response delay

of astrocytes were reduced in a dose-dependent manner, with a sharp fall-off between 0.9% and 1.2%. This suggests the possibility that a critical, high level of local neuronal activity is necessary to elicit astrocyte responses. Similar to the TBOA experiments at 630 nm (Fig. 3.3c), the divergence of the signal traces at low and high isoflurane occurred sharply at ~3 s (arrow in Fig. 3.5c, d), which corresponds well to the delay in astrocytes responses. Importantly, two different high isoflurane concentrations (blue and black curves) had similar effects on both global and mapping intrinsic optical imaging signals, as did two different low concentrations (red and green curves), suggesting a sharp nonlinear drop between low and high "states" around 1.0% (Fig. 3.5c, d). These effects on intrinsic signal components related to blood volume are consistent with the quasi-nonlinear effect of isoflurane on astrocyte calcium responses relative to neuronal responses.

In summary, manipulation of astrocyte responses demonstrates that in almost all aspects – including the temporal delay, spatially graded effect of TBOA injection, and dose-dependent effect of isoflurane – hemodynamic signals from local blood volume changes match astrocyte activation levels. Notably, regardless of whether the effect on neuronal responses is an increase (by TBOA) or a decrease (by isoflurane), the level of astrocyte response appears to determine the magnitude of blood volume increase, in both direct (546 nm) and indirect (630 nm) measurement. When the sharply tuned orientation selective responses of astrocytes are blocked, so that the coupling between neurons and the vasculature is broken, the mapping signal (which is highly orientation selective) is greatly decreased. These findings clearly demonstrate the key role of astrocytes as critical mediators of neurovascular coupling. The activation of neurons is necessary for obtaining strong functional imaging signals related to blood volume changes, but this influence is mediated by astrocytes and is thus secondary to astrocyte activation. The patterns of evoked blood volume signals follow those of astrocytes than neurons – a conclusion that helps elucidate some previous findings in the literature.

3.4 Conclusions and Outstanding Issues

3.4.1 Astrocytes and Neurovascular Coupling

Previously, it was believed that the blood volume component of intrinsic signals was regulated with low spatial precision, exceeding the size of individual neuronal functional modules in cortex, and therefore was less suitable for high resolution imaging than oximetric signals (Grinvald et al. 1999). However, high quality functional maps derived from blood volume changes have been obtained under green light in several intrinsic signal imaging experiments in auditory cortex (Versnel et al. 2002; Dinse et al. 1997), as well as in visual cortex as shown here. We conclude that a key variable is to keep astrocyte responses intact. Similarly, although there is still debate on whether an initial dip (from an early increase of dHb) exists in the fMRI BOLD signal (Buxton 2001), it is clear that the majority of the BOLD signal originates from the late stage hyperemia; thus the activation of astrocytes is essential for BOLD signal imaging. Intrinsic signal optical imaging differs slightly from BOLD signal imaging in that there are multiple sources for the intrinsic signal, and some components may not be regulated by astrocytes. For example, light scattering signals under red light (810 nm) decrease but still remain largely intact shortly after inactivation of astrocytes (our unpublished data), though the chronic effect remains unknown. It is notable, however, that the signal with green light is more than 40 times greater in absolute magnitude than that at 810 nm.

The coupling between neuronal activity and hemodynamic response magnitude is found to be linear only over a narrow range (Hewson-Stoate et al. 2005); strong nonlinearities are also seen, which are better described by a threshold or power law relationship (Sheth et al. 2004). These complexities can be explained at least partially by the nonlinear response properties of astrocytes, which have a higher threshold of activation than neurons (Schummers et al. 2008). Thus, the spatial precision and sharp tuning of astrocyte responses allow the spatially selective control of local blood volume of individual functional modules, and the thresholded responses of astrocytes help explain the nonlinear nature of neurovascular coupling. Therefore, the activity of astrocytes is critical for obtaining robust mapping signals for hemodynamic imaging, and manipulations (in addition to isoflurane and TBOA) that influence the functional state of astrocytes are likely to influence such imaging. It follows that maintaining astrocytes in a healthy condition is a critical step in functional brain imaging such as intrinsic signal imaging and fMRI of BOLD signals.

3.4.2 Neural Activity, Astrocyte Activity, and Hemodynamic Response Parameters

While understanding the role of astrocytes in neurovascular coupling is informative from a mechanistic point of view, a more detailed, quantitative, description of the role of astrocytes in neurovascular coupling will be essential to improve the interpretation of hemodynamic imaging data. The data described above are highly suggestive of a nonlinear, thresholded transformation between neuronal activity and astrocyte calcium responses. This transformation should be studied in more detail and parametrically. How much synaptic activity is required to elicit a measurable astrocyte response? Over what ranges is this relationship linear, and how can the nonlinearities be modeled? It is also noteworthy that calcium signaling in astrocyte processes may play a role in neurovascular coupling, which was not addressed with somatic measurements alone.

The transformation at the other end of the coupling – from astrocyte to vascular also warrants further quantitative study. How much calcium signal in an astrocyte is necessary to elicit a measurable hemodynamic response? How linear is this relationship? Ultimately, we may hope that with adequate characterization of astrocyte activation in relationship to both neural and vascular responses, we will obtain a quantitative understanding of the transformation from neural activity to the commonly measured parameters in hemodynamic imaging modalities.

3.4.3 Effects of Anesthesia on Astrocyte Responses

It is clear that anesthesia can alter both neuronal and astrocyte behavior. Our data described above, and those of others (Takano et al. 2006), suggest that astrocytes may be particularly susceptible to anesthetics. If we are to be able to apply our understanding derived from anesthetized preparations to awake, especially human, subjects, we will need to assure ourselves that the same principles apply. More work will need to be done in awake animal preparations in order to bridge this gap. Recent advances on multiple fronts (Greenberg et al. 2008; Goense and Logothetis 2008; Dombeck et al. 2007) promise to promote this effort in the near future.

References

Anderson CM, Nedergaard M (2003) Astrocyte-mediated control of cerebral microcirculation. Trends Neurosci 26:340–344 author reply 344–345

Anderson CM, Swanson RA (2000) Astrocyte glutamate transport: review of properties, regulation, and physiological functions. Glia 32:1–14

Bonhoeffer T, Grinvald A (1991) Iso-orientation domains in cat visual cortex are arranged in pinwheel-like patterns. Nature 353:429–431

Bushong EA, Martone ME, Jones YZ, Ellisman MH (2002) Protoplasmic astrocytes in CA1 stratum radiatum occupy separate anatomical domains. J Neurosci 22:183–192

Bushong EA, Martone ME, Ellisman MH (2003) Examination of the relationship between astrocyte morphology and laminar boundaries in the molecular layer of adult dentate gyrus. J Comp Neurol 462:241–251

Buxton RB (2001) The elusive initial dip. Neuroimage 13:953–958

Buxton RB, Wong EC, Frank LR (1998) Dynamics of blood flow and oxygenation changes during brain activation: the balloon model. Magn Reson Med 39:855–864

Cauli B, Tong XK, Rancillac A, Serluca N, Lambolez B, Rossier J, Hamel E (2004) Cortical GABA interneurons in neurovascular coupling: relays for subcortical vasoactive pathways. J Neurosci 24:8940–8949

Cornell-Bell AH, Finkbeiner SM, Cooper MS, Smith SJ (1990) Glutamate induces calcium waves in cultured astrocytes: long-range glial signaling. Science 247:470–473

Cox SB, Woolsey TA, Rovainen CM (1993) Localized dynamic changes in cortical blood flow with whisker stimulation corresponds to matched vascular and neuronal architecture of rat barrels. J Cereb Blood Flow Metab 13:899–913

De Saint Jan D, Westbrook GL (2005) Detecting activity in olfactory bulb glomeruli with astrocyte recording. J Neurosci 25:2917–2924

Dinse HR, Godde B, Hilger T, Reuter G, Cords SM, Lenarz T, Von Seelen W (1997) Optical imaging of cat auditory cortex cochleotopic selectivity evoked by acute electrical stimulation of a multi-channel cochlear implant. Eur J Neurosci 9:113–119

Dombeck DA, Khabbaz AN, Collman F, Adelman TL, Tank DW (2007) Imaging large-scale neural activity with cellular resolution in awake, mobile mice. Neuron 56:43–57

Dreier JP, Korner K, Gorner A, Lindauer U, Weih M, Villringer A, Dirnagl U (1995) Nitric oxide modulates the CBF response to increased extracellular potassium. J Cereb Blood Flow Metab 15:914–919

Duong TQ, Kim DS, Ugurbil K, Kim SG (2001) Localized cerebral blood flow response at submillimeter columnar resolution. Proc Natl Acad Sci USA 98:10904–10909

Fox PT, Raichle ME (1986) Focal physiological uncoupling of cerebral blood flow and oxidative metabolism during somatosensory stimulation in human subjects. Proc Natl Acad Sci USA 83:1140–1144

Fox PT, Mintun MA, Reiman EM, Raichle ME (1988) Enhanced detection of focal brain responses using intersubject averaging and change-distribution analysis of subtracted PET images. J Cereb Blood Flow Metab 8:642–653

Friston KJ, Mechelli A, Turner R, Price CJ (2000) Nonlinear responses in fMRI: the Balloon model, Volterra kernels, and other hemodynamics. Neuroimage 12:466–477

Frostig RD, Lieke EE, Ts'o DY, Grinvald A (1990) Cortical functional architecture and local coupling between neuronal activity and the microcirculation revealed by in vivo high-resolution optical imaging of intrinsic signals. Proc Natl Acad Sci USA 87:6082–6086

Genoud C, Quairiaux C, Steiner P, Hirling H, Welker E, Knott GW (2006) Plasticity of astrocytic coverage and glutamate transporter expression in adult mouse cortex. PLoS Biol 4:e343

Goense JB, Logothetis NK (2008) Neurophysiology of the BOLD fMRI signal in awake monkeys. Curr Biol 18:631–640

Greenberg DS, Houweling AR, Kerr JN (2008) Population imaging of ongoing neuronal activity in the visual cortex of awake rats. Nat Neurosci 11:749–751

Grinvald A, Lieke E, Frostig RD, Gilbert CD, Wiesel TN (1986) Functional architecture of cortex revealed by optical imaging of intrinsic signals. Nature 324:361–364

Grinvald A, Shoham D, Shmuel A, Glaser D, Vanzetta I, Shtoyerman E, Slovin H, Arieli A (1999) *In vivo* optical imaging of cortical architecture and dynamics. In: Modern techniques in neuroscience research (Windhorst U, Johansson H, eds), pp 893–969. New York: Springer.

Haber M, Zhou L, Murai KK (2006) Cooperative astrocyte and dendritic spine dynamics at hippocampal excitatory synapses. J Neurosci 26:8881–8891

Hamel E (2006) Perivascular nerves and the regulation of cerebrovascular tone. J Appl Physiol 100:1059–1064

Hawrylak N, Greenough WT (1995) Monocular deprivation alters the morphology of glial fibrillary acidic protein-immunoreactive astrocytes in the rat visual cortex. Brain Res 683:187–199

Haydon PG, Carmignoto G (2006) Astrocyte control of synaptic transmission and neurovascular coupling. Physiol Rev 86:1009–1031

Helmchen F, Denk W (2005) Deep tissue two-photon microscopy. Nat Methods 2:932–940

Hewson-Stoate N, Jones M, Martindale J, Berwick J, Mayhew J (2005) Further nonlinearities in neurovascular coupling in rodent barrel cortex. Neuroimage 24:565–574

Hirase H, Creso J, Buzsaki G (2004a) Capillary level imaging of local cerebral blood flow in bicuculline-induced epileptic foci. Neuroscience 128:209–216

Hirase H, Qian L, Bartho P, Buzsaki G (2004b) Calcium dynamics of cortical astrocytic networks in vivo. PLoS Biol 2:E96

Hirrlinger J, Hulsmann S, Kirchhoff F (2004) Astroglial processes show spontaneous motility at active synaptic terminals in situ. Eur J Neurosci 20:2235–2239

Iadecola C (2004) Neurovascular regulation in the normal brain and in Alzheimer's disease. Nat Rev Neurosci 5:347–360

Kuschinsky W, Wahl M (1978) Local chemical and neurogenic regulation of cerebral vascular resistance. Physiol Rev 58:656–689

Kwong KK, Belliveau JW, Chesler DA, Goldberg IE, Weisskoff RM, Poncelet BP, Kennedy DN, Hoppel BE, Cohen MS, Turner R (1992) Dynamic magnetic resonance imaging of human brain activity during primary sensory stimulation. Proc Natl Acad Sci USA 89:5675–5679

Logothetis NK, Pauls J, Augath M, Trinath T, Oeltermann A (2001) Neurophysiological investigation of the basis of the fMRI signal. Nature 412:150–157

Mathiesen C, Caesar K, Akgoren N, Lauritzen M (1998) Modification of activity-dependent increases of cerebral blood flow by excitatory synaptic activity and spikes in rat cerebellar cortex. J Physiol 512(Pt 2):555–566

Muller CM (1990) Dark-rearing retards the maturation of astrocytes in restricted layers of cat visual cortex. Glia 3:487–494

Muller CM (1992) Astrocytes in cat visual cortex studied by GFAP and S-100 immunocytochemistry during postnatal development. J Comp Neurol 317:309–323

Nakadate K, Imamura K, Watanabe Y (2001) Effects of monocular deprivation on the expression pattern of alpha-1 and beta-1 adrenergic receptors in the kitten visual cortex. Neurosci Res 40:155–162

Ngai AC, Ko KR, Morii S, Winn HR (1988) Effect of sciatic nerve stimulation on pial arterioles in rats. Am J Physiol 254:H133–H139

Nimmerjahn A, Kirchhoff F, Kerr JN, Helmchen F (2004) Sulforhodamine 101 as a specific marker of astroglia in the neocortex in vivo. Nat Methods 1:31–37

Niwa K, Lindauer U, Villringer A, Dirnagl U (1993) Blockade of nitric oxide synthesis in rats strongly attenuates the CBF response to extracellular acidosis. J Cereb Blood Flow Metab 13:535–539
Niwa K, Araki E, Morham SG, Ross ME, Iadecola C (2000) Cyclooxygenase-2 contributes to functional hyperemia in whisker-barrel cortex. J Neurosci 20:763–770
Ogawa S, Lee TM (1990) Magnetic resonance imaging of blood vessels at high fields: in vivo and in vitro measurements and image simulation. Magn Reson Med 16:9–18
Ogawa S, Lee TM, Kay AR, Tank DW (1990a) Brain magnetic resonance imaging with contrast dependent on blood oxygenation. Proc Natl Acad Sci USA 87:9868–9872
Ogawa S, Lee TM, Nayak AS, Glynn P (1990b) Oxygenation-sensitive contrast in magnetic resonance image of rodent brain at high magnetic fields. Magn Reson Med 14:68–78
Ohki K, Chung S, Kara P, Hubener M, Bonhoeffer T, Reid RC (2006) Highly ordered arrangement of single neurons in orientation pinwheels. Nature 442:925–928
Paulson OB, Newman EA (1987) Does the release of potassium from astrocyte endfeet regulate cerebral blood flow? Science 237:896–898
Perea G, Araque A (2005) Properties of synaptically evoked astrocyte calcium signal reveal synaptic information processing by astrocytes. J Neurosci 25:2192–2203
Petzold GC, Albeanu DF, Sato TF, Murthy VN (2008) Coupling of neural activity to blood flow in olfactory glomeruli is mediated by astrocytic pathways. Neuron 58:897–910
Porter JT, Mccarthy KD (1996) Hippocampal astrocytes in situ respond to glutamate released from synaptic terminals. J Neurosci 16:5073–5081
Rubio R, Berne RM (1975) Regulation of coronary blood flow. Prog Cardiovasc Dis 18:105–122
Schummers JR, Yu H, Sur M (2008) Tuned responses of astrocytes and their influence on hemodynamic signals in the visual cortex. Science 320:1638–1643
Sheth SA, Nemoto M, Guiou M, Walker M, Pouratian N, Toga AW (2004) Linear and nonlinear relationships between neuronal activity, oxygen metabolism, and hemodynamic responses. Neuron 42:347–355
Simard M, Arcuino G, Takano T, Liu QS, Nedergaard M (2003) Signaling at the gliovascular interface. J Neurosci 23:9254–9262
Stephan KE, Weiskopf N, Drysdale PM, Robinson PA, Friston KJ (2007) Comparing hemodynamic models with DCM. Neuroimage 38:387–401
Stosiek C, Garaschuk O, Holthoff K, Konnerth A (2003) In vivo two-photon calcium imaging of neuronal networks. Proc Natl Acad Sci USA 100:7319–7324
Swanson RA, Morton MM, Sagar SM, Sharp FR (1992) Sensory stimulation induces local cerebral glycogenolysis: Demonstration by autoradiography. Neuroscience 51:451–461
Takano T, Tian GF, Peng W, Lou N, Libionka W, Han X, Nedergaard M (2006) Astrocyte-mediated control of cerebral blood flow. Nat Neurosci 9:260–267
Takata N, Hirase H (2008) Cortical layer 1 and layer 2/3 astrocytes exhibit distinct calcium dynamics in vivo. PLoS ONE 3:e2525
Thompson JK, Peterson MR, Freeman RD (2003) Single-neuron activity and tissue oxygenation in the cerebral cortex. Science 299:1070–1072
Turner R, Le Bihan D, Moonen CT, Despres D, Frank J (1991) Echo-planar time course MRI of cat brain oxygenation changes. Magn Reson Med 22:159–166
Vanzetta I, Slovin H, Omer DB, Grinvald A (2004) Columnar resolution of blood volume and oximetry functional maps in the behaving monkey; implications for FMRI. Neuron 42:843–854
Vanzetta I, Hildesheim R, Grinvald A (2005) Compartment-resolved imaging of activity-dependent dynamics of cortical blood volume and oximetry. J Neurosci 25:2233–2244
Vaucher E, Tong XK, Cholet N, Lantin S, Hamel E (2000) GABA neurons provide a rich input to microvessels but not nitric oxide neurons in the rat cerebral cortex: a means for direct regulation of local cerebral blood flow. J Comp Neurol 421:161–171
Versnel H, Mossop JE, Mrsic-Flogel TD, Ahmed B, Moore DR (2002) Optical imaging of intrinsic signals in ferret auditory cortex: responses to narrowband sound stimuli. J Neurophysiol 88:1545–1558

Volterra A, Meldolesi J (2005) Astrocytes, from brain glue to communication elements: the revolution continues. Nat Rev Neurosci 6:626–640

Wang X, Lou N, Xu Q, Tian GF, Peng WG, Han X, Kang J, Takano T, Nedergaard M (2006) Astrocytic Ca^{2+} signaling evoked by sensory stimulation in vivo. Nat Neurosci 9:816–823

Yu H, Farley BJ, Jin DZ, Sur M (2005) The coordinated mapping of visual space and response features in visual cortex. Neuron 47:267–280

Chapter 4
Somatosensory: Imaging Tactile Perception

Li Min Chen, Robert Friedman, and Anna W. Roe

Abstract Optical imaging of intrinsic signals has extended our understanding of the functional organization of primary somatosensory cortex (SI) in primates. This chapter describes the findings which show that somatotopy, long a staple of somatosensory cortical functional organization, may not be as precise as the maps drawn from single and multiunit recordings. Optical maps of the tactile funneling illusion, which demonstrates a map of how tactile stimuli are perceived in SI rather than a map of skin topography, support the topographic representation in SI which is not a physical body map, but a perceptual map. Optical images of vibrotactile pressure, flutter, and vibration submodality domains overlaying the somatotopy reveal striking similarities and differences between other modality maps in somatosensory as revealed through electrophysiological methods and visual cortices as revealed through optical imaging methods. In an effort to extend these findings to functional MRI studies and somatic perception in humans, we compared in the same animal the maps acquired with high spatial resolution optical imaging in monkey SI with positive BOLD maps acquired at high filed (9.4T). We demonstrated that the positive BOLD fMRI maps were comparable to the fine scale OI maps in revealing both somatotopy and funneling. With the addition of high field human fMRI studies at 7 Tesla, this combination of approaches can investigate the relationship between somatotopic representation and sensory perception from the human to the modular domain and single unit level in primates.

L.M. Chen (✉)
Department of Radiology and Radiological Science, Institute of Imaging Science, Vanderbilt University, AA 1105 MCN, 1161 21st Avenue South, Nashville, TN, 37232, USA
e-mail: limin.chen@vanderbilt.edu

A.W. Roe (ed.), *Imaging the Brain with Optical Methods*,
DOI 10.1007/978-1-4419-0452-2_4, © Springer Science+Business Media, LLC 2010

4.1 Introduction

This chapter reviews some exciting new studies on primate somatosensory cortex using intrinsic signal optical imaging and high field fMRI methods. Previous electrophysiological approaches have provided a great deal of information regarding neural processes within primary somatosensory cortex (SI). However, there is limited understanding, besides the presence of a somatotopy and columnar organization, of the functional organization in SI. Optical imaging (OI) is a method well suited for revealing the functional organization of cortex. One only needs to view the highly structured functional maps observed in primary visual cortex to appreciate the role OI can play in revealing the functions of other cortical areas (review see Zepeda et al. (2004)). Viewing the functional organization of visual cortex with OI has provided significant insight into the integration and segregation of different visual perceptual channels and has inspired a large number of studies addressing these issues. The hope is that greater understanding of functional organizations in somatosensory cortex through imaging methods will similarly illuminate the mechanisms of SI in perception and behavior.

New World monkeys are an attractive species for optical imaging studies because their lissencephalic cortex allows unobstructed viewing of multiple cortical areas, especially Brodmann areas 3a, 3b, 1, and 2 (SI) (cf. for visual cortex: Malach et al. (1994); Xu et al. (2004); Roe et al. (2005)). The functional organization of SI in New World monkeys largely parallels with those in old world monkeys (cf. Sur et al. (1982); Pons et al. (1987b)). Stimulating a single digit produces activation in each of Areas 3b and 1 in a topographically predictable manner (cf. Chen et al. (2001); Chen et al. (2005); Friedman et al. (2008)). In the squirrel monkey, we have observed, with optical imaging, an orderly topographical map of the fingerpads within areas 3b and 1, consistent with the progression of maps determined with electrophysiological mapping procedures (e.g., Sur et al. (1982); Merzenich et al. (1987)). Activation sites were focal, measuring 0.5–1 mm in size. Furthermore, the physiology and anatomy of the somatosensory system have been extensively studied in New World monkeys, making them prime candidates for studies on the functional organization of cortex in sensation and behavior. For these reasons, the studies summarized in this chapter are mostly derived from studies on New World monkeys.

Ongoing efforts have been taken to develop optical imaging and functional MRI in awake behaving monkeys. To underscore the importance of these developing approaches, one of the challenges is to understand the modular basis of cognition. How do networks of modules within and across cortical areas produce our perceptions, memories, or emotions? Visualization of such modular activations during behavior is tantamount to watching the brain at work. Over the past decade, methods have been developed to visualize activations through implanted "windows on the brain" (e.g., Shtoyerman et al. (2000); Chen et al. (2002); Seidemann et al. (2002); Slovin et al. (2002)) as well as high-resolution fMRI (e.g., Van Essen et al. (2001); Tootell et al. (2003); Shmuel et al. (2006); Chen et al. (2007); Hadj-Bouziane et al. (2008); Tootell et al. (2008)). By applying these methods to the awake behaving animal, it is hoped that these methods can open new vistas in our understanding of the neural basis of cognitive function. Furthermore, the ability of these studies to forge a critical link between a large body of work on animal models and functional imaging in humans will be invaluable. There are a handful of optical imaging and fMRI studies on visual behaviors in the awake monkey (OI: Grinvald et al. (1991); Vnek et al. (1999); Shtoyerman et al. (2000); S1eidemann et al. (2002); Slovin et al. (2002); Siegel et al. (2003); fMRI: Van Essen et al. (2001); Vanduffel et al. (2001); Vanduffel et al. (2002); Tootell et al. (2003); Shmuel et al. (2006); Keliris et al. (2007); Tootell et al. (2008)). However, such studies on somesthesis in alert animal are still few (Chen et al. 2005). FMRI of SI in monkeys will provide a linkage between anatomical, electrophysiological, and optical imaging data on monkeys and the fMRI data of humans in somatic perception.

4.2 Methodology of Optical Imaging of Primary Somatosensory Cortex in New World Monkeys

4.2.1 The Somatosensory Optical Imaging Signal

Optical imaging has been used in the study of somatosensory function in rats (Sherrick 1964; Masino et al. 1993; Narayan et al. 1994; Masino and Frostig 1996; Goldreich et al. 1998; Sheth et al. 1998; Sheth et al. 2004), non-human primates (Tommerdahl et al. 1996; Tommerdahl et al. 1999; Chen et al. 2001; Shoham and Grinvald 2001; Chen et al. 2003, 2005; Tommerdahl et al. 2005a), and to some extent, in humans (Cannestra et al. 1998; Schwartz et al. 2004; Sato et al. 2005; Suh et al. 2005). In these studies, somatosensory stimulus-induced local decreases in the tissue reflectance of light are used to measure cortical responses. This intrinsic signal arises from a variety of physiological responses that accompany cortical activation; sources which include changes in blood volume due to local capillary recruitment, activity-dependent changes in the oxygen saturation level of hemoglobin (increase or decrease levels of deoxyhemoglobin), and changes in light scattering caused by signal arise from ion and water flux . The magnitude, latency, and timecourse of a reflectance change are dependent on the illumination wavelength used (Bonhoeffer and Grinvald 1996; Vanzetta et al. 2005). The intrinsic signal is correlated with both spiking and subthreshhold components of the neural response (Sheth et al. 2003; Thompson et al. 2003), for a recent review see Zepeda et al. (2004), and has been associated with the so-called "initial dip" that can be observed with the BOLD fMRI signal (e.g., Kim et al. (2000); Cannestra et al. (2001); Yacoub et al. (2001); Zarahn (2001); Fukuda et al. (2006b)). Although intrinsic imaging lacks the temporal resolution offered by voltage sensitive dyes, it is a high spatial resolution imaging method that does not require application of any external agents to the brain.

4.2.2 Relationship of Tactile Stimulation with the Optical Signal

In SI cortex, the typical intrinsic signal timecourse (under 600–630 nm illuminant, wavelengths at which the oximetry component is relatively large) peaks within 2–4 s with respect to the onset of a brief stimulus (Fig. 4.1; cf. Chen et al. 2001, 2003, 2005), which is similar in timecourse and magnitude of the intrinsic signal observed in response to a brief visual stimulus (Bonhoeffer and Grinvald 1996; Vnek et al. 1999; Roe et al. 2006). Longer illumination wavelengths (>750 nm) capture intrinsic signals dominated by changes in tissue light scattering or water diffusion. Imaging with infrared light generates similar somatotopic maps (e.g., 833 nm wavelength, cf. Tommerdahl et al. (1998); Tommerdahl et al. (1999); Tommerdahl et al. (2005b); Tommerdahl et al. (2006)), but could potentially generate different maps than those dominated by changes in oximetry. With respect to stimulus intensity dependent cortical responses in SI, studies have demonstrated a linear relationship between the amplitude of the optical signal and stimulus intensity

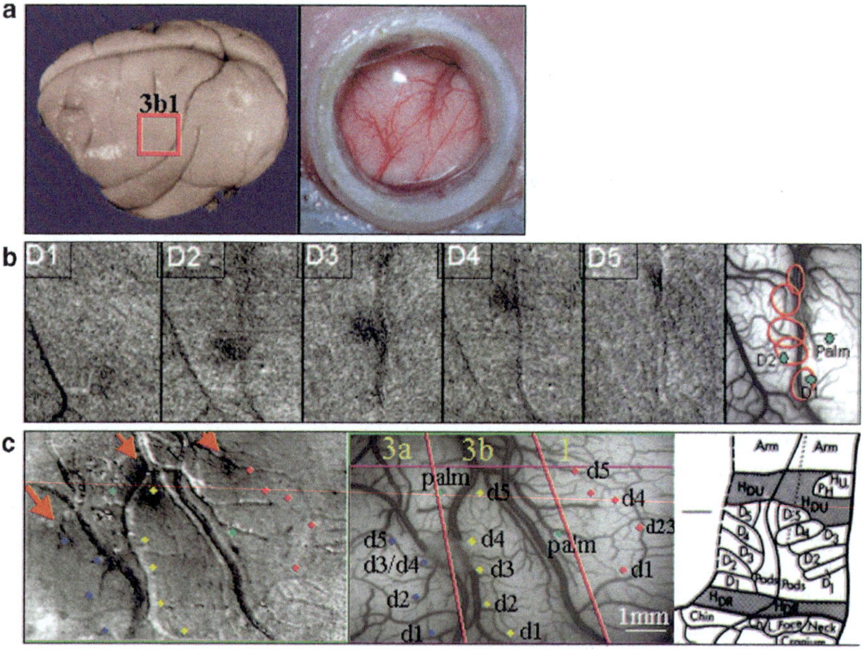

Fig. 4.1 Optical imaging of digit topography in SI of anesthetized squirrel monkey. (**a**) *Left*: Location of areas 3b and 1 in squirrel monkey brain. *Right*: Optical window over this area. (**b**) Five images obtained in response to indentation of digits D1–D5, respectively. Dark pixels indicate cortical activation. *Far right panel*: D1–D5 activation locations overlaid on blood vessel map (from Chen et al. 2001). (**c**) Imaging of Areas 3a, 3b, and 1 in response to indentation of digit D5 with a 0.2 mm dia. probe. D5 activation zones (*red arrows*) correlate well with the electrophysiological map (*dots*). *Middle Panel*: Topography and size of focal activations (0.5–1 mm) area consistent with published maps (Sur et al. 1982)

in area 3b (Chen et al. 2003; Simons et al. 2005; Friedman et al. 2008). Those graded stimulus intensities evoke graded percepts, which support the notion that cortical responses as estimated by population neural activity measures (versus single unit activity) correlate better with perception rather than responses of individual neurons. A comparable relationship is observed for stimulus duration (Simons et al. 2007). These observations support the idea that optical imaging can be used as a reliable tool to explore cortical functions of underlying perception.

4.3 Somatotopic Representation in Primary Somatosensory Cortex

4.3.1 Topography in Somatosensory Cortex

Orderly topographic sensory maps in primary somatosensory cortex (SI) serve as an anchor for our understanding of somatosensory cortical organization (Woolsey et al. 1942; Nelson et al. 1980; Sur et al. 1982; Pons et al. 1985; Pons et al. 1987a). Primate primary somatosensory cortex (SI) in the postcentral gyrus contains four complete topographic maps of the body surface that fall within the architectonically defined Brodmann's Areas 3a, 3b, 1, and 2 (Woolsey et al. 1942; Powell and Mountcastle 1959; Kaas et al. 1979; Nelson et al. 1980; Sur et al. 1982; Pons et al. 1985; Pons et al. 1987a). Areas 3b and 1 receive input primarily from cutaneous afferents whereas areas 3a and 2 receive input from deep afferents (muscle spindles and joints, e.g., Tanji and Wise (1981)). Other parietal areas, such as areas 5 and 7, also process somatosensory information (Murray and Mishkin 1984; Dong et al. 1994; Burton et al. 1997; Duhamel et al. 1998; Debowy et al. 2001). Somatotopic maps are also found laterally in SII and adjacent area PV (Burton et al. 1995; Krubitzer et al. 1995), and there are other somatosensory areas in insular cortex that receive cutaneous, nociceptive, and visceral information (Robinson and Burton 1980a, b; Schneider et al. 1993; Craig, 2003a, b).

4.3.2 Optical Imaging of Cortical Topography in Anesthetized Monkeys

A number of studies have employed the optical imaging approach to examine topographic representation in somatosensory cortex. Such studies have revealed organizations quite consistent with previous electrophysiological studies of SI topography (e.g., in rats: Masino et al. (1993); Masino and Frostig (1996); Goldreich et al. (1998); Sheth et al. (1998); in non-human primates and humans: Narayan et al. (1994); Cannestra et al. (1998); Schwartz et al. (2004)). In the squirrel monkey, studies of somatotopy have produced images of the body map (hand, leg, and foot: Tommerdahl et al. (1999)), of the distal fingerpads (Chen et al. 2001) and of the face and multiple sites along a digit (Chen et al. unpublished data). Finger topography has also been mapped in area 1 of the macaque monkey (Shoham and Grinvald 2001; Friedman et al. 2006). These images have been consistent with the previously described somatotopic maps discerned with electrophysiological recordings (e.g., Sur et al. (1982); Pons et al. (1985, 1987a, b); Nelson et al. (1980)).

Figure 4.1 shows optical imaging maps from areas 3b and 1 from the anesthetized squirrel monkey as shown in Fig. 4.1a. In Fig. 4.1b, focal activations roughly 0.5–1 mm in size are obtained following the stimulation of individual digit tips with indenting probes. Comparison of D1 (thumb) to D5 digit tip activations reveals an orderly map within Area 3b of the squirrel monkey. These maps are quite consistent with published maps (Fig. 4.1c right panel) obtained with electrophysiological methods (e.g., Sur et al. (1982); Merzenich et al. (1987)). Imaging a larger field of view reveals multiple representations, which include Areas 3a, 3b, and 1 (Fig. 4.1c). For example, stimulation of a single digit tip (D5) with a sharp probe (0.2 mm dia) elicits activations in Area 3a, 3b, and 1 respectively (Fig 4.1c, red arrows); the locations of these activations are consistent with electrophysiological recordings in this case (color dots). Thus, intrinsic signal imaging is a useful tool for quickly and clearly revealing tactile topography in primates.

4.3.3 Optical Imaging of Cortical Topography in Alert Monkeys

Our view of somatosensory topography is largely based on the data collected from the anesthetized animal. However, little is known about these topographies in the awake primate (McKenna et al. 1982; Iwamura et al. 1993; Blankenburg et al. 2003). The question remains as to whether the fundamental structure of topographic maps is the same in the anesthetized and the awake behaving animal. For example, does topography remain stable over time in the awake animal or is it dependent on behavioral context? Neither is much known regarding the relative activations of the different Brodmann areas 3a, 3b, 1 and 2 found within primary somatosensory cortex of the primate. Since different cortical areas are characterized by distinct stimulus preferences (Mountcastle and Powell 1959a, b; Hyvarinen et al. 1968; Costanzo and Gardner 1980; Carlson 1981; Sur et al. 1985; Iwamura et al. 1993), it is possible that different cortical areas become dominant under changing stimulus and behavioral contexts. Behavioral modulation of neuronal responses within SI has been confirmed by a number of studies (Nelson 1987; Nelson et al. 1991; Hsiao et al. 1993; Iwamura et al. 1995; Mesulam 1998; Burton and Sinclair 2000; Meftah el et al. 2002; Fitzgerald et al. 2006a, b; Thakur et al. 2006), however, little is known about the modulation of interareal dominance or of patterned area-specific functional organizations while an animal is performing a task. Indeed, what is the behavioral relevance of known functional organizations and topographies? Such studies would provide important constraints on understanding feedforward versus feedback influences in cortical processing (cf. Roe (2003)).

Our optical imaging study in somatosensory cortex revealed that activation patterns could differ significantly between anesthetized and awake preparations (Chen et al. 2005). Using intrinsic signal optical imaging of Areas 3b and 1, images response to vibrotactile and electrocutaneous stimuli presented to individual fingerpads in awake and anesthetized squirrel monkeys collected repeatedly for over a period of 2 years. Imaged somatotopic maps were stable over this period in both anesthetized and awake states as shown in Fig. 4.2. The image maps were also consistent with the electrophysiologically obtained topographic maps in Areas 3b and 1 recorded in the anesthetized state.

In the awake animal, optical signal sizes were greater in size and more variable than signals from an anesthetized animal, leading to larger activations (both in area and amplitude), and suggesting a less precise topography. Topographically, the map in the awake state appeared distinctly different and less precise from that in the anesthetized state. As shown in Fig. 4.3, in the awake state, the activations to D3 and D4 stimulation are highly overlapped. The significance of this finding remains to be explored. Larger activations and larger receptive field sizes are traditionally associated with less refined topography and perhaps decreased discriminability. However, the possibility remains that larger imaged activations reflect a substrate for greater cortical dynamics. Indeed, these findings are consistent with previous studies in the awake animal that have shown large receptive fields and dynamic changes in receptive field structure (such as increases in receptive field size, shifting "hot spots" or changes in spatial sensitivity profiles within receptive fields) or

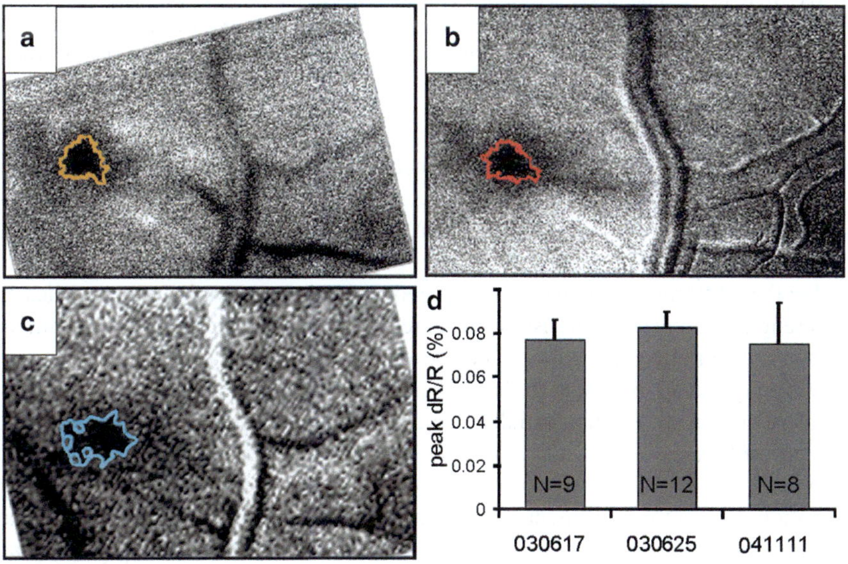

Fig. 4.2 Stability of topographic maps in the anesthetized monkey in SI (from Chen et al. 2005). Row images taken over three different imaging sessions (**a–c**). Color outlines derived from filtering and thresholding method were overlaid on the raw images to indicate the location of the greatest response. (**d**) Percentage peak amplitude changes derived from three imaging sessions

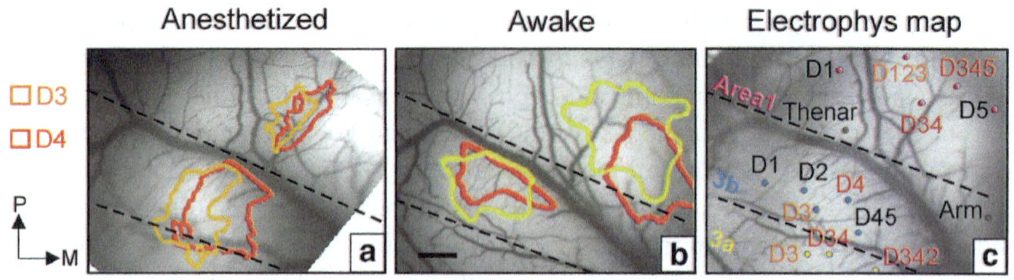

Fig. 4.3 Comparison of topographic activations in the anesthetized and awake states. Cortical response to vibrotactile stimulation of D3 (*orange*) and D4 (*red*) in the squirrel monkey. (**a**) Anesthetized activation is more prominent in Area 3b. (**b**) Awake activation is more prominent in Area 1. (**c**) Electrophysiological mapping of Areas 3a, 3b, and 1

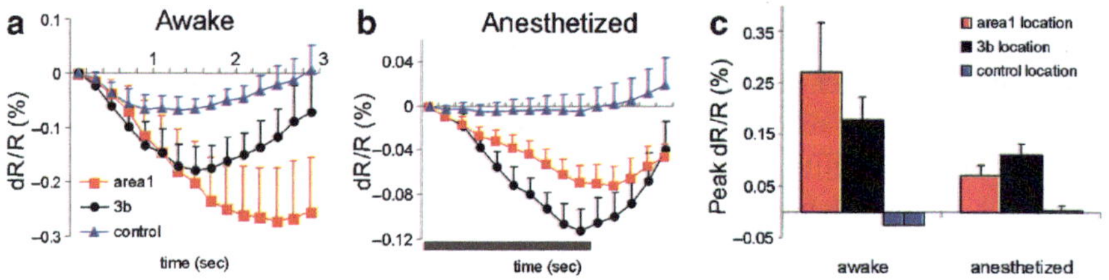

Fig. 4.4 Timecourse of optical signal to identical vibrotactile stimuli on the same digit in both awake and anesthetized conditions in one monkey. (**a, b**) Temporal profiles of the optical signal derived from the location of D4 activation in area 3b (*black line*), area 1 (*red line*), and control (*blue line*) locations in the awake (**a**) and anesthetized (**b**) states. The *gray bar* in (**b**) indicates the period of stimulus presentation. (**c**) Comparison of the absolute peak amplitude of response in area 3b (*black bar*), area 1 (*red bar*), and control (*blue bar*) locations in awake (*left*) and anesthetized (*right*) animals. Error bars represent SE

other modulation of cortical activity with behavioral context (Chapin and Lin 1984; Nelson 1987; Nicolelis et al. 1993; Schroeder et al. 1995). Thus, topographic organizations present in the anesthetized animal may be used in different ways in the awake animal.

Another difference between anesthetized and awake activation patterns lies in the strength of activations across cortical areas. Whereas in the anesthetized animal, strongest imaging signals are obtained from Area 3b, in the awake animal Area 1 activation dominate over that in Area 3b, suggesting that interareal interactions in the alert animal differ substantially from that in the anesthetized animal. In the anesthetized state, we generally observe stronger activations in Area 3b than in Area 1 (Fig. 4.4b, c). The opposite is observed in the awake state, where Area 1 activations are more prominent (Fig. 4.4a, c). However, one thing to note is that although the amplitude of optical signal in awake state was several times larger than in anesthetized state (Fig. 4.4c), the overall temporal profiles of the optical signal were similar in these two conditions. This observation suggests that anesthetics did not significantly affect the temporal characteristics of the hemodynamic response associated with the neural activities driven by the peripheral tactile stimuli.

Together, these results suggest significant awake versus anesthetized differences in somatosensory activations, probably due to differences in neural response.

4.3.4 Correlations of Optical Imaging and fMRI Maps

Many studies suggest that the optical intrinsic signal corresponds to the early negative BOLD signal (the so-called 'initial dip") (Malach et al. 1994; Toth et al. 1996; Cannestra et al. 2001; Pouratian et al. 2002; Sheth et al. 2003; Thompson et al. 2003). This early negative BOLD signal is more focal and is believed to be more closely related to the underlying neural activity. This signal, however, is quite small and not reliably detected with standard fMRI methods (Duong et al. 2000; Cannestra et al. 2001). The late positive BOLD is associated with a large influx of oxygenated blood, which is much less spatially specific. Therefore, the early signal is more spatially specific and leads to higher resolution maps than the late signal. Although alternative fMRI approaches, such as cerebral blood flow (CBF) and cerebral blood volume (CBV) methods, have been shown to be able to reveal submillimeter-sized columnar and laminar organizations of cortex (Duong et al. 2001; Fukuda et al. 2006a; Fukuda et al. 2006b; Harel et al. 2006b; Harel et al. 2006a) and retina (Cheng et al. 2006), their applications in humans have been limited because of the requirement of contrast agent administration in CBV measurements. Whether such spatial resolution can be obtained with positive BOLD signal, which is detected in most human fMRI studies, is unknown. To enable this, we deviated from common fMRI brain scan methodology. Instead of scanning in the traditional coronal, sagittal, or horizontal planes, we employed a surface coil coupled with a high field 9.4T magnet. This focuses the strongest signal over a small area of interest and at the same time permits imaging in a plane that is parallel to the cortical surface. This approach permits the viewing of different depths into cortex and also permits correlation with brain landmarks such as vascular patterns. We conducted somatosensory mapping experiments in anaesthetized squirrel monkeys with these fMRI methods. As shown in Fig. 4.5, BOLD fMRI is capable of revealing the same digit maps that were obtained with optical imaging. Positive BOLD revealed discrete and focal activation during vibrotactile stimulation of the fingerpads (Fig. 4.5a–c). The composite images (Fig. 4.5d, e) illustrates a well-organized somatotopy of D2 to D4, and the activation size and topography are consistent with previous studies (Fig. 4.5i). Changing the statistical threshold of the fMRI and optical activation maps led to relatively small changes in the area of activation and did not alter their locations. Furthermore, when the same animal is imaged with both optical imaging (Fig. 4.5f–h, j) and fMRI methods, the maps show a high degree of alignment (Chen et al. 2007). This gives us confidence that noninvasive fMRI methods (without the use of any injectable contrast agents!) can achieve high spatial resolution, and so can be used to map fine functional structures.

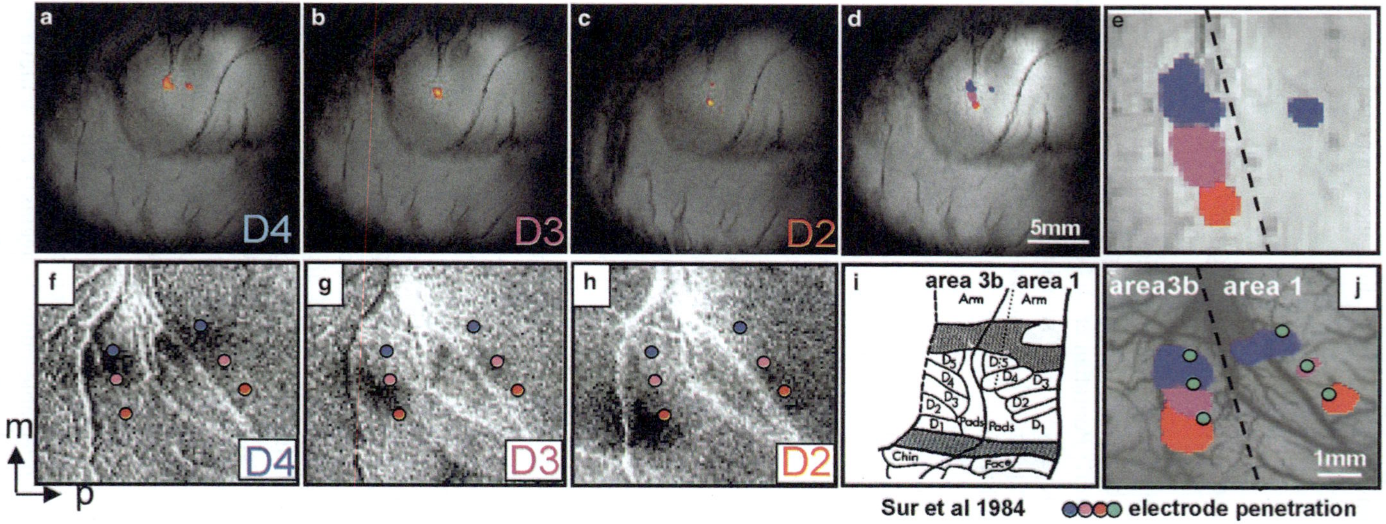

Fig. 4.5 Similar digit topography revealed by fMRI and Optical Imaging in the same animal. (**a–c**) Single-condition fMRI activation maps of vibrotactile stimulation of D4 (**a**), D3 (**b**), and D2 (**c**). (**d–e**) Composite map of digit topography and enlarged view (**e**). (**f–h**) Single-condition optical imaging activation maps of vibrotactile stimulation of D4 (**f**), D3 (**g**), and D2 (**h**). (**i**) Fingerpad topography in areas 3b and 1 closely matches the known somatopic map of squirrel monkey area 3b and area 1 (from Sur et al. 1982). (**j**) Composite activation map on blood vessel image. *m* medial; *p* posterior

4.4 Representation of Perception in Primary Somatosensory Cortex

The long-standing view that sensory topography in the somatosensory cortex reflects a "body map" is well supported. However, a recent study has called into question traditional views of somatotopic cortical maps (Chen et al., 2003). This study used the so-called "funneling illusion" to demonstrate that topographic maps in SI can reflect the perceived location of tactile stimulation rather than the location of physical stimulation on the skin. This finding suggested that, in contrast to previous views, the topographic map in Area 3b is a perceptual map rather than a physical one.

4.4.1 The Funneling Illusion

The funneling illusion is the illusory perception of skin stimulation at a single site central to a line of multiple stimulation sites (Sherrick 1964; Gardner and Spencer 1972a, b; Gardner and Tast 1981; Hashimoto et al. 1999). Inputs at lateral sites are "funneled" centrally so that perceived intensity at the central site is greater than that to stimulation at the middle site alone. With two-point stimulation, a funneled sensation is produced at a central location that is without direct stimulation. This illusion has been reported on the forearm, palm, and fingers. When two digit tips are simultaneously stimulated, subjects report sensations such as "a mound or trapezoid centered over a finger," "a mound or trapezoid bridging fingers," or "a mound bridging fingers with 2 lateral humps" (Chen et al. 2003 supplemental material). Thus, the funneling illusion is characterized by a perception of spatial mislocalization and increased tactile intensity. The neural basis for this illusion is thought to involve a complex integration of inhibitory interactions. Previous studies have shown that the funneling illusion is encoded in primary somatosensory cortex (SI), and not peripherally at the skin (Gardner and Spencer 1972a). Responses of SI neurons to three-point skin stimulation have demonstrated that a broad distribution of cortical neurons is recruited (Gardner and Spencer 1972a; Gardner and Costanzo 1980).

4.4.2 Two-Point Stimulation Produces Cortical Merging in Area 3b

The central representation of this phenomenon was not expected to be in primary somatosensory cortex. Since the illusion involves integration across multiple digits, it was unlikely to occur in Area 3b where receptive fields are confined to single digits. However, a recent study has demonstrated that this integration does indeed occur in early somatosensory areas. Figure 4.6 illustrates that stimulation of either D2 alone

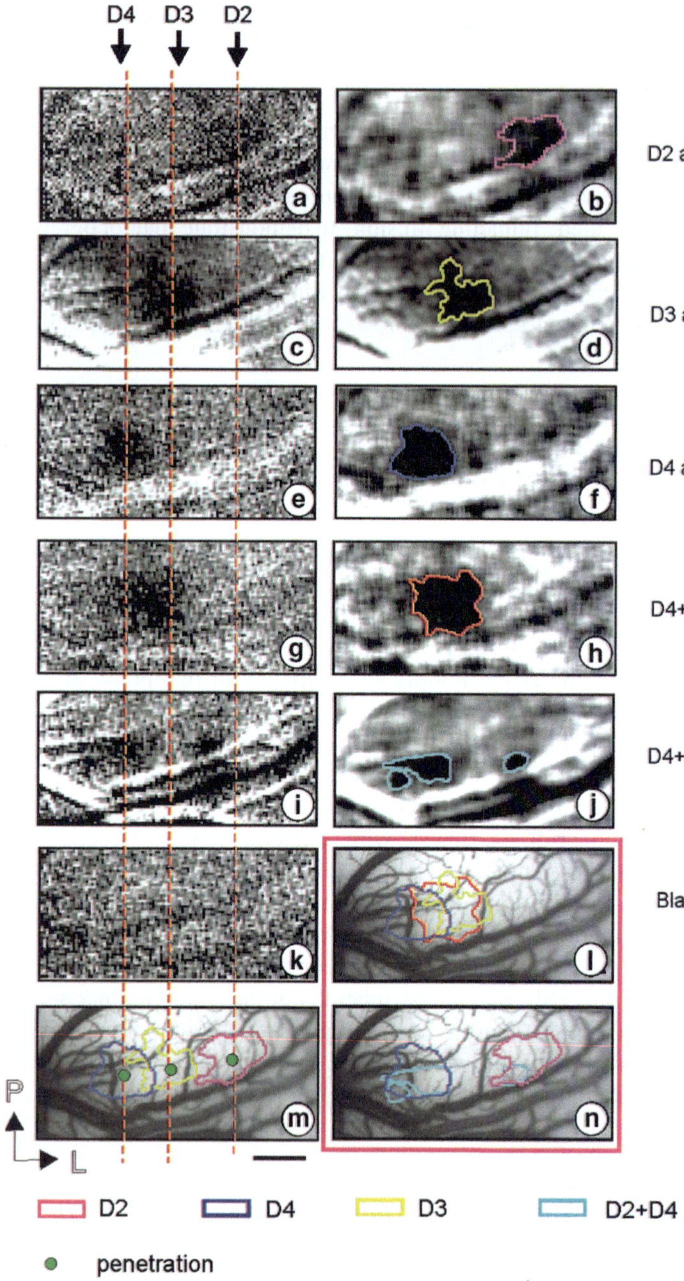

Fig. 4.6 Cortical activations by 1- and 2-digit stimulation in Area 3b. Raw images of D2, D3, D4, D3+D4, D2+D4 stimulation (**a, c, e, g, i**); low-passed images with threshold outlines (**b, d, f, h, j**). Digit activation centers are indicated by *red-dashed lines*. (**k**) Blank condition. (**l–n**) Overlay of outlines of adjacent digit (**l**), single digit (**m**), and nonadjacent digit (**n**) activations. (**m**) Penetrations show lateral to medial topographic locations of D2, D3 and D4 (*green dots*). Scale bar: 1 mm

(a, b), D3 alone (c, d) or D4 alone (e, f) elicits single focal mm-sized activations. However, stimulation of adjacent digits D4+D3 (g, h) produces a single activation site whose center lies between the D3 and D4 sites (l). This merged activation is not observed for paired stimulation of nonadjacent digits D4+D2 (i, j) (n). Thus, stimulation at each location alone activates spatially distinct locations in the cortex, but paired stimulation of adjacent skin sites produces a single centrally located peak of activation (Gardner and Spencer 1972a, b; Gardner and Costanzo 1980).

4.4.3 Intensity of Funneling Percept

The second notable characteristic of multipoint stimulation (two or more points) is the increased intensity of the funneled sensation in comparison with the stimulation of a single point. Gardner and Costanzo (1980) demonstrated that responses of cortical neurons to multipoint stimulation were often similar in magnitude to single point stimulation and were always less than the sum of component single point responses. From this, they suggested that the increased intensity is not due to greater neuronal firing rates, but rather due to the recruitment of a broader distribution of neurons. Such broader distribution can be reflected in either greater *area* of activation or greater *amplitude* of imaged signal.

In our imaging studies (Chen et al. 2003), we found that both the area and amplitude of activation produced by paired digit stimulation (both adjacent and nonadjacent) are *smaller* than the sum of the single digit activation areas. This reduction was observed in all digit pairs examined (both adjacent and nonadjacent). Thus, two finger stimulation leads to a *reduction* in activation area (for both adjacent and nonadjacent digit pairs).

Electrophysiological recording of single units also show similar reductions in activation. As shown in Fig. 4.7, two units, (one with rapidly adapting response properties (RA) and one with slowly adapting response properties (SA), were recorded with receptive fields on the D4 fingerpad. Stimulation of D4 alone (left column) revealed an early transient component and a weak late component (indicated by arrows) in both of the RA and SA cells' responses. Stimulation of either adjacent D3 or nonadjacent D2 alone produced no initial transient and relatively weak late response. Simultaneous stimulation of two adjacent fingerpads (right column, D4+D3) reduced the amplitude of the initial transient. This reduction of the initial transient was not observed with simultaneous stimulation of nonadjacent fingerpads (right, D4+D2). Thus, adjacent digit stimulation leads to weakening of digit activation as evidenced by decreased area and amplitude of optical signal, late component suppression in optical signal (not shown), and decline in size of single unit response. Schematic model is illustrated in Fig. 4.8.

We suggest that the decreased cortical activation observed during 2-digit stimulation may be due to inhibitory interdigit interactions. Lateral inhibition is a well-described phenomenon throughout the somatosensory system. Somatosensory afferent input produces both inhibition as well as excitation of central neurons,

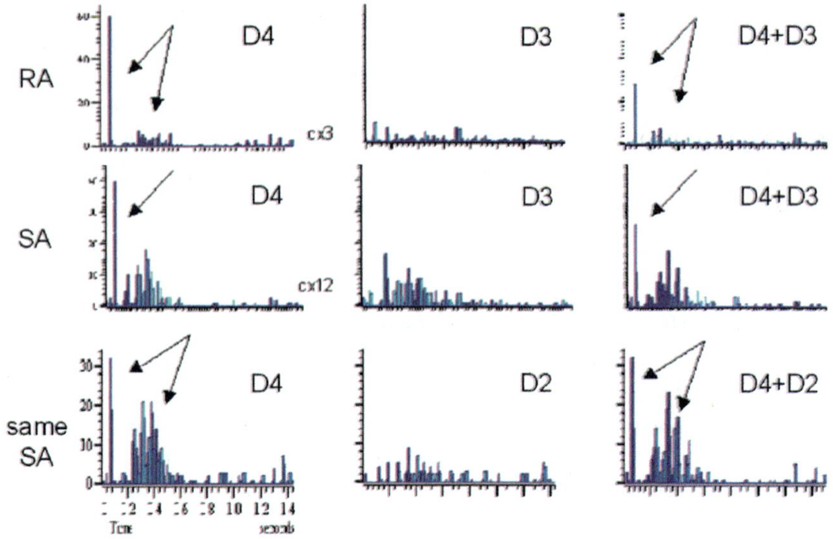

Fig. 4.7 Two units recorded on D4 fingerpad. *Top row*: RA cell response to D4, D3, and D4+D3 stimulation. *Middle row*: SA cell response to D4, D3, and D4+D3 stimulation. *Bottom row*: same SA cell response to D4, D2, and D4+D2 stimulation. 20 trials, bin width 100 ms. *Arrows* specify early and late components of response

Fig. 4.8 Summary of cortical merging. (**a**) Adjacent digit stimulation results in merging to a single central site, and increased signal amplitude at the merged location concomitant with decreased amplitudes nearby. (**b**) Nonadjacent digit stimulation results in decreased signal area and amplitude at each of the stimulation sites

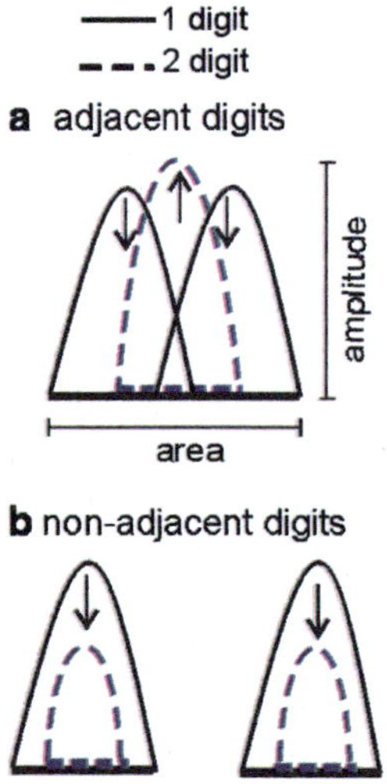

including the dorsal horn, the dorsal column nuclei, the ventrobasal complex, and the somatosensory cortex. These inhibitory processes are thought to be mediated by local feedforward or feedback circuits within the target nuclei. The spatial profile of inhibition is often larger than that of excitation, producing a "center-surround" receptive field organization (e.g., Janig et al. (1979)); this sharpening has also been referred to as "coning." Stimulus induced inhibition has longer latencies than excitation and can last for 100 ms or more (e.g., Laskin and Spencer (1979)). These lateral inhibitory processes may serve to limit the spatial extent of discharge zones and enhance stimulus contrast, thereby improving spatial and form discrimination.

In sum, simultaneous digit stimulation produces a reduced activation at the sites of digit representation. This reduction is coupled with increased response amplitude in the merged zone (where no actual stimulation occurs); this increased response at the merged zone is comparable in magnitude to that of single digit activations. In other words, even though no physical stimulus occurs at the merged site, the response is similar in size to that of an actual single digit stimulus. We speculate that a change in the balance of interaction between neurons at the merged site (increased amplitude) and those at nearby sites (decreased amplitudes) serves to heighten the perceived intensity and sharpen the focus (decreased area) during the funneling illusion (Fig. 4.6). The fact that activation of nearby cortical sites leads to single merged activations suggests that spatial percepts are strongly dictated by central representations. Indeed, physical perception can occur where no physical stimulus occurred. This study further suggests that, under certain contextual situations, receptive fields of neurons in Area 3b can span more than a single digit. The extent of these contextual influences is likely to be determined by mechanisms dependent on intracortical distance, center and surround interactions, and cortical feedback.

4.4.4 Tactile Funneling Illusion Revealed by High-Resolution fMRI

We have shown that fMRI maps of individual digits were similar to the maps revealed by optical imaging (Fig. 4.5). We then asked whether BOLD fMRI was capable of achieving *submillimeter* resolution, as we demonstrated in the tactile funneling illusion with optical imaging (Chen et al. 2003). This tactile funneling illusion became our test bed for high spatial mapping with fMRI because the center of the observed funneling activation was only 0.5 mm away from the D3 and D4 activation locations. Could BOLD fMRI detect this small submillimeter shift in activation? We, therefore, repeated this stimulation paradigm in anesthetized squirrel monkeys in the fMRI. Consistent with our optical imaging results (Fig. 4.9a–d), we found that simultaneous stimulation of D3 and D4 produced a single central focal cortical activation located roughly 0.5 mm away from the individual D3 and D4 activations (Fig. 4.5e–h). This suggests that even submillimeter spatial resolutions can be achieved with BOLD fMRI technology. Further examination of the similarity between the fMRI and OIS maps obtained in the same animal indicated that the activation patterns obtained with the two methods are same.

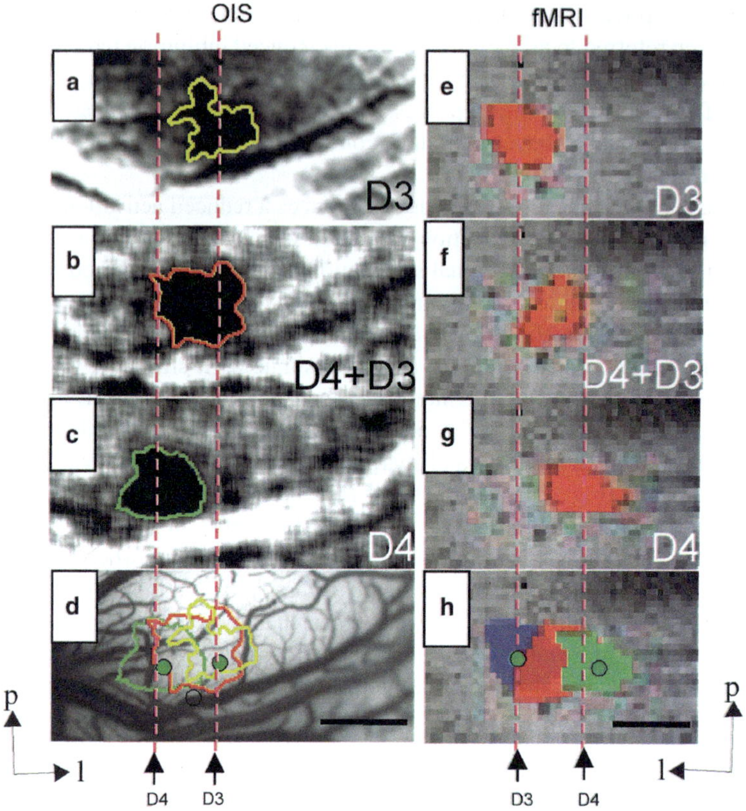

Fig. 4.9 Tactile funneling illusory response revealed by both optical imaging in cortical area 3b of the anesthetized squirrel monkey. (**a–c**) Optical imaging revealed activations in response to single (D3 **a**; and D4 **c**) and paired digits stimulation (D3+D4, **b**). (**d**) Composite image with color outlines showing the relative activation centre shift in paired digits condition. (**e–g**) FMRI revealed activations in response to single (D3 **e**, **a**; and D4 **g**) and paired digits stimulation (D3+D4, **f**). (**h**) Composite image with color patches showing the relative activation centre shift in paired digits condition. *Dotted red lines* indicate the centre locations of individual digit activations. *Green dots* indicate the locations of D3 and D4 identified by electrophysiology. *P* posterior, *L* lateral. Scale bar: 1 mm

4.5 Modality Representation in SI

4.5.1 "Labeled Lines" in Touch

Anatomical, physiological, and psychophysical evidence suggests "labeled lines" of modality-specific cutaneous information (Verrillo 1966a, b; Talbot et al. 1968; Verrillo and Bolanowski 1986). Psychophysically, vibrotactile stimuli produce three distinct sensations on the skin: (1) A pressure sensation is induced by stimuli below 4 Hz, (2) A flutter sensation is evoked by stimulation in the 4–40 Hz range, and (3) a vibration sensation is evoked by higher frequencies (40–200 Hz) (Johansson et al. 1982). These pressure, flutter, and vibratory sensations are mediated by slowly adapting (SA), rapidly adpating (RA), and pacinian (PC) receptors, respectively (Mountcastle et al. 1972; LaMotte and Mountcastle 1975; Cohen and Vierck 1993). These modalities remain largely separate in their central projections. For example, direct electrical stimulation of single identified low threshold mechanoreceptive afferents (SA, RA, or PC) can evoke only one type of perception (pressure, flutter, or vibration, respectively) (Torebjork and Ochoa 1980; Vallbo et al. 1981). Psychophysical studies fail to find vibrotactile masking and adaptation between stimulus frequencies that produce pressure (0.5 Hz), flutter (20 Hz), and vibratory sensations (200 Hz) (Gescheider et al. 1979; Gescheider et al. 1985; Bolanowski et al. 1988). In addition, frequency-specific electrical stimulation of a cortical RA-dominated site in Area 3b mimics the effect of stimulating RA receptors of the skin (Romo et al. 1998). Remarkably, even the transfer of tactile learning from one digit to another is modality-specific (Harris et al. 2001). These studies suggest a marked degree of separation in the experiences of pressure, flutter, and vibration, mediated by separate populations of receptors which remain separate in their central projections, and perhaps even to higher cortical areas involved in tactile learning and memory (Romo et al. 2000). Thus, both psychophysical and neurophysiological studies suggest some degree of modality-specific functional segregation in somatosensory cortex.

4.5.2 Presence of Interdigitated Multiple Maps

Anatomical and physiological evidence also suggests parallel modality-specific pathways, from periphery through the dorsal column nuclei, to the thalamus, and into early somatosensory cortical areas. Dykes et al. (1981) have described the segregation of RA, SA, and PC responses in the VPL and VPI. Jones and colleagues (Jones et al. 1982) have suggested that "rods" of topography and modality-specific cells project to similar modality specific bands in Area 3b, and perhaps also Area 1. Connections between Area 3b and 1 are topographically homotopic, with feedforward projections being more robust than feedback (Jones and Powell 1969; Jones et al. 1978; Cusick et al. 1985; Burton and Fabri 1995; Burton et al. 1995). Using 2-deoxyglucose labeling methods combined with anatomical tracer injections, Juliano et al. (1990) suggest that excitatory information is transmitted from area 3b to area 1 in a way that connects clusters of cells with similar response properties.

However, there is limited evidence as to whether different tactile features form multiple functional domains within each of Areas 3a, 3b, 1, 2. Electrophysiological mapping studies describe zones of neurons with SA, RA, and PC mechanoreceptor responses within Area 3b (Paul et al. 1972; Sur et al. 1981; Sretavan and Dykes 1983; Sur et al. 1984). Based on densely spaced electrode penetrations, Sur et al. (1981, 1984) found a segregation of SA and RA cells in the middle layers of Area 3b and suggested that these are organized in irregular antero-posterior "bands."

This ground-breaking work was the first to suggest the presence of multiple maps in single cortical areas in SI.

Our recent studies (Chen et al. 2001; Friedman et al. 2004) using optical imaging methods revealed networks of vibrotactile domains in SI. Vibrotactile stimulation of the digit fingerpads at frequencies that produce the sensations of pressure (1 Hz), flutter (30 Hz), and vibration (200 Hz) was used in the anesthetized squirrel monkey. These stimuli produced characteristic SA-dominated, RA-dominated, or PC-dominated responses, respectively. Intrinsic signal optical maps were obtained in response to each of these stimuli. A vector summation method was used to determine a pixel-by-pixel weighted response to the pressure, flutter, and vibration stimuli (similar to that used for visual cortical orientation maps, methodology details are described in Friedman et al. (2004). Clusters of pixels with saturated color indicate that one vector magnitude dominates the other two. Three examples of such pixel-wise SA/RA/PC vector summation are illustrated in Fig. 4.10a–c. Pixel locations with a dominant SA response appear bright red, those with a dominant RA response appear bright green, and those with a dominant PC response appear bright blue. Patches of cortex that are coded white indicate areas with strong response to each of the pressure, flutter, and vibratory stimuli. These vibrotactile domains were typically 200–300 μm in size in both Area 3b and Area 1 (Fig. 4.10). These results suggest the information channels related to pressure, flutter, and vibration retain some separation not only in area 3b but also in area 1.

Imaging using near-infrared signals has also shown a fine topography consistent with previous electrophysiological methods (Tommerdahl et al. 2002). Further research using 25 Hz flutter and 200 Hz vibratory stimuli on the palm found that flutter stimulation led to focal topographic activation in area 3b whereas vibration only led to delayed reduced activation or an inhibition in areas 3a and 3b that was supported by a reduced activation to paired stimulation (Tommerdahl et al. 1999). The apparent contradiction between a lack of activation to 200 Hz vibration in studies by Tommerdahl et al. (2005a, b) and the focal activation observed by Chen et al. (2001) and Friedman et al. (2004) in areas 3b and 1 has been reconciled by the finding that the responses to vibration are site dependent (i.e., digits vs. palm). Further research has shown that the extent and magnitude of cortical response to flutter stimuli is duration-dependent (Chiu et al. 2005; Simons et al. 2005; Simons et al. 2007), with longer (2–5 s) stimulation leading to suppression in surrounding regions. This finding parallels the improved ability to spatially localize following presentation of longer adapting stimulation (Tannan et al. 2006).

Other imaging studies performed by Tommerdahl and associates have revealed that longer tactile stimulation periods produce longer signal timecourses. When stimulus duration increased from 0.5, 1.0, 2.0, and 5.0 s, the responses got stronger and more focal and became surrounded by suppressed response regions (Tommerdahl et al. (2005a, b); Tommerdahl et al. 2006). Interestingly, the magnitude of suppression was not uniform and usually was strongest along medial to posterior axis to the maximally activated regions (Simons et al. 2007). Authors of this study suggest that these suppressed cortical regions represent skin sites that are normally costimulated during tactile exploration. This could implicate that the function of connections that link

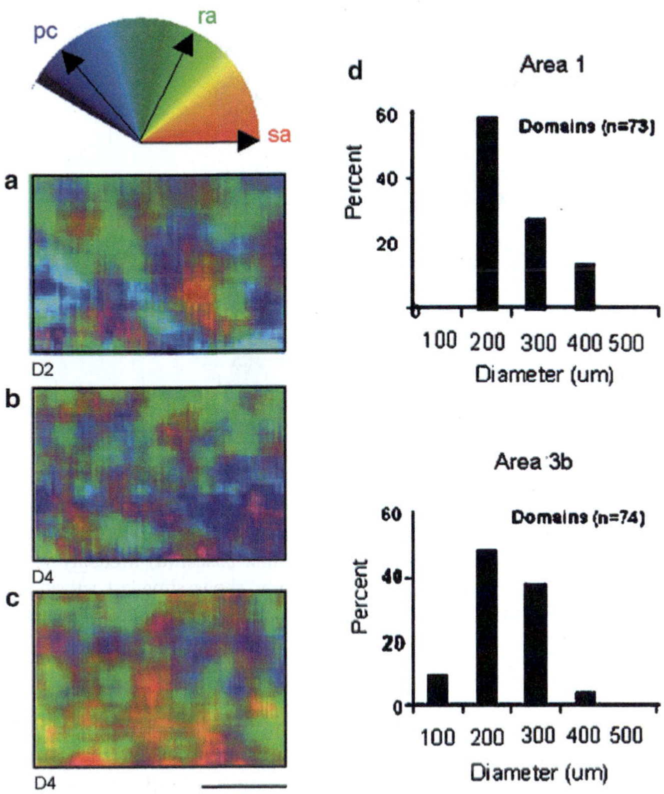

Fig. 4.10 Modality maps in SI. *Top*: Vibrotactile frequency-specific responses are plotted in different directions of color space (*red*: Sa, 1 Hz; *green*: Ra, 30 Hz; *blue*: Pc, 200 Hz). (**a–c**) Three cases: vector maps obtained through pixel by pixel vector summation. Domains dominated by single colors indicate regions preferentially responsive to a single vibrotactile frequency. Scale bar: 1 mm. (**d**) Size distribution of vibrotactile domains in Area 1 (*left*) and Area 3b (*right*)

columns in neighboring, which are somatotopically distinct regions play a crucial role in tactile acuity. In sum, the research of Tommerdahl and associates in anterior parietal cortex has revealed complex spatial and temporal responses to relatively simple tactile stimuli that might reflect the sensations in humans to the same kinds of stimuli (Tommerdahl et al. (2005a, b); Tannan et al. 2006).

4.5.3 Relationship of Vibrotactile Domains with Somatotopy

It is evident that the modality-specific response extends beyond the classically defined topographic map as revealed by simple indentation stimuli. As shown in Fig. 4.11b, an indentation stimulus (which activates all three receptors types) to digit D2 produces a fairly focal activation in both Area 3b and Area 1 (outline in

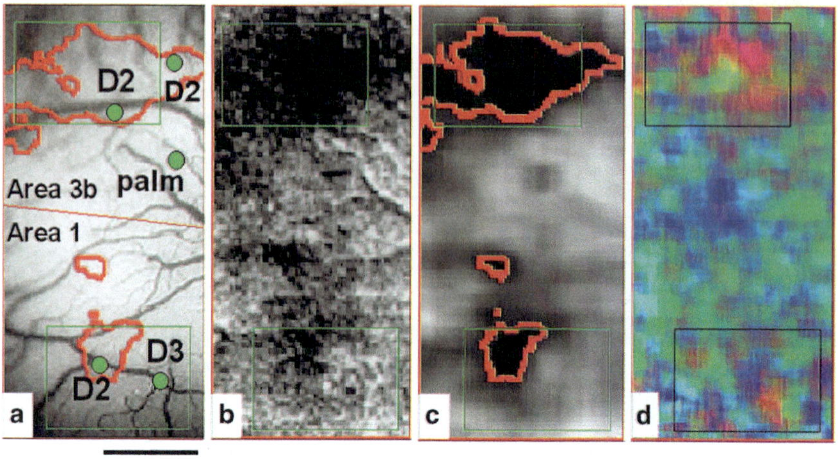

Fig. 4.11 Relationship of topography and vibrotactile response. D2 activation in Area 3b and 1. (**a**) Vessel map. D2 activations in Area 3b (*above*) and Area 1 (*below*). (**b–c**) Raw and filtered image of D2 indentation. Red outline delineates strongest activation zones. (**d**) Modality vector map (*red*, *green*, and *blue* response for SA, RA, and PC, respectively) shows strongest activation in centers of D2 activation and weaker activation outside of D2 zones. Scale bar: 1 mm (Friedman et al. 2004)

red in Fig. 4.11c). In response to pressure, flutter, and vibration stimulation, the vector summation map reveals that the strongest responses (most saturated red, green, and blue regions) correspond with the topographic digit locations. However, clustered responses, though weaker, are also evident away from the location of D2 representation (outside the boxes). This additional nontopographic activation is reminiscent of the finding in visual cortex in which complete orientation maps are obtained even though only a single eye is stimulated (Blasdel 1992). Thus, modality maps and topographic maps exhibit some degree of independence. Whether this extended nontopographic region of activation is due to spiking and/or subthreshhold activity remains to be determined.

4.6 A New Model of Functional SI Organization

The finding that somatosensory cortical domains are roughly 200–300 μm in size strengthen the view that modularity is a common organizational feature of cortical representation. Cortical domains of similar size have been described in multiple cortical areas [in V1 and V2 (see Roe 2003 for review), and V4 (Felleman et al. 1997), IT (Tsunoda et al. 2001), Area 7 (Siegel et al. 2003), and in prefrontal areas (Kritzer and Goldman-Rakic 1995). These findings suggest an update of previous views of SI organization, which were based primarily on electrophysiological recordings. Previously, each cortical "hypercolumn" was thought to contain segregated

Sa and Ra columns innervated predominantly by thalamocortical fibers of a single vibrotactile modality (Sur et al. 1981). Optical imaging evidence now suggests a modification of this view (Fig. 4.12). As shown by reconstruction of single thalamocortical arbors (Garraghty et al. 1989), inputs to SI have multiple arbors (200–300 μm in size) that span several mm of cortex (see also (Jones et al. 1982). These and other corticocortical arbors are likely to give rise to clustered activations that have been observed in 2-deoxyglucose studies (Juliano et al. 1990; cf. Burton and Fabri 1995). Such arbors could give rise either to an array of discrete clusters or, by varying arbor overlap, continuous modality maps. Thus, regions dominated by single SA, RA, or PC inputs would give rise to SA (red), RA (blue), or PC (green) domains. Regions of some overlap would appear as magenta or yellow-green colors (not depicted). And, regions of high SA, RA, and PC overlap could underlie the so-called "hot spots" commonly described in SI receptive fields (in optical images, these locations would appear as black, gray, or white domains); these would be well activated by broadband stimuli such as skin indentation. Other arbors extend to nontopographic locations away from the hot spot and establish locally some degree of modality-specific dominance (cf. Fig. 4.10). Thus, not unlike the way horizontal iso-orientation networks in V1 give rise to resulting orientation map structure, the observed maps result from overlapping horizontal networks of patchy, modality-specific dominance. In sum, in the revised view, each digit representation is served by collections of interdigitating SA, RA, and PC columns. The effect of topographic stimulation is, therefore, no longer so discrete and can, under certain stimulation conditions, have nontopographic consequences.

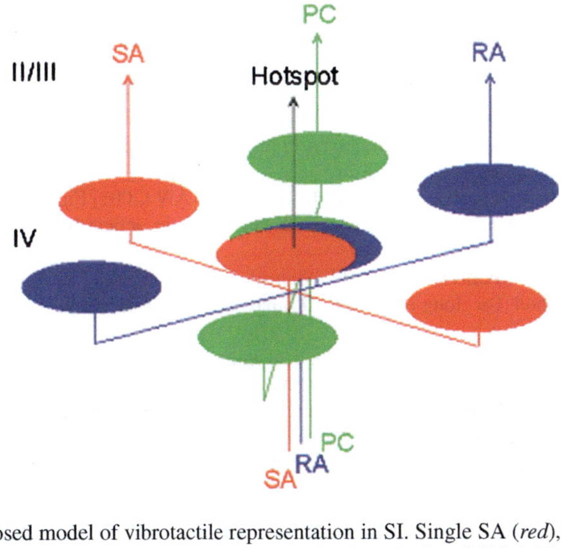

Fig. 4.12 Proposed model of vibrotactile representation in SI. Single SA (*red*), RA (*blue*), or PC (*green*) fibers terminate in layer IV, each with several arbors (*colored disks*). These arbors project in turn to superficial layers II/III (*vertical arrows*). Thus, some cortical columns are dominated by a single SA, RA, or PC input, while others have mixed input (central overlapping disks)

Acknowledgments Supported by NINDS (AWR), NIDCR (LMC), and Packard Foundation.

References

Blankenburg F, Ruben J, Meyer R, Schwiemann J, Villringer A (2003) Evidence for a rostral-to-caudal somatotopic organization in human primary somatosensory cortex with mirror-reversal in areas 3b and 1. Cereb Cortex 13:987–993

Blasdel GG (1992) Orientation selectivity, preference, and continuity in monkey striate cortex. J Neurosci 12:3139–3161

Bolanowski SJ Jr, Gescheider GA, Verrillo RT, Checkosky CM (1988) Four channels mediate the mechanical aspects of touch. J Acoust Soc Am 84:1680–1694

Bonhoeffer T, Grinvald A (1996) Optical imaging based on intrinsic signals: The methodology. Academic, London

Burton H, Fabri M (1995) Ipsilateral intracortical connections of physiologically defined cutaneous representations in areas 3b and 1 of macaque monkeys: projections in the vicinity of the central sulcus. J Comp Neurol 355:508–538

Burton H, Sinclair RJ (2000) Tactile-spatial and cross-modal attention effects in the primary somatosensory cortical areas 3b and 1–2 of rhesus monkeys. Somatosens Mot Res 17:213–228

Burton H, Fabri M, Alloway K (1995) Cortical areas within the lateral sulcus connected to cutaneous representations in areas 3b and 1: a revised interpretation of the second somatosensory area in macaque monkeys. J Comp Neurol 355:539–562

Burton H, Sinclair RJ, Hong SY, Pruett JR Jr, Whang KC (1997) Tactile-spatial and cross-modal attention effects in the second somatosensory and 7b cortical areas of rhesus monkeys. Somatosens Mot Res 14:237–267

Cannestra AF, Black KL, Martin NA, Cloughesy T, Burton JS, Rubinstein E, Woods RP, Toga AW (1998) Topographical and temporal specificity of human intraoperative optical intrinsic signals. Neuroreport 9:2557–2563

Cannestra AF, Pouratian N, Bookheimer SY, Martin NA, Beckerand DP, Toga AW (2001) Temporal spatial differences observed by functional MRI and human intraoperative optical imaging. Cereb Cortex 11:773–782

Carlson M (1981) Characteristics of sensory deficits following lesions of Brodmann's areas 1 and 2 in the postcentral gyrus of Macaca mulatta. Brain Res 204:424–430

Chapin JK, Lin CS (1984) Mapping the body representation in the SI cortex of anesthetized and awake rats. J Comp Neurol 229:199–213

Chen LM, Friedman RM, Ramsden BM, LaMotte RH, Roe AW (2001) Fine-scale organization of SI (area 3b) in the squirrel monkey revealed with intrinsic optical imaging. J Neurophysiol 86:3011–3029

Chen LM, Heider B, Williams GV, Healy FL, Ramsden BM, Roe AW (2002) A chamber and artificial dura method for long-term optical imaging in the monkey. J Neurosci Methods 113:41–49

Chen LM, Friedman RM, Roe AW (2003) Optical imaging of a tactile illusion in area 3b of the primary somatosensory cortex. Science 302:881–885

Chen LM, Friedman RM, Roe AW (2005) Optical imaging of SI topography in anesthetized and awake squirrel monkeys. J Neurosci 25:7648–7659

Chen LM, Turner GH, Friedman RM, Zhang N, Gore JC, Roe AW, Avison MJ (2007) High-resolution maps of real and illusory tactile activation in primary somatosensory cortex in individual monkeys with functional magnetic resonance imaging and optical imaging. J Neurosci 27:9181–9191

Cheng H, Nair G, Walker TA, Kim MK, Pardue MT, Thule PM, Olson DE, Duong TQ (2006) Structural and functional MRI reveals multiple retinal layers. Proc Natl Acad Sci USA 103:17525–17530

Chiu JS, Tommerdahl M, Whitsel BL, Favorov OV (2005) Stimulus-dependent spatial patterns of response in SI cortex. BMC Neurosci 6:47

Cohen RH, Vierck CJ Jr (1993) Relationships between touch sensations and estimated population responses of peripheral afferent mechanoreceptors. Exp Brain Res 94:120–130

Costanzo RM, Gardner EP (1980) A quantitative analysis of responses of direction-sensitive neurons in somatosensory cortex of awake monkeys. J Neurophysiol 43:1319–1341

Craig AD (2003a) Pain mechanisms: labeled lines versus convergence in central processing. Annu Rev Neurosci 26:1–30

Craig AD (2003b) A new view of pain as a homeostatic emotion. Trends Neurosci 26:303–307

Cusick CG, Steindler DA, Kaas JH (1985) Corticocortical and collateral thalamocortical connections of postcentral somatosensory cortical areas in squirrel monkeys: a double-labeling study with radiolabeled wheatgerm agglutinin and wheatgerm agglutinin conjugated to horseradish peroxidase. Somatosens Res 3:1–31

Debowy DJ, Ghosh S, Ro JY, Gardner EP (2001) Comparison of neuronal firing rates in somatosensory and posterior parietal cortex during prehension. Exp Brain Res 137:269–291

Dong WK, Chudler EH, Sugiyama K, Roberts VJ, Hayashi T (1994) Somatosensory, multisensory, and task-related neurons in cortical area 7b (PF) of unanesthetized monkeys. J Neurophysiol 72:542–564

Duhamel JR, Colby CL, Goldberg ME (1998) Ventral intraparietal area of the macaque: congruent visual and somatic response properties. J Neurophysiol 79:126–136

Duong TQ, Kim DS, Ugurbil K, Kim SG (2000) Spatiotemporal dynamics of the BOLD fMRI signals: toward mapping submillimeter cortical columns using the early negative response. Magn Reson Med 44:231–242

Duong TQ, Kim DS, Ugurbil K, Kim SG (2001) Localized blood flow response at submillimeter columnar resolution. Proc Natl Acad Sci USA 98:10904–10909

Dykes RW, Sur M, Merzenich MM, Kaas JH, Nelson RJ (1981) Regional segregation of neurons responding to quickly adapting, slowly adapting, deep and Pacinian receptors within thalamic ventroposterior lateral and ventroposterior inferior nuclei in the squirrel monkey (Saimiri sciureus). Neuroscience 6:1687–1692

Felleman DJ, Xiao Y, McClendon E (1997) Modular organization of occipito-temporal pathways: cortical connections between visual area 4 and visual area 2 and posterior inferotemporal ventral area in macaque monkeys. J Neurosci 17:3185–3200

Fitzgerald PJ, Lane JW, Thakur PH, Hsiao SS (2006a) Receptive field properties of the macaque second somatosensory cortex: representation of orientation on different finger pads. J Neurosci 26:6473–6484

Fitzgerald PJ, Lane JW, Thakur PH, Hsiao SS (2006b) Receptive field (RF) properties of the macaque second somatosensory cortex: RF size, shape, and somatotopic organization. J Neurosci 26:6485–6495

Friedman RM, Chen LM, Roe AW (2004) Modality maps within primate somatosensory cortex. Proc Natl Acad Sci USA 101:12724–12729

Friedman RM, Chen LM, Winberry JE, Roe AW (2006) Intrinsic and voltage dye imaging of digit activation in somatosensory area 1 of the macaque monkey. Soc Neurosci Abstr 32:446.14

Friedman RM, Chen LM, Roe AW (2008) Responses of Area 1 in anesthetized squirrel monkeys to single and dual site stimulation of the digits. J Neurophysiol 100(6):3185–3196.

Fukuda M, Moon CH, Wang P, Kim SG (2006a) Mapping iso-orientation columns by contrast agent-enhanced functional magnetic resonance imaging: reproducibility, specificity, and evaluation by optical imaging of intrinsic signal. J Neurosci 26:11821–11832

Fukuda M, Wang P, Moon CH, Tanifuji M, Kim SG (2006b) Spatial specificity of the enhanced dip inherently induced by prolonged oxygen consumption in cat visual cortex: implication for columnar resolution functional MRI. Neuroimage 30:70–87

Gardner EP, Costanzo RM (1980) Spatial integration of multiple-point stimuli in primary somatosensory cortical receptive fields of alert monkeys. J Neurophysiol 43:420–443

Gardner EP, Spencer WA (1972a) Sensory funneling. II. Cortical neuronal representation of patterned cutaneous stimuli. J Neurophysiol 35:954–977

Gardner EP, Spencer WA (1972b) Sensory funneling. I. Psychophysical observations of human subjects and responses of cutaneous mechanoreceptive afferents in the cat to patterned skin stimuli. J Neurophysiol 35:925–953

Gardner EP, Tast JM (1981) Psychophysical measurements of perceived intensity of single-point and multiple-point cutaneous stimuli in humans and subhuman primates. J Neurophysiol 46:479–495

Garraghty PE, Pons TP, Sur M, Kaas JH (1989) The arbors of axons terminating in middle cortical layers of somatosensory area 3b in owl monkeys. Somatosens Mot Res 6:401–411

Gescheider GA, Frisina RD, Verrillo RT (1979) Selective adaptation of vibrotactile thresholds. Sens Processes 3:37–48

Gescheider GA, Sklar BF, Van Doren CL, Verrillo RT (1985) Vibrotactile forward masking: psychophysical evidence for a triplex theory of cutaneous mechanoreception. J Acoust Soc Am 78:534–543

Goldreich D, Peterson BE, Merzenich MM (1998) Optical imaging and electrophysiology of rat barrel cortex. II. Responses to paired-vibrissa deflections. Cereb Cortex 8:184–192

Grinvald A, Frostig RD, Siegel RM, Bartfeld E (1991) High-resolution optical imaging of functional brain architecture in the awake monkey. Proc Natl Acad Sci USA 88:11559–11563

Hadj-Bouziane F, Bell AH, Knusten TA, Ungerleider LG, Tootell RB (2008) Perception of emotional expressions is independent of face selectivity in monkey inferior temporal cortex. Proc Natl Acad Sci USA 105:5591–5596

Harel N, Ugurbil K, Uludag K, Yacoub E (2006a) Frontiers of brain mapping using MRI. J Magn Reson Imaging 23:945–957

Harel N, Lin J, Moeller S, Ugurbil K, Yacoub E (2006b) Combined imaging-histological study of cortical laminar specificity of fMRI signals. Neuroimage 29:879–887

Harris IM, Harris JA, Caine D (2001) Object orientation agnosia: a failure to find the axis? J Cogn Neurosci 13:800–812

Hashimoto I, Yoshikawa K, Kimura T (1999) Sensory funneling of liminal multiple- point air-puff stimulation produces dramatic reduction in reaction time but relatively invariant P300 somatosensory evoked potential. Neuroreport 10:3201–3205

Hsiao SS, O'Shaughnessy DM, Johnson KO (1993) Effects of selective attention on spatial form processing in monkey primary and secondary somatosensory cortex. J Neurophysiol 70:444–447

Hyvarinen J, Sakata H, Talbot WH, Mountcastle VB (1968) Neuronal coding by cortical cells of the frequency of oscillating peripheral stimuli. Science 162:1130–1132

Iwamura Y, Tanaka M, Sakamoto M, Hikosaka O (1993) Rostrocaudal gradients in the neuronal receptive field complexity in the finger region of the alert monkey's postcentral gyrus. Exp Brain Res 92:360–368

Iwamura Y, Tanaka M, Hikosaka O, Sakamoto M (1995) Postcentral neurons of alert monkeys activated by the contact of the hand with objects other than the monkey's own body. Neurosci Lett 186:127–130

Janig W, Spencer WA, Younkin SG (1979) Spatial and temporal features of afferent inhibition of thalamocortical relay cells. J Neurophysiol 42:1450–1460

Johansson RS, Landstrom U, Lundstrom R (1982) Responses of mechanoreceptive afferent units in the glabrous skin of the human hand to sinusoidal skin displacements. Brain Res 244:17–25

Jones EG, Powell TP (1969) Connexions of the somatic sensory cortex of the rhesus monkey. II. Contralateral cortical connexions. Brain 92:717–730

Jones EG, Coulter JD, Hendry SH (1978) Intracortical connectivity of architectonic fields in the somatic sensory, motor and parietal cortex of monkeys. J Comp Neurol 181:291–347

Jones EG, Friedman DP, Hendry SH (1982) Thalamic basis of place- and modality-specific columns in monkey somatosensory cortex: a correlative anatomical and physiological study. J Neurophysiol 48:545–568

Juliano SL, Friedman DP, Eslin DE (1990) Corticocortical connections predict patches of stimulus-evoked metabolic activity in monkey somatosensory cortex. J Comp Neurol 298:23–39

Kaas JH, Nelson RJ, Sur M, Lin CS, Merzenich MM (1979) Multiple representations of the body within the primary somatosensory cortex of primates. Science 204:521–523
Keliris GA, Shmuel A, Ku SP, Pfeuffer J, Oeltermann A, Steudel T, Logothetis NK (2007) Robust controlled functional MRI in alert monkeys at high magnetic field: effects of jaw and body movements. Neuroimage 36:550–570
Kim DS, Duong TQ, Kim SG (2000) High-resolution mapping of iso-orientation columns by fMRI. Nat Neurosci 3:164–169
Kritzer MF, Goldman-Rakic PS (1995) Intrinsic circuit organization of the major layers and sublayers of the dorsolateral prefrontal cortex in the rhesus monkey. J Comp Neurol 359:131–143
Krubitzer L, Clarey J, Tweedale R, Elston G, Calford M (1995) A redefinition of somatosensory areas in the lateral sulcus of macaque monkeys. J Neurosci 15:3821–3839
LaMotte RH, Mountcastle VB (1975) Capacities of humans and monkeys to discriminate vibratory stimuli of different frequency and amplitude: a correlation between neural events and psychological measurements. J Neurophysiol 38:539–559
Laskin SE, Spencer WA (1979) Cutaneous masking. II. Geometry of excitatory and inhibitory receptive fields of single units in somatosensory cortex of the cat. J Neurophysiol 42:1061–1082
Malach R, Tootell RB, Malonek D (1994) Relationship between orientation domains, cytochrome oxidase stripes, and intrinsic horizontal connections in squirrel monkey area V2. Cereb Cortex 4:151–165
Masino SA, Frostig RD (1996) Quantitative long-term imaging of the functional representation of a whisker in rat barrel cortex. Proc Natl Acad Sci USA 93:4942–4947
Masino SA, Kwon MC, Dory Y, Frostig RD (1993) Characterization of functional organization within rat barrel cortex using intrinsic signal optical imaging through a thinned skull. Proc Natl Acad Sci USA 90:9998–10002
McKenna TM, Whitsel BL, Dreyer DA (1982) Anterior parietal cortical topographic organization in macaque monkey: a reevaluation. J Neurophysiol 48:289–317
Meftah el M, Shenasa J, Chapman CE (2002) Effects of a cross-modal manipulation of attention on somatosensory cortical neuronal responses to tactile stimuli in the monkey. J Neurophysiol 88:3133–3149
Merzenich MM, Nelson RJ, Kaas JH, Stryker MP, Jenkins WM, Zook JM, Cynader MS, Schoppmann A (1987) Variability in hand surface representations in areas 3b and 1 in adult owl and squirrel monkeys. J Comp Neurol 258:281–296
Mesulam MM (1998) From sensation to cognition. Brain 121(Pt 6):1013–1052
Mountcastle VB, Powell TP (1959a) Neural mechanisms subserving cutaneous sensibility, with special reference to the role of afferent inhibition in sensory perception and discrimination. Bull Johns Hopkins Hosp 105:201–232
Mountcastle VB, Powell TP (1959b) Central nervous mechanisms subserving position sense and kinesthesis. Bull Johns Hopkins Hosp 105:173–200
Mountcastle VB, LaMotte RH, Carli G (1972) Detection thresholds for stimuli in humans and monkeys: comparison with threshold events in mechanoreceptive afferent nerve fibers innervating the monkey hand. J Neurophysiol 35:122–136
Murray EA, Mishkin M (1984) Relative contributions of SII and area 5 to tactile discrimination in monkeys. Behav Brain Res 11:67–83
Narayan SM, Santori EM, Toga AW (1994) Mapping functional activity in rodent cortex using optical intrinsic signals. Cereb Cortex 4:195–204
Nelson RJ (1987) Activity of monkey primary somatosensory cortical neurons changes prior to active movement. Brain Res 406:402–407
Nelson RJ, Sur M, Felleman DJ, Kaas JH (1980) Representations of the body surface in postcentral parietal cortex of Macaca fascicularis. J Comp Neurol 192:611–643
Nelson RJ, Li B, Douglas VD (1991) Sensory response enhancement and suppression of monkey primary somatosensory cortical neurons. Brain Res Bull 27:751–757
Nicolelis MA, Lin RC, Woodward DJ, Chapin JK (1993) Dynamic and distributed properties of many-neuron ensembles in the ventral posterior medial thalamus of awake rats. Proc Natl Acad Sci USA 90:2212–2216

Paul RL, Merzenich M, Goodman H (1972) Representation of slowly and rapidly adapting cutaneous mechanoreceptors of the hand in Brodmann's areas 3 and 1 of Macaca mulatta. Brain Res 36:229–249
Pons TP, Garraghty PE, Cusick CG, Kaas JH (1985) The somatotopic organization of area 2 in macaque monkeys. J Comp Neurol 241:445–466
Pons TP, Garraghty PE, Friedman DP, Mishkin M (1987a) Physiological evidence for serial processing in somatosensory cortex. Science 237:417–420
Pons TP, Wall JT, Garraghty PE, Cusick CG, Kaas JH (1987b) Consistent features of the representation of the hand in area 3b of macaque monkeys. Somatosens Res 4:309–331
Pouratian N, Sicotte N, Rex D, Martin NA, Becker D, Cannestra AF, Toga AW (2002) Spatial/temporal correlation of BOLD and optical intrinsic signals in humans. Magn Reson Med 47:766–776
Powell TP, Mountcastle VB (1959) The cytoarchitecture of the postcentral gyrus of the monkey Macaca mulatta. Bull Johns Hopkins Hosp 105:108–131
Robinson CJ, Burton H (1980a) Organization of somatosensory receptive fields in cortical areas 7b, retroinsula, postauditory and granular insula of M. fascicularis. J Comp Neurol 192:69–92
Robinson CJ, Burton H (1980b) Somatic submodality distribution within the second somatosensory (SII), 7b, retroinsular, postauditory, and granular insular cortical areas of M. fascicularis. J Comp Neurol 192:93–108
Roe AW (2003) Modular complexity of area 2 in the macaque monkey. CRC, New York
Roe AW, Fritsches K, Pettigrew JD (2005) Optical imaging of functional organization of V1 and V2 in marmoset visual cortex. Anat Rec A Discov Mol Cell Evol Biol 287:1213–1225
Roe A, Lu H, Kamitani Y, Tong F (2006) Orientation-selective information in the late positive optical intrinsic signal: implications for fMRI. Human Brain Mapping, Florence, Italy
Romo R, Hernandez A, Zainos A, Salinas E (1998) Somatosensory discrimination based on cortical microstimulation. Nature 392:387–390
Romo R, Hernandez A, Zainos A, Brody CD, Lemus L (2000) Sensing without touching: psychophysical performance based on cortical microstimulation. Neuron 26:273–278
Sato K, Nariai T, Tanaka Y, Maehara T, Miyakawa N, Sasaki S, Momose-Sato Y, Ohno K (2005) Functional representation of the finger and face in the human somatosensory cortex: intraoperative intrinsic optical imaging. Neuroimage 25:1292–1301
Schneider RJ, Friedman DP, Mishkin M (1993) A modality-specific somatosensory area within the insula of the rhesus monkey. Brain Res 621:116–120
Schroeder CE, Seto S, Arezzo JC, Garraghty PE (1995) Electrophysiological evidence for overlapping dominant and latent inputs to somatosensory cortex in squirrel monkeys. J Neurophysiol 74:722–732
Schwartz TH, Chen LM, Friedman RM, Spencer DD, Roe AW (2004) Intraoperative optical imaging of human face cortical topography: a case study. Neuroreport 15:1527–1531
Seidemann E, Arieli A, Grinvald A, Slovin H (2002) Dynamics of depolarization and hyperpolarization in the frontal cortex and saccade goal. Science 295:862–865
Sherrick CE Jr (1964) Effects of Double Simultaneous Stimulation of the Skin. Am J Psychol 77:42–53
Sheth BR, Moore CI, Sur M (1998) Temporal modulation of spatial borders in rat barrel cortex. J Neurophysiol 79:464–470
Sheth S, Nemoto M, Guiou M, Walker M, Pouratian N, Toga AW (2003) Evaluation of coupling between optical intrinsic signals and neuronal activity in rat somatosensory cortex. Neuroimage 19:884–894
Sheth SA, Nemoto M, Guiou M, Walker M, Pouratian N, Hageman N, Toga AW (2004) Columnar specificity of microvascular oxygenation and volume responses: implications for functional brain mapping. J Neurosci 24:634–641
Shmuel A, Augath M, Oeltermann A, Logothetis NK (2006) Negative functional MRI response correlates with decreases in neuronal activity in monkey visual area V1. Nat Neurosci 9:569–577
Shoham D, Grinvald A (2001) The cortical representation of the hand in macaque and human area S-I: high resolution optical imaging. J Neurosci 21:6820–6835

Shtoyerman E, Arieli A, Slovin H, Vanzetta I, Grinvald A (2000) Long-term optical imaging and spectroscopy reveal mechanisms underlying the intrinsic signal and stability of cortical maps in V1 of behaving monkeys. J Neurosci 20:8111–8121

Siegel RM, Raffi M, Phinney RE, Turner JA, Jando G (2003) Functional architecture of eye position gain fields in visual association cortex of behaving monkey. J Neurophysiol 90:1279–1294

Simons SB, Tannan V, Chiu J, Favorov OV, Whitsel BL, Tommerdahl M (2005) Amplitude-dependency of response of SI cortex to flutter stimulation. BMC Neurosci 6:43

Simons SB, Chiu J, Favorov OV, Whitsel BL, Tommerdahl M (2007) Duration-dependent response of SI to vibrotactile stimulation in squirrel monkey. J Neurophysiol 97:2121–2129

Slovin H, Arieli A, Hildesheim R, Grinvald A (2002) Long-term voltage-sensitive dye imaging reveals cortical dynamics in behaving monkeys. J Neurophysiol 88:3421–3438

Sretavan D, Dykes RW (1983) The organization of two cutaneous submodalities in the forearm region of area 3b of cat somatosensory cortex. J Comp Neurol 213:381–398

Suh M, Shariff S, Bahar S, Mehta AD, Schwartz TH (2005) Intrinsic optical signal imaging of normal and abnormal physiology in animals and humans–seeing the invisible. Clin Neurosurg 52:135–149

Sur M, Wall JT, Kaas JH (1981) Modular segregation of functional cell classes within the post-central somatosensory cortex of monkeys. Science 212:1059–1061

Sur M, Nelson RJ, Kaas JH (1982) Representations of the body surface in cortical areas 3b and 1 of squirrel monkeys: comparisons with other primates. J Comp Neurol 211:177–192

Sur M, Wall JT, Kaas JH (1984) Modular distribution of neurons with slowly adapting and rapidly adapting responses in area 3b of somatosensory cortex in monkeys. J Neurophysiol 51:724–744

Sur M, Garraghty PE, Bruce CJ (1985) Somatosensory cortex in macaque monkeys: laminar differences in receptive field size in areas 3b and 1. Brain Res 342:391–395

Talbot WH, Darian-Smith I, Kornhuber HH, Mountcastle VB (1968) The sense of flutter-vibration: comparison of the human capacity with response patterns of mechanoreceptive afferents from the monkey hand. J Neurophysiol 31:301–334

Tanji J, Wise SP (1981) Submodality distribution in sensorimotor cortex of the unanesthetized monkey. J Neurophysiol 45:467–481

Tannan V, Whitsel BL, Tommerdahl MA (2006) Vibrotactile adaptation enhances spatial localization. Brain Res 1102:109–116

Thakur PH, Fitzgerald PJ, Lane JW, Hsiao SS (2006) Receptive field properties of the macaque second somatosensory cortex: nonlinear mechanisms underlying the representation of orientation within a finger pad. J Neurosci 26:13567–13575

Thompson JK, Peterson MR, Freeman RD (2003) Single-neuron activity and tissue oxygenation in the cerebral cortex. Science 299:1070–1072

Tommerdahl M, Delemos KA, Vierck CJ Jr, Favorov OV, Whitsel BL (1996) Anterior parietal cortical response to tactile and skin-heating stimuli applied to the same skin site. J Neurophysiol 75:2662–2670

Tommerdahl M, Delemos KA, Favorov OV, Metz CB, Vierck CJ Jr, Whitsel BL (1998) Response of anterior parietal cortex to different modes of same-site skin stimulation. J Neurophysiol 80:3272–3283

Tommerdahl M, Delemos KA, Whitsel BL, Favorov OV, Metz CB (1999) Response of anterior parietal cortex to cutaneous flutter versus vibration. J Neurophysiol 82:16–33

Tommerdahl M, Favorov O, Whitsel BL (2002) Optical imaging of intrinsic signals in somatosensory cortex. Behav Brain Res 135:83–91

Tommerdahl M, Favorov OV, Whitsel BL (2005a) Effects of high-frequency skin stimulation on SI cortex: mechanisms and functional implications. Somatosens Mot Res 22:151–169

Tommerdahl M, Simons SB, Chiu JS, Tannan V, Favorov O, Whitsel B (2005b) Response of SII cortex to ipsilateral, contralateral and bilateral flutter stimulation in the cat. BMC Neurosci 6:11

Tommerdahl M, Simons SB, Chiu JS, Favorov O, Whitsel BL (2006) Ipsilateral input modifies the primary somatosensory cortex response to contralateral skin flutter. J Neurosci 26:5970–5977

Tootell RB, Tsao D, Vanduffel W (2003) Neuroimaging weighs in: humans meet macaques in "primate" visual cortex. J Neurosci 23:3981–3989

Tootell RB, Devaney KJ, Young JC, Postelnicu G, Rajimehr R, Ungerleider LG (2008) fMRI mapping of a morphed continuum of 3D shapes within inferior temporal cortex. Proc Natl Acad Sci USA 105:3605–3609

Torebjork HE, Ochoa JL (1980) Specific sensations evoked by activity in single identified sensory units in man. Acta Physiol Scand 110:445–447

Toth LJ, Rao SC, Kim DS, Somers D, Sur M (1996) Subthreshold facilitation and suppression in primary visual cortex revealed by intrinsic signal imaging. Proc Natl Acad Sci USA 93:9869–9874

Tsunoda K, Yamane Y, Nishizaki M, Tanifuji M (2001) Complex objects are represented in macaque inferotemporal cortex by the combination of feature columns. Nat Neurosci 4:832–838

Vallbo AB, Hulliger M, Nordh E (1981) Do spindle afferents monitor joint position in man? A study with active position holding. Brain Res 204:209–213

Van Essen DC, Lewis JW, Drury HA, Hadjikhani N, Tootell RB, Bakircioglu M, Miller MI (2001) Mapping visual cortex in monkeys and humans using surface-based atlases. Vision Res 41:1359–1378

Vanduffel W, Fize D, Mandeville JB, Nelissen K, Van Hecke P, Rosen BR, Tootell RB, Orban GA (2001) Visual motion processing investigated using contrast agent-enhanced fMRI in awake behaving monkeys. Neuron 32:565–577

Vanduffel W, Tootell RB, Schoups AA, Orban GA (2002) The organization of orientation selectivity throughout macaque visual cortex. Cereb Cortex 12:647–662

Vanzetta I, Hildesheim R, Grinvald A (2005) Compartment-resolved imaging of activity-dependent dynamics of cortical blood volume and oximetry. J Neurosci 25:2233–2244

Verrillo RT (1966a) Effect of spatial parameters on the vibrotactile threshold. J Exp Psychol 71:570–575

Verrillo RT (1966b) Vibrotactile thresholds for hairy skin. J Exp Psychol 72:47–50

Verrillo RT, Bolanowski SJ Jr (1986) The effects of skin temperature on the psychophysical responses to vibration on glabrous and hairy skin. J Acoust Soc Am 80:528–532

Vnek N, Ramsden BM, Hung CP, Goldman-Rakic PS, Roe AW (1999) Optical imaging of functional domains in the cortex of the awake and behaving monkey. Proc Natl Acad Sci USA 96:4057–4060

Woolsey C, Marshall W, Bard P (1942) Representation of cutaneous tactile sensibility in the cerebral cortex of the monkey as indicated by evoked potentials. Bull Johns Hopkins Hosp 70:399–441

Xu X, Collins CE, Kaskan PM, Khaytin I, Kaas JH, Casagrande VA (2004) Optical imaging of visually evoked responses in prosimian primates reveals conserved features of the middle temporal visual area. Proc Natl Acad Sci USA 101:2566–2571

Yacoub E, Shmuel A, Pfeuffer J, Van De Moortele PF, Adriany G, Ugurbil K, Hu X (2001) Investigation of the initial dip in fMRI at 7 Tesla. NMR Biomed 14:408–412

Zarahn E (2001) Spatial localization and resolution of BOLD fMRI. Curr Opin Neurobiol 11:209–212

Zepeda A, Arias C, Sengpiel F (2004) Optical imaging of intrinsic signals: recent developments in the methodology and its applications. J Neurosci Methods 136:1–21

Chapter 5
How Images of Objects Are Represented in Macaque Inferotemporal Cortex

Manabu Tanifuji, Takayuki Sato, Go Uchida, Yukako Yamane, and Kazushige Tsunoda

Abstract Visual object recognition is a simple and easy task in our daily life. However, the mechanisms for recognizing objects are not at all simple nor easy. To understand neural mechanisms of object recognition, we have investigated representation of object images in macaque inferior temporal cortex that is the area essential for object recognition. Optical intrinsic signal imaging has revealed that object images are represented by the combinatorial code at the columnar level, where each column represents a visual feature of object images. The visual features represented by columns include local features as well as global features representing spatial arrangements of local features. Here, columns are supposed to be functional units for object representation. However, difference in object selectivity among nearby cells does not support the concept of columns as the functional units. Quantitative analysis of object responses of single cells and population activity revealed that each cell in a columnar region is characterized by cell specific property and property common across the cells in the columnar region, suggesting two different levels (single cell and columnar level) of object representation. Possible role of these two levels of object representation will be discussed.

5.1 Introduction

In order to understand neural mechanisms of object recognition, we have investigated representation of object images in inferotemporal (IT) cortex of macaque monkeys. This area of brain is essential for the recognition of objects by their visual images, and neurons in this area are known to respond to images of complex objects (Gross 1994; Gross et al. 1979; Desimone et al. 1984; Bruce et al. 1981; Perrett

M. Tanifuji (✉)
Laboratory for Integrative Neural Systems, RIKEN Brain Science Institute, Wako, Saitama, 351-0198, Japan
e-mail: tanifuji@riken.go.jp

A.W. Roe (ed.), *Imaging the Brain with Optical Methods*,
DOI 10.1007/978-1-4419-0452-2_5, © Springer Science+Business Media, LLC 2009

et al. 1982; Tanaka et al. 1991). Although some of these neurons specifically respond to certain objects such as faces, many IT neurons also well respond to visual features that are less complex than object images (Desimone et al. 1984; Bruce et al. 1981; Perrett et al. 1982; Tanaka et al. 1991). Since none of these visual features are specific enough, object representation takes the combined activation of multiple neurons to represent a particular object image in IT cortex. Investigations of object representation require analyses of activities of multiple neurons evoked by an object image.

Neurons responding to particular objects or visual features are not randomly distributed in IT cortex. Tanaka and colleagues found that neurons tuned to similar features were clustered together and formed a column in IT cortex (Feature column) (Fujita et al. 1992). Optical intrinsic signal imaging (OISI), that has columnar level resolution, is an appropriate technique to investigate object representation with multiple neurons (Grinvald et al. 1999). In fact, OISI revealed that an object image activates multiple spots (activity spots) in IT cortex, and that distribution patterns of activity spots are different from object to object (Tsunoda et al. 2001; Yamane et al. 2006). In this chapter, we present our studies with OISI that revealed what and how activity spots (corresponding to feature columns) are used to represent object images in IT cortex (Tsunoda et al. 2001; Yamane et al. 2006; Wang et al. 1996; Wang et al. 1998).

OISI is still limited in spatial resolution. It does not allow us to resolve visual responses at a level of single cells. Strictly speaking, object representation revealed by OISI is at the level of columns, and, in principle, columnar level representation could be different from representation at the level of single cells. In addition, recent studies have reported that object selectivity of nearby cells is largely different; these observations seemingly contradict to the column hypothesis (Tamura et al. 2005; Kreiman et al. 2006; Sato et al. 2009). Our recent study suggests that this apparent contradiction is due to the response property specific to neurons that obscure the response property common across the cells in a columnar region (Sato et al. 2009). The critical question is how to relate object representation at columnar level and a possible representation by combinations of activities of single cells in close vicinity. In the last part of the chapter, we discuss this issue. The above study also suggests that, in contrast to the columnar organizations in V1, the columnar structure in IT cortex may not uniformly cover entire IT (Sato et al. 2009).

IT cortex is subdivided into multiple areas. Our investigations focus on dorsal part of area TEa, that is the most anterior part of IT cortex where OISI is still applicable. Furthermore, the results shown here were based on the experiments with anesthetized monkeys. Thus, we cannot directly relate object representation observed in our studies to recognition in waking animals, but studies with anesthetized animals have an advantage that they can reveal object representation unbiased toward particular behavioral contexts.

To begin with, basic characteristics of intrinsic signals in IT cortex and physiological studies related to the columnar organizations are briefly reviewed, since studies on these two issues provide basis for our studies with OISI.

5.2 Optical Intrinsic Signal Imaging (OISI) in IT Cortex

Activation of neurons elicits changes in reflection of light from neural tissues. The changes (intrinsic signals) reflect three secondary physiological mechanisms associated with neural activation: deoxygenation level of hemoglobin (Vanzetta and Grinvald 1999), blood volume changes in capillaries (Fukuda et al. 2005; Vanzetta et al. 2004), and microscopic changes in tissue structure that cause changes in light scattering (MacVicar and Hockman 1991; Tsunoda et al. 2004). Relative contribution of these three components to intrinsic signals depends on the wavelength of light used for illuminating the cortex. In OISI from area TE, we used wavelengths at around 610 nm, where deoxygenation of hemoglobin is the dominant component. The typical time courses of the signals were not very different from those in V1: the signals reached a peak of increase in absorption at about 1–2 s after the stimulus onset, crossed the baseline absorption level at around 2–4 s. after the stimulus onset, and then showed absorption decreases that lasted for several seconds (Fig. 5.1). The initial increase in absorption corresponds to deoxyhemoglobin increase due to oxygen consumption by neurons activated by the stimulus and late decrease in absorption corresponds to replacement of deoxyhemoglobin with oxyhemoglobin due to increased blood flow triggered by the neural activation.

To obtain spatial patterns of intrinsic signals in area TE, we took images of cortical surface during 2.5 s period starting from 0.5 s after the stimulus onset by a CCD

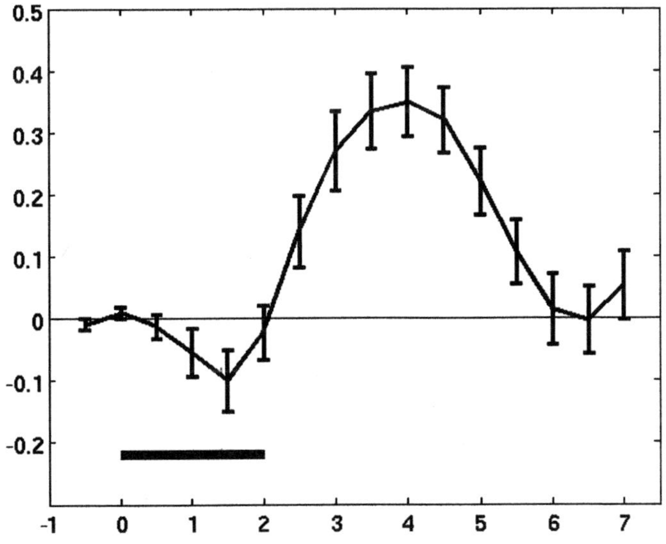

Fig. 5.1 The time course of intrinsic signals in area TE. Horizontal and vertical axes represent time (s) and reflection changes (%), respectively. *Horizontal bar*, the period for visual stimulation

camera, averaged them, and subtracted an image taken just before the stimulus onset from the averaged image (Tsunoda et al. 2001; Yamane et al. 2006). Thus, spatial patterns of intrinsic signals in area TE mainly reflect deoxygenation of hemoglobin induced by increased neural activity. Though OISI is typically used to visualize activation at the level of columns, raw intrinsic signals were not confined to columnar regions (global signals) (Fig. 5.2c). The global signals were, however, locally modulated. The local peaks of modulation (activity spots) were extracted by removing a low spatial frequency component of intrinsic signals (filtered image, Fig. 5.2d). Activity spots were then demarcated by drawing contours crossing half height between peak absorption of the spot and background in the filtered image (Fig. 5.2b). The magnitudes of the absorption changes in the filtered intrinsic signals were well correlated with the average of multi-unit activities (avgMUA) recorded from the same spot. For eleven activity spots, for example, the mean and standard deviation of correlation coefficient was 0.61 ± 0.17, and this value is statistically significant ($p < 0.05$). As shown later, avgMUA gives a good estimate of the neural activity at the columnar level. Thus, the high spatial frequency component

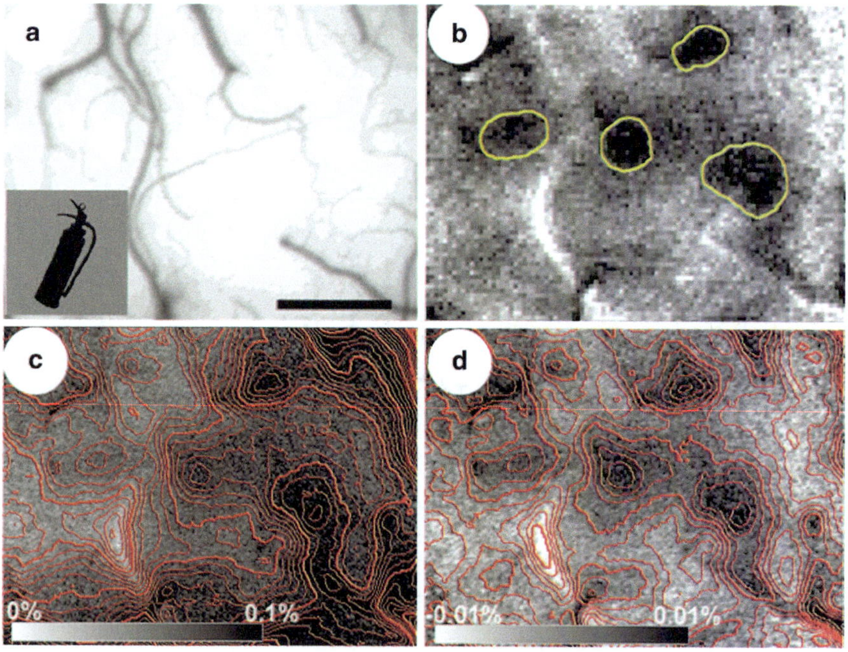

Fig. 5.2 A spatial pattern of intrinsic signals in IT. (**a**) Surface view of the exposed portion of IT cortex. Scale bar, 1 mm. (**b**) Activity spots outlined at the level of half the peak absorption value. (**c**) the spatial pattern of raw intrinsic signals without filtering. (**d**) spatial pattern of activity spots obtained by filtering raw intrinsic signals Scale bars in (**c**) and (**d**), percent changes in absorption increase. (modified from Tsunoda et al. (2001))

of intrinsic signals as in Fig 5.2b,d is in a good agreement with firing activity of population of neurons as seen in intrinsic signals in V1.

What about the low spatial frequency component of the raw intrinsic signals indicates? One possibility is that this low spatial frequency component (global signals) reflects synaptic potentials elicited by visual stimuli. To examine this possibility, we stained the cortex with the voltage sensitive dye RH1692 and compared the dye signal with intrinsic signals recorded from the same region (Homma and Tanifuji 2003). It is known that the voltage sensitive dye stains cellular membrane and that changes in flourescence of the dye reflect synaptic potentials rather than spiking activities (Grinvald et al. 1999). We found that the dye signals also revealed low and high spatial components, and local peaks of the dye signal coincide well with activity spots revealed by OISI. Since the low spatial frequency component of voltage sensitive dye signals was similar to that of the intrinsic signals, we think that raw intrinsic signals may also reflect synaptic potentials, and that intrinsic signals outside of the regions of activity spots reflect synaptic potentials below the threshold of neuronal firing. Similar argument was made for the global and local modulation of intrinsic signals observed in cat striate cortex (Das and Gilbert 1995). Interestingly, the low spatial frequency component is not uniformly distributed in space. For example, more absorption is seen in the right half than in the left half (Fig. 5.2c). Similar regional specificity was also observed in activation patterns recorded by voltage sensitive dye imaging with various object images. In this way, the distribution of the low frequency components was stimulus dependent. Some representations of object images may exist at the level of synaptic potentials as well as those at columnar levels.

5.3 Evidence for the Columnar Organization with Respect to the Critical Features in Area TE

Columnar organizations in the area TE were systematically investigated by Tanaka and colleagues (Tanaka et al. 1991; Fujita et al. 1992). They first determined a visual feature critical for each cell (Tanaka et al. 1991), and then investigated columnar organizations with respect to the critical features (Fujita et al. 1992). Here, the critical feature for a cell means the simplest visual feature that activates the cell equally well as the best object stimulus. To find such a critical feature, first, they searched for the most effective stimulus for each cell among more than a hundred of three-dimensional object stimuli. These stimuli included stuffed animals, plastic fruits and vegetables, and experimenter's hand and body. These stimuli were presented to the monkeys from various viewpoints to maximize the chance of finding the most effective image. Then, they generated modifications of the most effective stimulus and examined responses evoked by the simplified stimuli. If the cell responded to one of the simplified stimuli equally well as the original object stimulus, this new stimulus was further simplified. This procedure was repeated until the experimenter could not simplify the stimulus without significant reduction of the evoked responses. Figure 5.3 shows a representative case of the stimulus simplification procedure according to their method, where we found a combination of the circle and rectangle was essential for the maximal activation of a cell. This cell was recorded from the activity spot that was expected to represent spatial relationship among object parts (discussed later in detail).

Tanaka and colleagues examined many TE neurons in this way and identified visual features essential for activating individual cells (Tanaka et al. 1991; Kobatake and Tanaka 1994). Many of these features are the combinations of simple shapes,

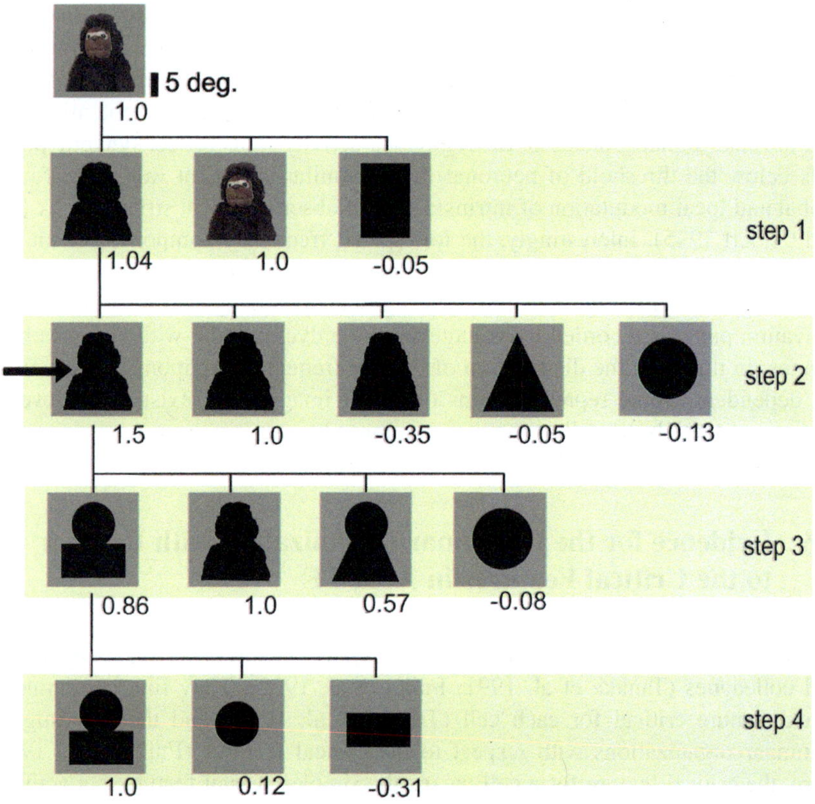

Fig. 5.3 Systematic stimulus simplification of an object image. The most effective object stimulus (*top* picture) was simplified step by step to determine the simplest stimulus that maximally activates the cell. Step 1 shows that neuronal activity elicited by the most effective object and its silhouettes are the same. The number below each picture indicates the response amplitudes normalized to the response to the reference stimulus. Step 2 examines the effect of the "sharpness" of the corner at the junction of the *upper* and *lower* parts (*arrow*) and shows that the silhouette with the sharpest corners (*leftmost* picture) is the most effective stimulus. Step 3 shows that activities elicited by the two *leftmost* stimuli are not significantly different. Step 4 shows that neither the *upper* nor *lower* part activates the cell. In this case, the critical feature was determined as the combination of a circle and a rectangle (*leftmost* picture in step 4). (Modified from Yamane et al. (2006))

colors, luminance gradient/contrast, and textures. These features are more complex than the optimal visual stimuli for cells in areas V1, V2, and V4, but still less complex than natural objects (Kobatake and Tanaka 1994).

Fujita et al. (1992) systematically examined the columnar organization in the area TE with respect to the critical features. First, they penetrated electrodes perpendicular to the cortical surface and determined a critical feature of one neuron within the track. Then, they prepared a set of visual stimuli including this critical feature and its modifications and examined the response selectivity of other neurons within the track using the set of visual stimuli. This experiment showed that the most effective stimulus for the cells in the track is the critical feature or visual features similar to it. They also examined neuronal responses along a tangential electrode penetration and found that neurons with similar responsiveness were localized within the range about 0.4 mm along the tangential track. These results suggest a columnar organization in the IT cortex with respect to the critical features of the cells. However, we often encounter nearby cells spaced only for about 150 μm apart but having different selectivity for object images. We will discuss later in this chapter what makes this difference in the selectivity of the critical features and object images.

Many investigations suggest that neurons in IT cortex were plastic. For example, the above group trained monkeys for a specific set of visual stimuli with a delayed match-to-sample task (Kobatake et al. 1998). They found that, compared with naïve monkeys, IT cortex of the trained monkeys contained more neurons tuned to a trained stimulus: the response to the best of the trained stimuli was higher than the responses to any other object images. Because we cannot keep tracking responses of a single neuron during training that takes more than a month, these results may not indicate that training changed visual responses of IT cells. Neurons that were previously not visually driven may turn out to be responsive to the trained stimulus. In either case, the columnar organization in IT cortex, if it exists, seems not static but could be modified by extensive experience of the monkeys.

5.4 Object Representation by Combinations of Activity Spots in Area TE

OISI revealed that an object image activates multiple activity spots (Figs. 5.2 and 5.4). The spatial patterns of these activity spots are different from object to object. Some of the spots were co-activated by different objects, while other spots were activated only by one object (Fig. 5.4a, b). Assuming that each spot represents a visual feature, activation of the spots specific to a single object is likely to indicate that the other objects lack the visual feature represented by these spots. For example, the spots only activated by stimulus 1 may be related to horizontal red/blue stripes only seen in stimulus 1 (Fig. 5.4b). Comparison between the distribution patterns of activity spots and those produced by systematically simplified stimuli revealed that this was indeed the case (Fig. 5.4c). Here, a "black cat" (c-**1**) was

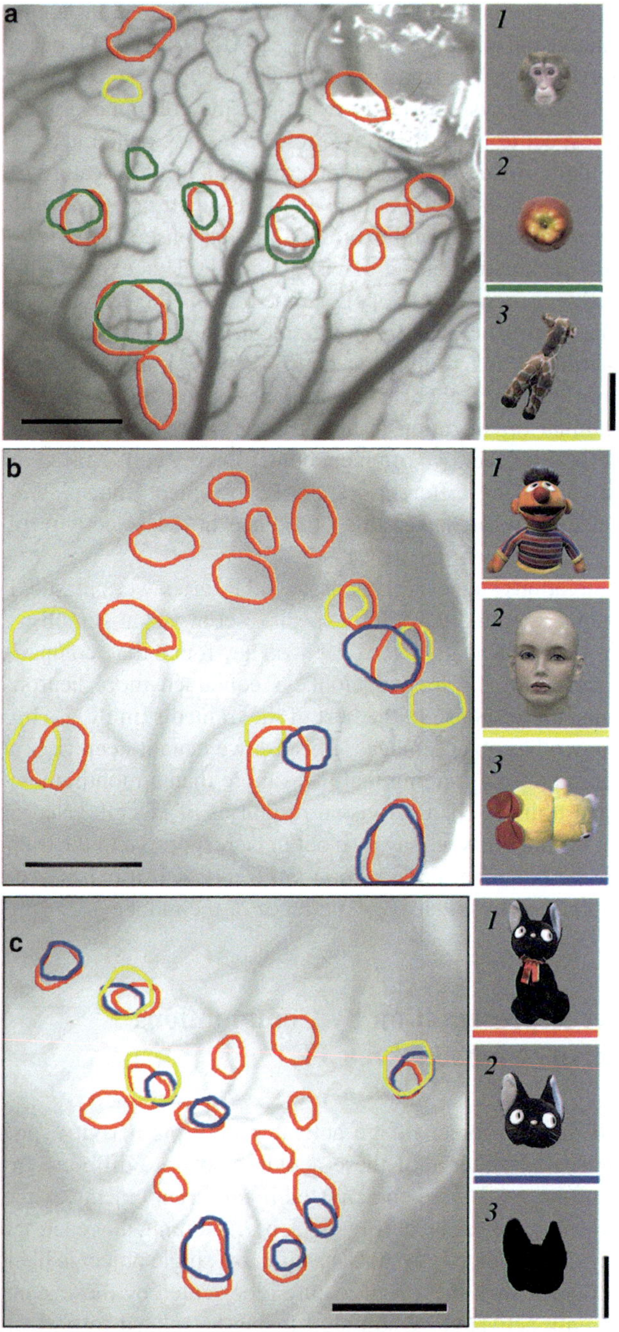

Fig. 5.4 Distribution of activity spots elicited by various object images (**a**,**b**) and simplified images (**c**). Contours of the activity spots were drawn as in Fig. 5.2. Horizontal scale bars, 1.0 mm. Vertical scale bars, 10°. (Modified from Tsunoda et al. (2001))

simplified to the "head" (c-**2**), and then to the "silhouette of the head" (c-**3**). The original image (c-**1**) elicited 14 spots, but presenting the "head" (c-**2**) elicited only eight spots of the original fourteen. The "silhouette" (c-**3**) only activated three (arrows) of the eight spots elicited by the "head" (c-**2**). Simplified stimuli lacking some of the visual features of the original image activated only a subset of the spots elicited by the stimuli before simplification. We examined 12 pairs of activation patterns elicited by the original and the simplified stimuli, and found that five pairs (42%) showed similar results.

Interestingly, in addition to the disappearance of spots, there were also cases where new spots emerged by the simplification of an object (Fig. 5.5). For example, in Fig. 5.5a, spots A and B disappeared but spot C appeared when stimulus **1** was simplified to stimulus **3**. Similarly in Fig. 5.5b, spot A was only activated by the simplified stimulus. Among the twelve pairs, the emergence of spots by simplification was observed in seven pairs (58%) (two cases showed only emergence and five cases showed both emergence and disappearance).

To identify the visual feature represented by each spot and also to find reasons for emergence of activity spots with stimulus simplification, we recorded single cellular activities from neurons in spots shown in Fig. 5.5a (Fig. 5.6). Cells in spots A and B were significantly activated by the "handle and hose" in isolation (Fig. 5.6a-1, b-1). This result is consistent with optical response patterns elicited by stimuli 1 and 3 (Fig. 5.5a). In addition, the cells in spot A were activated by the "handle" (Fig. 5.6a-2) having protrusions, but not by the "hose" (Fig. 5.6a-3). Furthermore, other stimuli with sharp protrusions, such as a "hand" (Fig. 5.6a-4) and "cat's head" (Fig. 5.6a-5), also activated the cells. These cells seemed to require a generic visual feature, "sharp protrusions," for activation. In contrast, the cells in spot B were activated by the "hose" (Fig. 5.6b-3), but neither by the "handle" (Fig. 5.6b-2) nor by a "line segment" (Fig. 5.6b-4). We also found no activation was elicited by a segment of a circle (Data not shown). The critical feature of these cells was, thus, an "asymmetric arc."

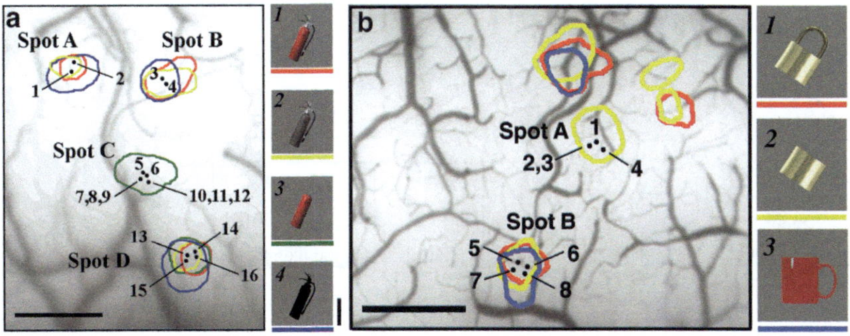

Fig. 5.5 Stimulus simplification causes appearance of new spots. The numbers indicate electrode penetration sites. (see Tsunoda et al. 2001 for electrode recordings in b) (Modified from Tsunoda et al. (2001))

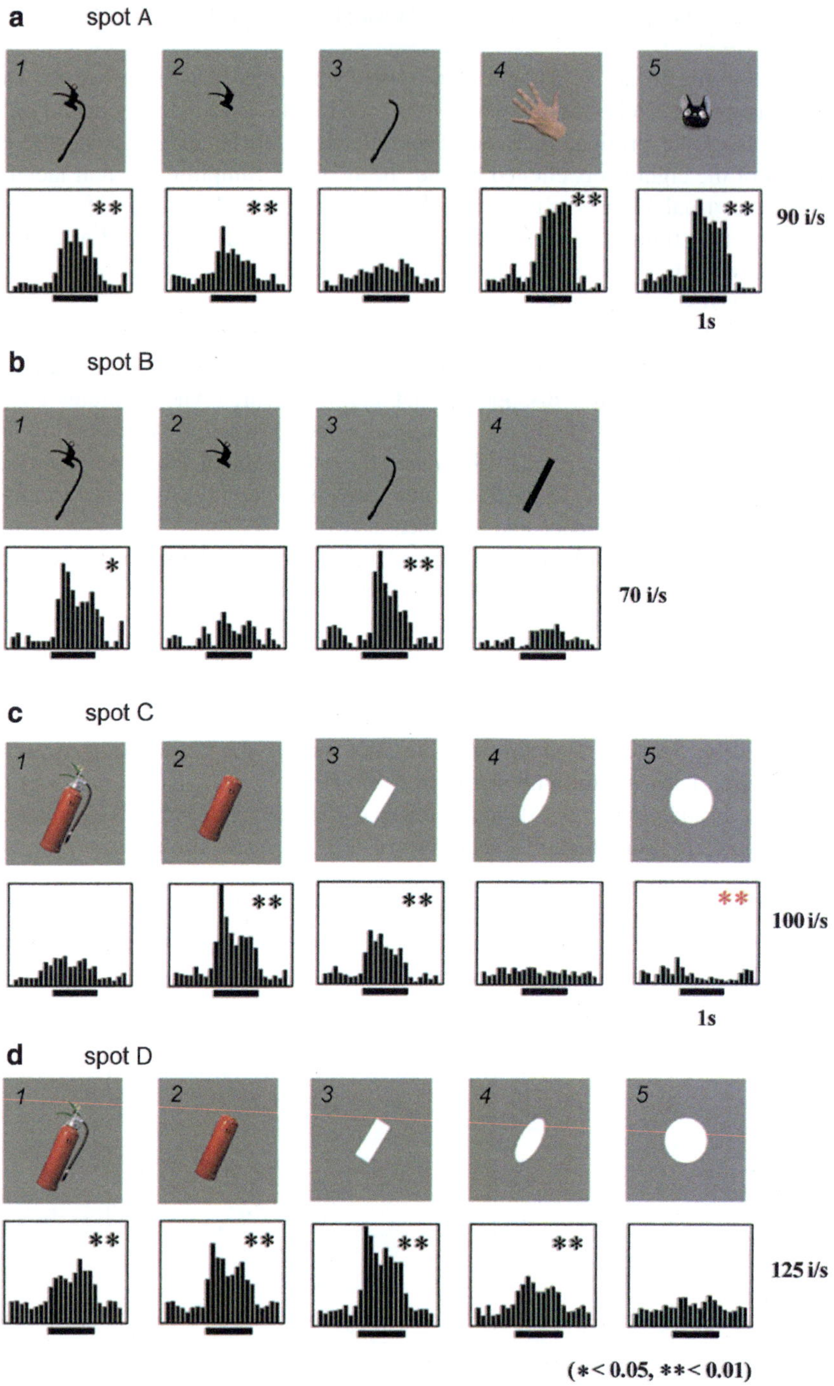

Fig. 5.6 Visual responsiveness of representative cells in spots A–D in Fig. 5.5a. Each histogram (PSTH) shows neuronal responses elicited by the above stimulus. Red asterisks indicate significant inhibition ($p<0.01$). *$p<0.05$, **$p<0.01$. Scale bars, 1-s periods of stimulation. (Modified from Tsunoda et al. (2001))

OISI revealed that the "cylinder" produced significant responses in both spots C and D (Figs. 5.5a-3), while other stimuli having "the cylinder" as a part (Figs. 5.5a-1, a-2 and a-4) only activated spot D but not spot C. The neural responses of cells in these spots were consistent with the imaging results: the cells in spot C was activated by the "cylinder" but not by the original "fire extinguisher" (Figs. 5.6c-1 and c-2), and a cell in spot D was significantly activated by both stimuli (Fig. 5.6d-1, d-2). The feature critical for the cells in spot D was a "rectangular shape" (Fig. 5.6d-3), but the cells also responded significantly to an "ellipse" (Fig. 5.6d-4). Since there were no responses to a "circle" (Fig. 5.6d-5), an "elongated structure" seemed to be necessary for activation. The simplest visual feature for the cells in spot C was also a "rectangular shape" (Fig. 5.6c-3). In contrast to the cells in spot D, however, none of the cells was activated by an "ellipse" (Fig. 5.6c-4). In addition, all of these cells were inhibited by the "circle" (Fig. 5.6c-5). These results suggest that the response properties of the cells in spot C (Fig. 5.5a) were determined by the balance between excitatory and inhibitory inputs, that is, the excitatory inputs were given by a feature related to a rectangular shape and the inhibitory inputs were given by a feature related to a circular shape. This explanation would account for the lack of activation by "the fire extinguisher," where the hose having a circular shape was attached with the cylinder having a rectangular shape. In general, these results suggest that some of the spots representing a particular feature were inactive when other features were presented together with the feature. This could explain that active spots appeared following simplification of stimuli in some cases.

So far, we have revealed two factors involved in representation of object images in the area TE. First, some of visual features represented by activity spots are local features such as protrusion and asymmetric arc. We refer "local features" to as visual features that occupy part of an object image and are distinguishable from other parts of an object image by their particular shapes, colors, or textures. Second, specific object representation is not made by combinations only of active spots but also of inactive spots. In the case of representation of original fire extinguisher, activation of spots A and B represents the presence of protrusion and curvature in the object, respectively. Activation of spot D indicates that entire structure is elongated, but no activation in spot C further indicates the structure has to be relatively elliptic. In such a way, combinations of inactive as well as active columns help representation of object images to be more specific.

5.5 Representation of Configurational Information Appeared in Object Images

Specific representation of object images by combinations of local features requires mechanisms to represent information about the spatial arrangement of "local features" or about spatial arrangement of parts including local features ("configurational information"). In the previous section, we showed that spot C in Fig. 5.5a was not activated when the hose is attached to the side of the cylinder and makes the entire shape elliptical. This spot, however, would be activated if the hose were secured above the handle where the rectangular shape of cylinder was exposed. From the viewpoint of representing "configurational information," we may consider that activity in spot C carries information about the position of the hose relative to the cylinder, although the way to represent "configurational information" about the relative position of hose and cylinder is indirect.

We searched for the neural substrates that explicitly represent "configurational information." Particularly, we explored representation of a particular spatial relationship, that is, "on top of." This is a typical spatial relationship appeared in object images. For example, the head is above the body, the lampshade is on top of the base, and a pineapple is separated into upper part leaves and lower part fruit. To find candidate spots related to representation of the spatial relationship, we searched for the spots that showed a specific pattern of activation: activation by the object consisting of two parts and the same object with a gap introduced between parts of the object, but no activation by a part alone (Yamane et al. 2006). Among four hemispheres, we found three spots showing this specific pattern of activation (Fig. 5.7). These spots should not simply represent local features of an upper or lower part of the objects because either part is not essential for activation. Moreover, activation by the stimulus with an introduced gap indicates that local features appearing at the junction of two parts, such as sharp connecting corners, are also not essential. Thus, these spots were good candidates that could represent spatial relationship, "on top of." For further characterization of these spots, we made single cellular recordings from these spots, and found that three unique response properties. First, the cells in these spots preferred object images consisting of vertically aligned two parts with a small number of exception (Fig. 5.8). The critical features for these cells determined by the stimulus simplification procedure were also the combinations of vertically aligned two parts except two (Fig. 5.9b-7, b-8). Second, these cells were less sensitive to color, texture, and local shapes of either part. Thus, there were no changes in the responses after removing color and texture during the stimulus simplification procedure (Figs. 5.3 and 5.9). The changes in shapes of the parts did not significantly alter responses of these cells. For example, a neuron, whose critical feature was determined as a combination of a circle and a rectangle, was also significantly activated by a combination of a circle and an ellipse. Third, these cells were highly selective to a particular spatial arrangement of the upper part and the lower part (Fig. 5.10). Most of the cells were maximally activated when the upper and lower parts were vertically aligned or tilted only for 45° from vertical arrangement. We cannot explain the selectivity to a particular spatial alignment of upper and lower parts by changes in retinotopic positions of the upper parts that occurred incidentally during the spatial rearrangements of the parts because the receptive field of these cells covered even larger area in the visual field.

These response properties enable these spots to respond to two-part objects regardless of the local features embedded in either part, but only when the parts are aligned vertically (Fig. 5.8). These results, as well as neurons in spot C in

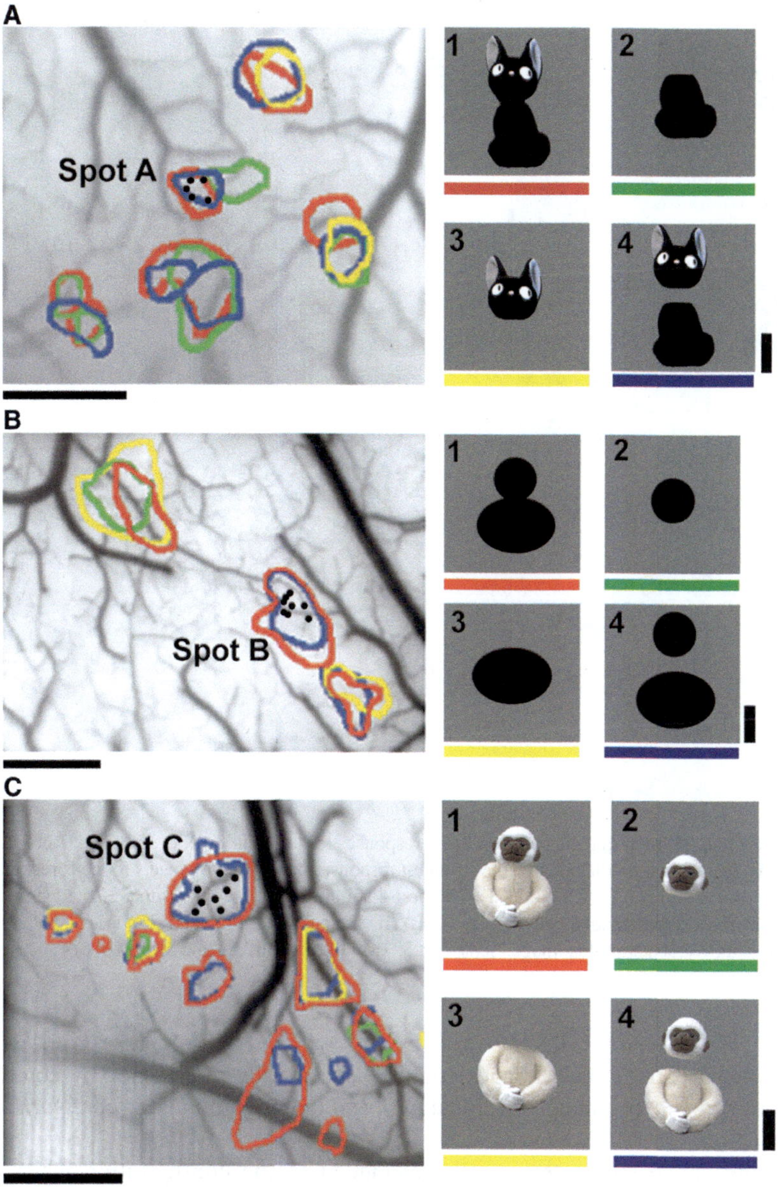

Fig. 5.7 Distribution of activity spots elicited by the stimulus set designed for searching spots related to spatial arrangement between *upper* and *lower* parts. Three spots A, B, and C were identified. *Black dots*, electrode penetration sites (Modified from Yamane et al. (2006))

Fig. 5.5a, suggest that neurons in area TE do not necessarily represent local features but also "configurational information" of the object images. As shown below, we consider that face neurons also represent configurational information

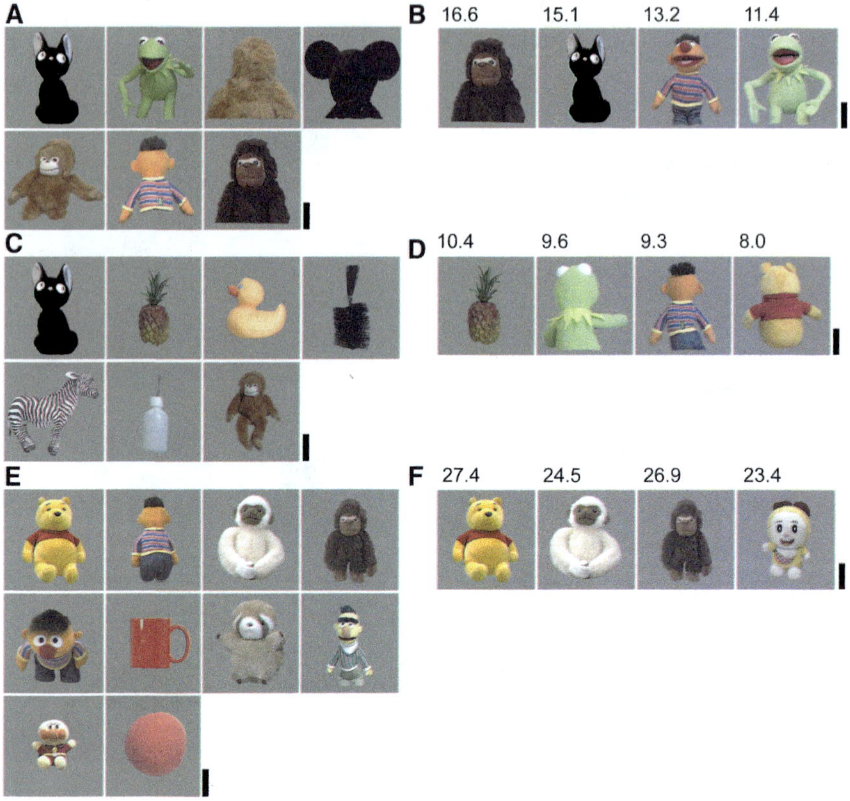

Fig. 5.8 Object stimuli that elicited significant responses for the cells in spots A, B, and C in Fig. 5.7. (**A, C, E**) Object images that elicited the strongest responses out of the 96 objects for cells recorded in spots A (**A**), B (**C**) and C (**E**). (**B, D, F**) The best 4 stimuli out of 96 objects that elicited significant responses for a representative cell in spots A (**B**), B (**D**), and C (**F**). Evoked responses (spikes/s) are indicated above each stimulus image. Scale bar, 5° (Modified from Yamane et al. (2006))

about facial parts. Object images could be specifically represented by combinations of spots representing "local features" and those representing "configurational information."

In the above study, these cells responded to the configuration where the upper part was above the lower part, but not to the configuration where the upper part is below the lower part (Fig. 5.10). This result means that, in some ways, the cells distinguished the upper part and the lower part of objects. In this respect, the cells were not entirely insensitive to local features. At least, the cells could distinguish upper and lower parts. These cells may be sensitive to difference of area of parts. Alternatively, some combination of a curvature in the upper part and in the lower part may be the critical factor (Brincat and Connor 2004).

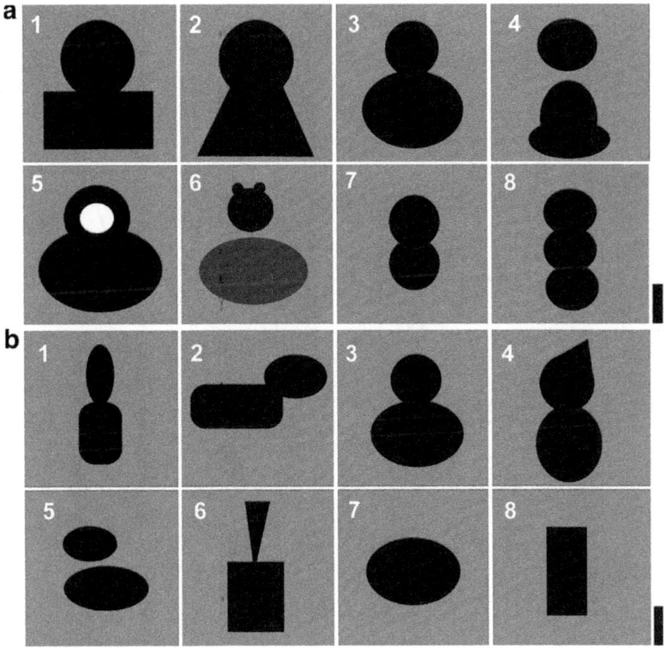

Fig. 5.9 The critical features of the cells in spots A(a) and B(b) related to representation of spatial relationship among parts. The critical features were not investigated for spot C. (Modified from Yamane et al. (2006))

5.6 Face Neurons in Area TE as Ones that Represent Facial Configuration

"Face neurons" is the neurons that respond to "faces," but these responses cannot be explained by specific responses to a part of "face" (Fig. 5.11). For example, a face without eyes did not activate the cell, but there was no activation by "eyes" alone (Fig. 5.11b right most, c). Furthermore, previous studies have shown that "face" with scrambled facial parts do not activate these neurons (Bruce et al. 1981). There are two characteristic properties of "face neurons." First, many of them are tuned to images of faces from a particular vantage point (Fig. 5.11a)(Perrett et al. 1982; Perrett et al. 1991). Second, these cells are less sensitive to difference of individual faces (Perrett et al. 1984; Baylis et al. 1985; Yamane et al. 1988; Young and Yamane 1992). These response properties suggest that the face neurons represent not specific faces but facial configuration.

Intrinsic signal imaging showed that there are spots specifically activated by faces (Fig. 5.12) (Wang et al. 1996; Wang et al. 1998). Thus, face neurons, as well as neurons specifically responding to visual features, are clustered together. Furthermore, activation patterns produced by images of faces from different vantage points revealed that the peaks of activity spots shift along the cortical surface as the

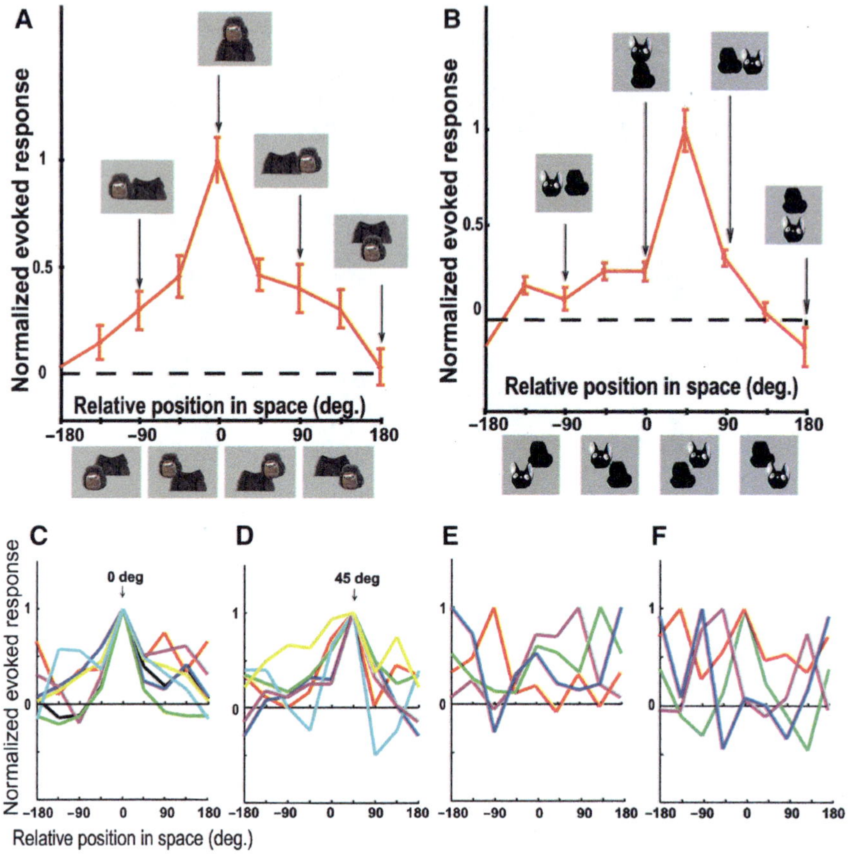

Fig. 5.10 Selectivity of cells to different spatial arrangements of the *upper* and the *lower* parts of object images. The normalized evoked responses (vertical axis) were plotted against the difference in the spatial arrangement of the parts (horizontal axis). The difference in spatial arrangement is defined by the angle between a line connecting the centers of the two parts of the best object stimuli and that of each rearranged stimulus. The pictures of stimuli corresponding to each angle are shown below the plot and also in the insets. The panels (**A**) and (**B**) show responses of representative cells in spots A and B in Fig. 5.7, respectively. (**C-F**) Tuning curves for other cells in others in spots A, B, and C with a single peak at 0°. (**C**), 45°. (**D**) and other angles (**E**), and tuning curves with multiple peaks (**F**). For simplicity, only the mean values of responses are plotted. (Yamane et al. (2006))

face rotates from the left profile to the right profile through the front face. This representation of faces from different vantage points in close vicinity may be important for view-independent recognition of faces.

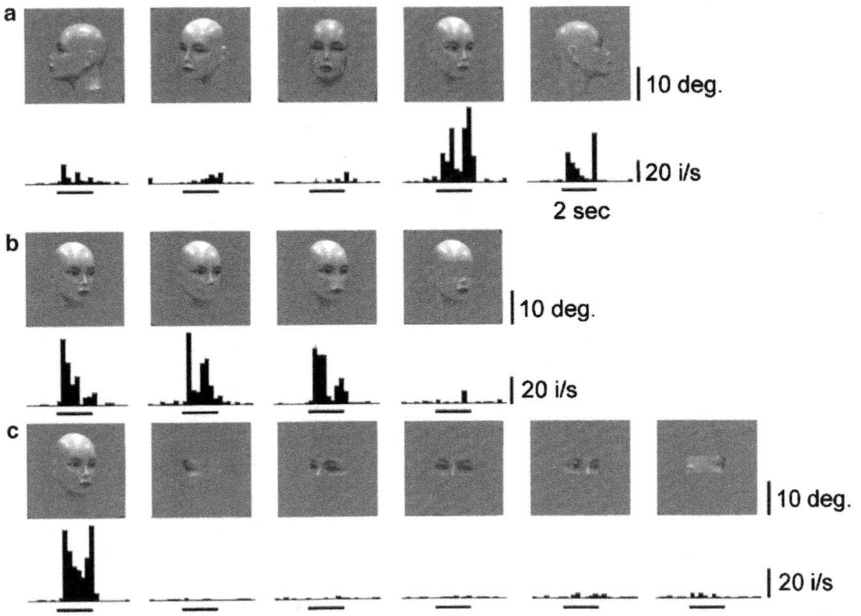

Fig. 5.11 Responses of a face neuron. The recording consists of three sessions. In the first session (**a**), the selectivity of the neuron for different views of a face is demonstrated. In the second and third sessions (**b**, **c**), selectivity of the same neuron for a face and facial parts were examined. Please note that no activation by either a face without eyes (**b**) or an eye alone (**c**) was observed. (Cited from Fukuda and Tanifuji, unpublished observation.)

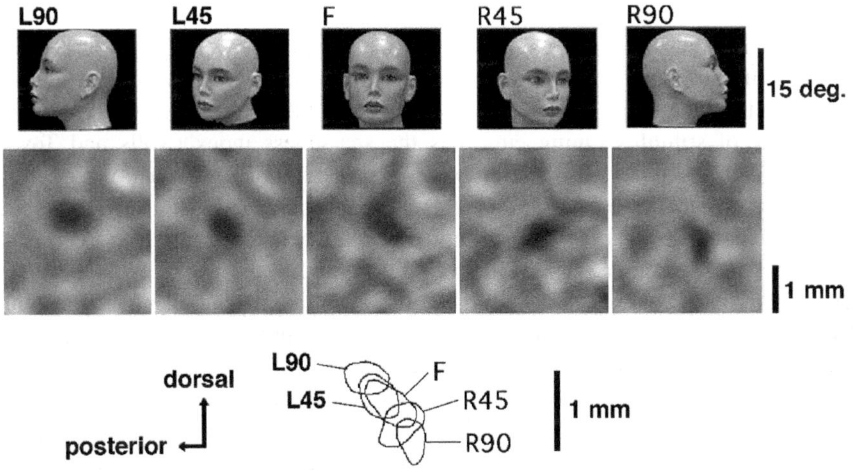

Fig. 5.12 Systematic shift in distribution of activity spots with rotation of the face. Images of the same cortical area (*middle panels*) obtained from five different views of the same mannequin face (*top panels*). The contours of the active spots are superimposed at the *bottom* (Modified from Wang et al. (1996))

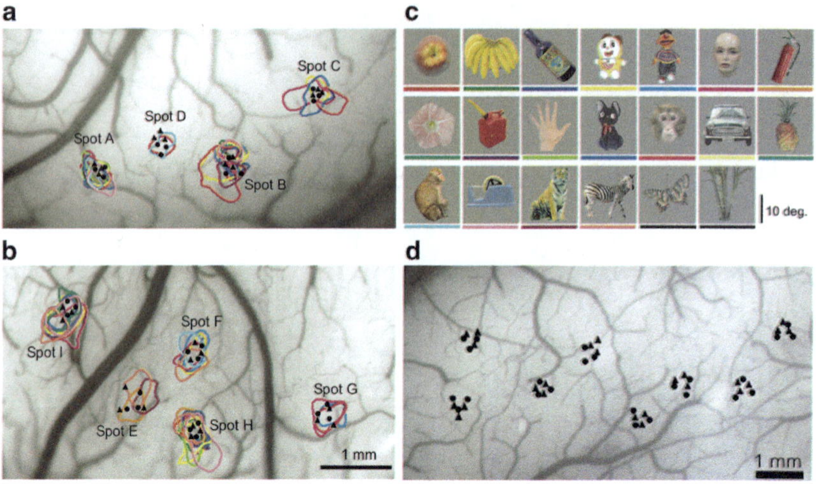

Fig. 5.13 The sites for electrophysiological recordings of single cells and MUs. Electrode penetration sites are given in filled circles and triangles. (**a, b**) The activity spots were shown by colored contours indicating area of activation by object images shown in (**c**). Contours were drawn as in Fig. 5.2. (**c**) Stimulus images for IOSI. The line type and the color are matched to the contours in **a**, and **b**. (**d**) The recording sites in the hemisphere where no IOSI was conducted beforehand. (Modified from Sato et al. (2009))

5.7 Object Representation at Different Levels: Columns and Single Cells Within a Column

Recently, several studies have quantitatively examined object selectivity of IT cells and found that IT cells were tuned to a relatively small number of object images (Tamura et al. 2005; Kreiman et al. 2006; Sato et al. 2009). One measure to quantify specificity of stimulus tuning curves is the sparseness index (Rolls and Tovee 1995). We obtained the value of the sparseness index for TE neurons for randomly chosen 80 object images as 0.19 ± 0.18 ($n = 218$). This value, 0.19, means that the neuron responds to only 19% of stimuli, if we approximate neuronal responses to stimuli in an all-or-none fashion (0 or 1). Furthermore, these studies have shown that selectivity of nearby cells was largely different from one another (Tamura et al. 2005; Kreiman et al. 2006; Sato et al. 2009). For example, five object images that evoked the strongest visual responses were different for two neurons even if these neurons were sampled from the same activity spot and spaced only for 150 μm apart (Fig. 5.13, 5.14). The correlation of object selectivity of nearby two cells within the same activity spots was as low as 0.15 in the mean value of the correlation coefficient, and the number of cell pairs that had significant correlation was only 28.5% ($p < 0.05$) (Fig. 5.15a) (Sato et al. 2009). The difference in object selectivity of nearby cells was not an artifact due to trial-by-trial variability of neuronal responses. The correlation coefficient expected from the trial-by-trial variability

Fig. 5.14 The best five stimuli of two nearby cells spaced 150 μm apart. These two cells were isolated from spot A in Fig. 5.13. The number above each stimulus was the evoked response by the stimulus (spikes/s). The correlation coefficient of selectivity of these two cells for 100 object images was 0.07

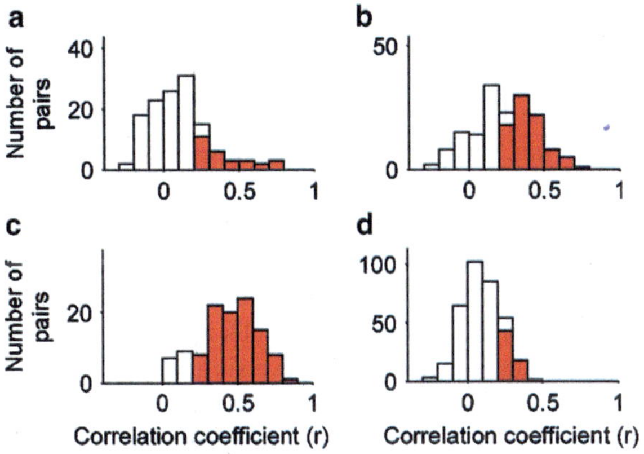

Fig. 5.15 Similarity in object selectivity between pairs quantified by correlation coefficient of evoked responses to 80 object stimuli. The pairs are made of single isolated cells (**a**), and MUs (**b**). Proportions of pairs that had significant correlation were 28.5% (**a**) and 60.0% (**b**). In (**c**), the correlation coefficient was calculated for the pair of MUs and the avgMU obtained from the same columns. Proportion of MUs that had significant correlation in object selectivity with avgMUs was 85.2%. On the other hand, in (**d**), we calculated correlation coefficient for the piars of MUs and avgMU obtained from different columns, Proportion of MUs that had significant correlation was only 16.4%. Thus, the common property represented by the avgMU was unique to the activity spot where the avgMU was calculated and was different from other spots. Horizontal axis, the value of correlation coefficient. The columns with red color represent group of pairs that has statistically significant correlation (p<0.05). (Modified from Sato et al. (2009))

was around 0.4 in correlation coefficient, and this value is higher than the value of similarity in object selectivity of nearby isolated cells. Thus, although evidence suggests a columnar organization in IT cortex (Fujita et al. 1992), object selectivity of nearby cells seemingly contradicts to the column hypothesis in IT cortex.

Furthermore, these findings raise a possibility that object images are specifically represented even by combinations of single cells in a local region that have different response property.

To resolve this contradiction, we analysed object responses of single cells and multi-units (MUs) recorded from activity spots (Fig. 5.13), and explained the reason for the difference in object selectivity of nearby cells that cell-specific response property obscured the common response property across the cells within a columnar region (Sato et al. 2009). In accordance with this explanation, the number of pairs of multiunits (MUs) (60%) showing significant correlation in object selectivity was higher than that of single neuron pairs (28.5%) probably because the variation caused by cell-specific responses were averaged across the cells and the common property became apparent in MU responses (Fig. 5.15a, b). Furthermore, we found that this common property (quantified as the average of MU activities recorded from the same activity spot (avg MU activity)) was different from activity spot to activity spot (Fig. 5.15c, d). Thus, although there is cell-to-cell variability in object responses, the columnar structure is maintained since these cells share a common response property. The relationship between the common property and cell-specific property is demonstrated in Fig. 5.16 by plotting object responses of individual cells against descending order of object responses of the avgMU activity. Furthermore, the PCA analysis of MUs recorded from multiple activity spots revealed that one activity spot was characterized by one or a few common properties in object responses (Fig. 5.17). As we mentioned earlier, analysis of the critical features of individual cells first revealed the columnar organization in area TE (Fujita et al. 1992). We consider that the procedure of stimulus simplification made them find the visual feature that represents the common property across the cells within a columnar region, although the authors of the paper did not intended to search for the common property across the cells. Recently, it has been shown that nearby cells in V1 also respond differently to natural images although these cells should have similar orientation preference (Yen et al. 2007). Thus, the differentiation between the response property specific at a single cell level and the response property at the columnar level may be a universal characteristic in cerebral cortices.

The critical question particularly in area TE is how to relate object representation by combinations of columns and a possible representation by combinations of single cells in a local region. Taking into account trial-by-trial variation of neuronal activity, we consider that representation at the columnar level has primary importance. Neuronal firing approximately follows Poisson distribution, and thus, trial-by-trial variability in neuronal responses is higher for the good stimuli that evoked strong mean responses than the stimuli that evoked weak mean responses. Thus, in general, it is difficult to extract information about the stimulus from activity in a single trial of a single neuron. Since we recognize an object instantaneously without trial-by-trial averaging, one plausible way to extract reliable sensory information is to take an ensemble average of responses of nearby neurons. Our results above point out that this ensemble average reflects tuning specificity of feature columns, and the specificity is generally different from single cells. Accordingly, we consider that recognition of object images presented at a

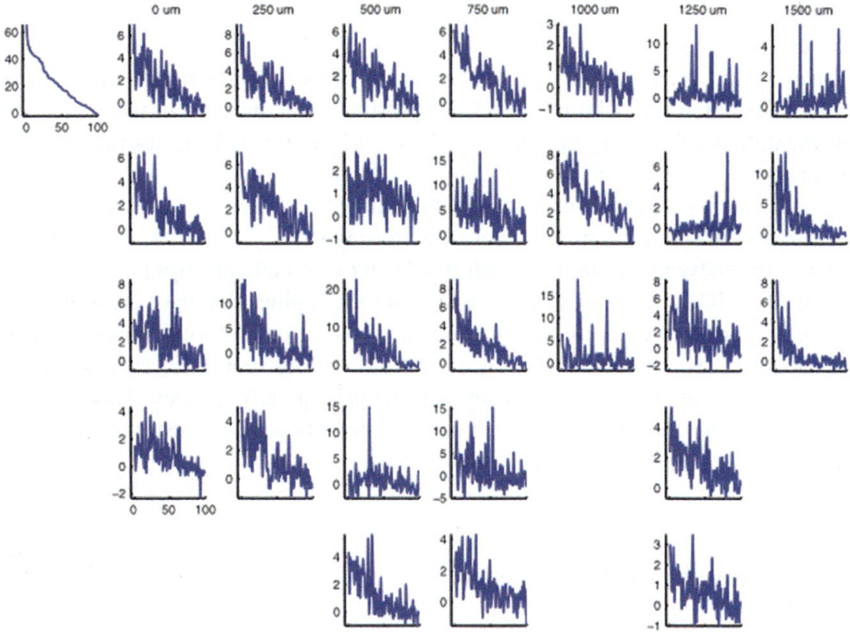

Fig. 5.16 The common response property revealed in tuning curves of the individual cells. Evoked responses of each cells were plotted against object stimuli rank ordered by the avgMU responses. The graph at the *left upper* corner represents the tuning curve of averaged MUs and the rests represent tuning curves of single cells at different depths. Depth of cells in each column is indicated at the *top*. Horizontal axes are rank-ordered according to the magnitude of evoked responses of averaged MUs to the 100 object stimuli in descending order. Vertical axes represent mean firing rate (spikes/s). As you see in single cell tuning curves, though the object that elicited the strongest responses was different from cell to cell, tuning curves of these cells have general tendency that rightward objects evoked weaker responses than leftward objects. (Modified from Sato et al. (2009))

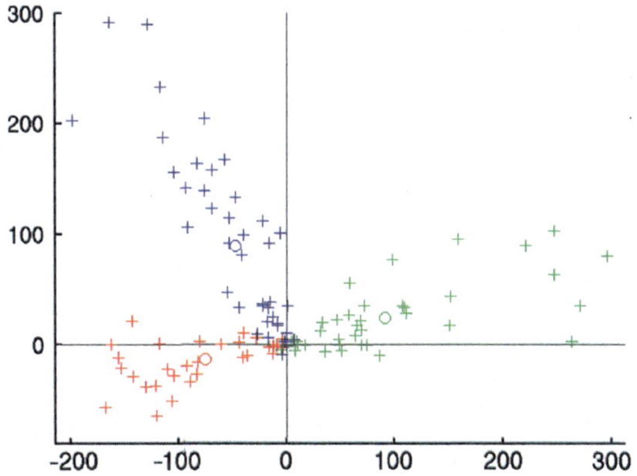

Fig. 5.17 One response property for each spots is approximated in comparison with difference in response property across activity spots. Object Responses of MUs of three spots (crosses; different color indicates different spots) are plotted in a multiple dimensional space made of 80 object images, and projection onto the 2-dimensional plane that includes responses of avgMU in three activity spots (open circles) is shown. (Modified from Sato et al. (2009))

glance is achieved based on representation of object images not at the level of single cells but at the columnar level. It is still an open question in what occasion representation of object images at the level of single cells is useful (but see Summery and Discussion).

Finally, unlike the columnar organizations in V1, we propose that the columnar organization may not cover entire cortex uniformly (Sato et al. 2009). In one hemisphere, we analyzed similarity in object selectivity of cells not from activity spots identified by IOSI but from arbitrary assigned local regions that have about the size of activity spots (Fig. 5.13d). If the columnar organization uniformly covers the cortex, the results would be the same as those from activity spots. However, we found that correlation in object selectivity among MU pairs greatly reduced. This observation suggests that in part of the cortex common property across the cells are not obvious than that in the activity spots.

5.8 Summary and Discussion

In this chapter, we have described object representation in IT cortex revealed by OISI: object images were represented by combinations of active and inactive feature columns, and these feature columns carry information about both local visual features as well as global features such as configuration of object parts. Though we showed representation of these two types of features in separate experiments, we do not intend to emphasize dichotomy between local and global features. It could be the case that some columns represent a visual feature that carries partly global and partly local information although such columns have not been explored yet. The visual features shown here, that is, protrusions, curvature, and rectangular shape for local features and the vertical alignment of parts and facial configuration for global features, are representatives that are easily described with a simple concept. As mentioned in an earlier part of the chapter, plasticity of IT cells make them adjust their response property to the new environment that monkeys react to. Thus, IT cells could tune to the visual features that are useful in certain behavior regardless of whether the represented features are local, global, or include both. Thus, in general, we may not easily describe visual features represented in IT cortex. For example, it would be difficult to characterize the visual features useful for Japanese to recognize Kanji characters at a glance and for Arabic to recognize Arabic characters at a glance.

Because of trial-by-trial variability in neuronal responses, we have discussed that object representation at the level of columns may have primary importance. Then, why do individual cells have a relatively high object tuning property that is different from cell to cell? One thought would be that variability in object responses among cells is necessary for shaping a response property essential for object representation at the columnar level. Alternatively, however, the object representation at the level of single cells may play an essential role in object recognition under certain circumstances. For example, when we inspect an object image, we gaze at the same portion of the image multiple times (Yarbus 1967).

Repeated fixations may play a role in making trial-by-trial averaging of responses at the level of single cells. Then, object recognition may have two stages: the first stage is to capture an object image at the level of feature columns by using ensemble averaging, and for the following careful inspection of the object image, we use representation at the level of single cells with trial-by-trial averaging through repeated fixations. At present, there is no evidence supporting this two stage model. At least, we need to investigate fundamental problems related to solving trial-by-trial variation in neuronal responses. In the case of ensemble averaging, it is essential whether the trail-by-trail variations are uncorrelated across neurons or not. On the other hand, in the case of solving trial-by-trial variations with repeated fixations, a certain kind of short term memory systems is required for holding sensory responses obtained in different time.

As described above, columnar organizations have a potential role in instantaneously and reliably detecting visual features at the columnar level. Thus, to maximize efficiency of plastic changes in IT cortex, plasticity may not only shape response property of individual cells depending on experience but also it may make clusters of these cells and generate new feature columns that are adjusted to the new environment. As described above, we suggest that columnar organization for visual features do not cover the entire region of IT, the rest of the region would be reserved area for generating new feature columns where neurons represent a common visual feature that is useful in certain behavior.

Since the results shown here were obtained from anesthetized monkeys, investigations of dynamic properties of object representation with behaving monkeys are required to relate object representation to recognition. For example, we have examined representation of a single object presented in the visual field. This experimental condition is very much far from natural conditions where many objects are presented simultaneously. Thus, one unsolved problem is how to deal with multiple objects in the visual field. One possible mechanism is proposed by Reynolds et al. (1999). They found that, regardless of whether the visual stimulus was presented in isolation or together with the other stimulus, visual responses of single V4 cell were the same when monkey pays attention to the stimulus. If this is the case, when multiple objects were presented to a monkey, spatial patterns of activity would change from one pattern to another depending on which object the monkey pays attention to. To address this kind of questions, investigations on representation by combinations of feature columns in behaving animals are required. So far in the literature, most of the investigations with behaving monkeys are based on physiological recordings from a single neuron at a time. Since the results presented here revealed that object representation is made of combinations of feature columns, we think that recordings from population activity particularly at the columnar level is essential. However, because of the slow time course of intrinsic signals (Fig. 5.1), OISI may not be the appropriate technique to address object recognition with behaving monkeys in which representation of object images supposed to be updated in a range of a few hundred milliseconds. Novel techniques with high temporal resolution as well as spatial resolution are required. Recently developed densely arranged multiple electrode arrays would be one of the such techniques (deCharms et al. 1999; Miyakawa et al. 2007).

References

Baylis GC, Rolls ET, Leonard CM (1985) Selectivity between faces in the responses of a population of neurons in the cortex in the superior temporal sulcus of the monkey. Brain Res 342(1):91–102

Brincat SL, Connor CE (2004) Underlying principles of visual shape selectivity in posterior inferotemporal cortex. Nat Neurosci 7:880–886

Bruce C, Desimone R, Gross CG (1981) Visual properties of neurons in a polysensory area in superior temporal sulcus of the macaque. J Neurophysiol 46(2):369–384

Das A, Gilbert CD (1995) Long-range horizontal connections and their role in cortical reorganization revealed by optical recording of cat primary visual cortex. Nature 375:780–784

deCharms RC, Blake DT, Merzenich MM (1999) A multielectrode implant device for the cerebral cortex. J Neurosci Methods 93:27–35

Desimone R, Albright TD, Gross CG, Bruce C (1984) Stimulus-selective properties of inferior temporal neurons in the macaque. J Neurosci 4(8):2051–2062

Fujita I, Tanaka K, Ito M, Cheng K (1992) Columns for visual features of objects in monkey inferotemporal cortex. Nature 360(6402):343–346

Fukuda M, Rajagopalan UM, Homma R, Matsumoto M, Nishizaki M, Tanifuji M (2005) Localization of activity-dependent changes in blood volume to submilimeter-scale functional domains in cat visual cortex. Cereb Cortex 15(6):823–833

Grinvald A, Shoham D, Shmuel A, Glaser D, Vanzetta I, Shtoyerman E, Slovin H, Wijnbergen C, Hildesheim R, Arieli A (1999) In-vivo optical imaging of cortical architecture and dynamics. In: Windhorst U and Johansson H (eds) Modern techniques in neuroscience research. Springer, Berlin Heidelberg New York pp. 893–970

Gross CG (1994) How inferior temporal cortex became a visual area. Cereb Cortex 5:455–469

Gross CG, Bender DB, Gerstein GL (1979) Activity of inferior temporal neurons in behaving monkeys. Neuropsychology 17:215–229

Homma R, Tanifuji M (2003) Comparison of functional MAPs in macaque area TE revealed by *in vivo* optical imaging with voltage-sensitive dye and intrinsic signal imaging. Abstr. viewer/ Itinerary Planner. Society for Neuroscience, Washington, DC. 818.21

Kobatake E, Tanaka K (1994) Neuronal selectivities to complex object features in the ventral visual pathway of the macaque cerebral cortex. J Neurophysiol 71(3):856–867

Kobatake E, Wang G, Tanaka K (1998) Effects of shape-discrimination training on the selectivity of inferotemporal cells in adult monkeys. J Neurophysiol 80(1):324–330

Kreiman G, Hung CP, Kraskov A, Quiroga RQ, Poggio T, DiCarlo JJ (2006) Object selectivity of local field potentials and spikes in the macaque inferior temporal cortex. Neuron 49:433–445

MacVicar BA, Hockman D (1991) Imaging of synaptically evoked intrinsic optical signals in hippocampal slices. J Neurosci 11:1458–1469

Miyakawa N, Vidal-Naquet M, Blake D, Merzenich M, Tanifuji M (2007). Activities from combination of columns in macaque area TE can encode object identity across viewing angles, Abstr. viewer/Itinerary Planner. Society for Neuroscience, San Diego. Online, 554.11.

Perrett DI, Rolls ET, Caan W (1982) Visual neurones responsive to faces in the monkey temporal cortex. Exp Brain Res 47(3):329–342

Perrett DI, Oram MW, Harries MH, Bevan R, Hietanen JK, Benson PJ, Thomas S (1991) Viewer-centred and object-centred coding of heads in the macaque temporal cortex. Exp Brain Res 86(1):159–173

Perrett DI, Smith PA, Potter DD, Mistlin AJ, Head AS, Milner AD, Jeeves MA (1984) Neurones responsive to faces in the temporal cortex: studies of functional organization, sensitivity to identity and relation to perception. Hum Neurobiol 3(4):197–208

Reynolds JH, Chelazzi L, Desimone R (1999) Competitive mechanisms subserve attention in macaque areas V2 and V4. J Neurosci 19:1736–1753

Rolls ET, Tovee MJ (1995) Sparseness of the neuronal representation of stimuli in the primate temporal visual cortex. J Neurophysiol 73:713–726

Sato T, Uchida G, Tanifuji M (2009) Cortical columnar organization is reconsidered in intorial tempral cortex. Coreb Corlex 19:1870–1880
Tamura H, Kaneko H, Fujita I (2005) Quantitative analysis of functional clustering of neurons in the macaque inferior temporal cortex. Neurosci Res 52:311–322
Tanaka K, Saito H, Fukada Y, Moriya M (1991) Coding visual images of objects in the inferotemporal cortex of the macaque monkey. J Neurophysiol 66(1):170–189
Tsunoda K, Oguchi Y, Hanazono G, Tanifuji M (2004) Mapping cone- and rod-induced retinal responsiveness in macaque retina by optical imaging. Invest Ophthalmol Vis Sci 45(10):3820–3826
Tsunoda K, Yamane Y, Nishizaki M, Tanifuji M (2001) Complex objects are represented in macaque inferotemporal cortex by the combination of feature columns. Nat Neurosci 4(8):832–838
Vanzetta I, Grinvald A (1999) Increased cortical oxidative metabolism due to sensory stimulation: implications for functional brain imaging. Science 286(5444):1555–1558
Vanzetta I, Solvin H, Omer DB, Grinvald A (2004) Columnar resolution of blood volume and oximetry functional maps in the behaving monkey; implications for fMRI. Neuron 42(5):843–854
Wang G, Tanaka K, Tanifuji M (1996) Optical imaging of functional organization in the monkey inferotemporal cortex. Science 272(5268):1665–1668
Wang G, Tanifuji M, Tanaka K (1998) Functional architecture in monkey inferotemporal cortex revealed by in vivo optical imaging. Neurosci Res 32(1):33–46
Yarbus AL (1967) Eye movements and vision. Plenum Press, New York
Yamane S, Kaji S, Kawano K (1988) What facial features activate face neurons in the inferotemporal cortex of the monkey? Exp Brain Res 73(1):209–214
Yamane Y, Tsunoda K, Matsumoto M, Phillips AN, Tanifuji M (2006) Representation of the spatial relationship among object parts by neruons in macaque inferotemporal cortex. J Neurophysiol 96:3147–3156
Yen S-C, Baker J, Gray CM (2007) Heterogeneity in the Responses of adjacent neurons to natural stimuli in cat striate cortex. J Neurophysiol 97:1326–1341
Young MP, Yamane S (1992) Sparse population coding of faces in the inferotemporal cortex. Science 256(5061):1327–1331

Chapter 6
Optical Imaging of Short–Term Working Memory in Prefrontal Cortex of the Macaque Monkey

Anna W. Roe

Abstract Prefrontal cortex is an area critical for cognitive functions such as planning, decision-making, and reasoning. Working memory is a key aspect to the execution of these functions and has been strongly associated with prefrontal function. This chapter reviews the functional organization of a prefrontal area, area 46, that has been associated with working memory in monkeys. Anatomical and optical imaging studies indicate the presence of a clustered organization within area 46, similar in nature to clustered organizations found in sensory cortical areas. Although the relationship of these clusters to working memory function is unknown, optical imaging studies suggest a spatial organization for mnemonic function. This 'spatial memory map' is topographically consistent with electrophysiologically established maps for visual and eye movement response. Interestingly, in trials in which response suppression is required, optical imaging reveals a possible suppressive signal; lack of this signal may underly the perseveration seen in diseases such as schizophrenia. In sum, I suggest that clustered organization in prefrontal cortex provides a scaffold upon which visual, mnemonic, and motor response are organized.

6.1 Introduction

Since a number of cognitive functions such as planning, decision-making, and reasoning require working memory, understanding the neural basis of cognitive function centers on the organization and encoding of working memory. Disruption of such processes has been associated with cognitive dysfunction (such as dissociation and perseveration) characteristic of mental diseases such as schizophrenia. Thus, understanding the neural basis underlying the working memory is important for a broad range of cognitive functions and for the development of clinical therapies that may lead to treatment of prefrontal dysfunctions.

A.W. Roe (✉)
Department of Psychology, Vanderbilt University, Nashville, TN, USA
e-mail: anna.roe@vanderbilt.edu

A.W. Roe (ed.), *Imaging the Brain with Optical Methods*,
DOI 10.1007/978-1-4419-0452-2_6, © Springer Science+Business Media, LLC 2009

6.2 Prefrontal Delay Period Activity Encodes Short–Term Working Memory

A significant literature suggests that the dorsolateral prefrontal cortex (dlPFC) or area 46 plays an important role in attention and working memory (Goldman-Rakic 1987; Petrides 1994; Fuster 1997; Inoue et al. 2004). It is well known that lesions to the dlPFC impair short-term working memory and the ability to suppress responses. In the macaque monkey (Fig. 6.1a), area 46 surrounds the principle sulcus and lies anterior to the arcuate sulcus (Fig. 6.1b, Walker 1940). (Note that this area has also been subdivided, based on histological and connectional evidence, into area 46 and area 9/46, Petrides and Pandya (1999) (Fig. 6.1c). A number of studies have examined the role of dlPFC in short-term working memory using delay match to sample paradigm. In the oculomotor delay response task (Fig. 6.2), monkeys are presented a cue (a), remember the cued location during the delay period (b), and, upon extinction of the fixation dot, subsequently perform an eye saccade to the remembered location (c). During the delay period, activity of many single neurons in area 46 is elevated, often exhibiting a sustained increase in spike firing. Furthermore, the delay period activity of single neurons has also been shown to be spatially tuned; that is, the delay activity is only elevated for the memory of certain spatial locations and not others (Fig. 6.3, Funihashi et al. 1989). Evidence further suggests that, analogous to population coding in motor cortex (Georgopoulos et al. 1988), location or direction of remembered location results from vector summation of many single unit responses (Takeda and Funihashi 2004). Furthermore, Takeda and Funihashi show that, in monkeys trained to perform a saccade 90° from the remembered location, during the delay period the population response shifts from the remembered location to the intended saccadic location, thereby suggesting a dynamic encoding of such population response. In sum, dlPFC delay period activity, both at the single cell and the population level, is consistent with the short-term storage of information and is regarded as a neuronal correlate of a short-term working memory. Such short-term storage is probably dynamic in nature.

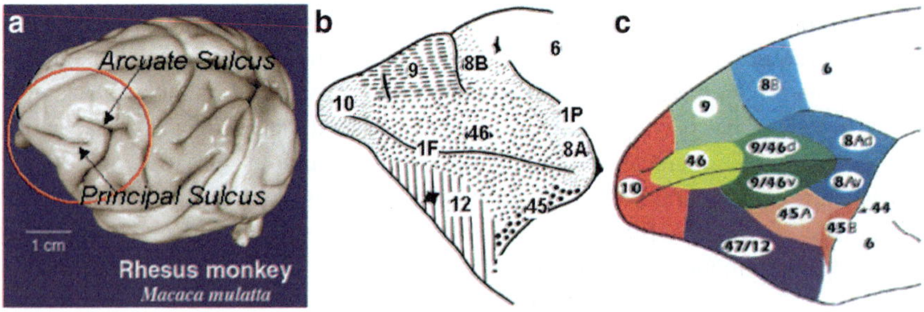

Fig. 6.1 Location of dorsolateral prefrontal cortex in macaque monkey. (**a**) Location of arcuate and principal sulci. (**b**) Area 46 surrounds the principle sulcus and lies anterior to the arcuate sulcus (adapted from Walker 1940). (**c**) Based on histological and connectional study, this area is also referred to as 46 and 9/46 (Petrides and Pandya 1999)

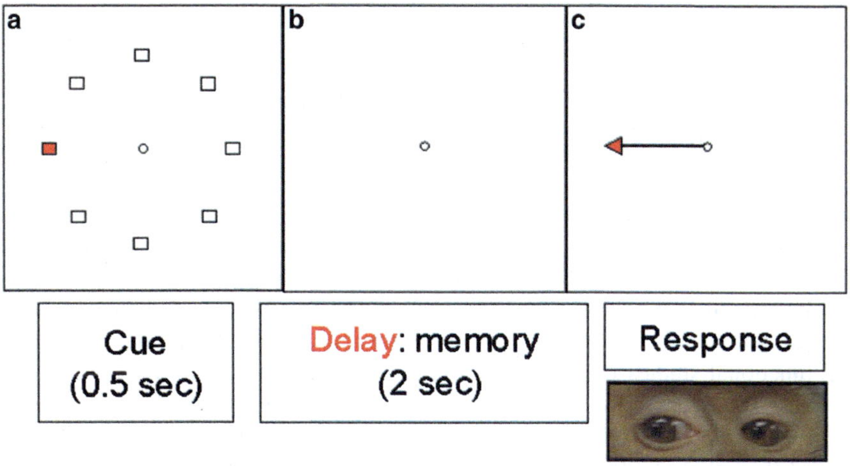

Fig. 6.2 Oculomotor delay response task. Monkeys are trained to visually fixate a central fixation dot (*small circle*) on a monitor continuously throughout this task. They are presented a cue (*red square* in **a**) at one of several possible locations (locations indicated by *empty squares*), the cue disappears and they are required to remember the cued location during the delay period (**b**). Upon extinction of the fixation dot, they subsequently perform an eye saccade to the remembered location (**c**). Correct performance is rewarded with a drop of juice

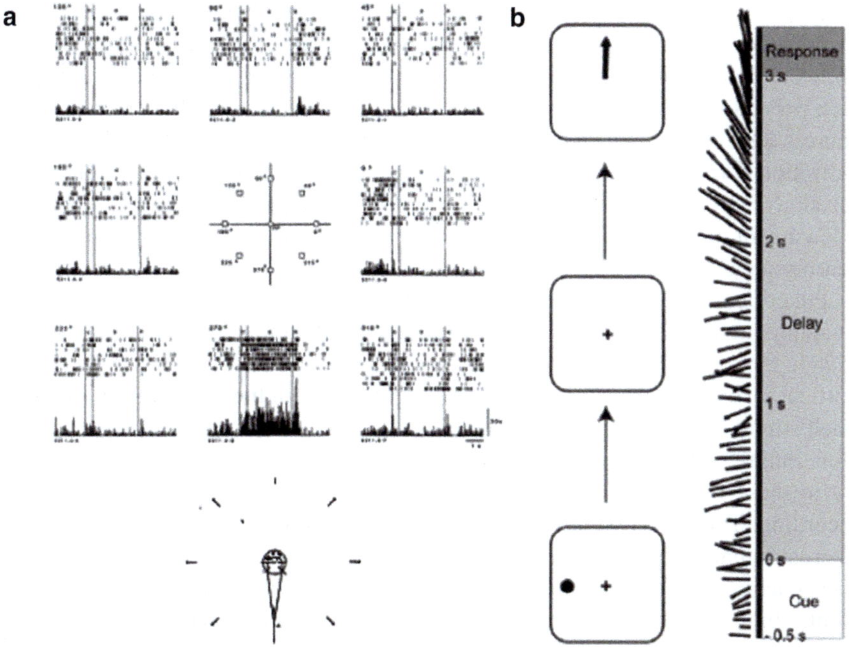

Fig. 6.3 Memory fields are spatially tuned. Delay period activity of prefrontal neuron in a monkey performing oculomotor delay response task. **a** Post-stimulus time histograms (*PSTH*) for activity during oculomotor delay response task to each of 8 directions. Prominent delay period activity is seen only for the DOWN direction (bottom *PSTH*). Resulting delay period tuning curve is shown *below* (from Funihashi et al. 1989). This indicates neurons in area 46 maintain memory trace of stimulus at specific spatial locations. (**b**) Shifting spatial selectivity during a mnemonic task via population vector coding of delay period activity. Direction of vector shifts during delay period reflecting initial representation of visual spatial code to one representing the encoding of intended saccade direction (visual input to motor output) (from Takeda and Funihashi 2004)

6.3 Does Prefrontal Cortex Contain Clustered Functional Organization?

One approach for studying prefrontal activity is optical imaging. This approach is useful only if activity within prefrontal areas is organized in a modular or clustered fashion. A number of cortical areas exhibit functional organizations consisting of repeated modules, typically 200–500 µm in size (terms such as domains, clusters, patches, puffs, and columns have also been used). Evidence from both anatomical and functional studies indicate the presence of clustered organization in sensory, motor, and association cortices (in V1 and V2 (for review see Roe 2003), V4 (Ghose and Ts'o 1997; Tanigawa et al. 2008; Roe 2008), IT (Tsunoda et al. 2001), Area 7 (Siegel et al. 2003; Raffi and Siegel 2005)). Whether regions such as prefrontal cortex contain clustered organizations is controversial (cf. Goldman-Rakic 1987, 1999). However, given the breadth of areas from which clustered organization has been observed, one could argue that the basic organization of prefrontal cortex is not dissimilar from other cortical areas.

Some evidence does point to the presence of clustered organization in prefrontal cortex. In dlPFC, tracer injections lead to a characteristic local network of labeled patches (Lund et al. 1993; Kritzer and Goldman-Rakic 1995). The pattern of label in principal sulcus following injections into other areas such as inferior parietal cortex appear patchy or columnar in nature (Cavada and Goldman-Rakic 1989). In fact, parietally-derived patches are observed to interdigitate with those from callosal sources (Goldman-Rakic and Schwartz 1982). 2-Deoxyglucose labeling of prefrontal activation during a spatial mnemonic task also suggest a columnar or patchy organization (Friedman and Goldman-Rakic 1994). These patches measure on average a few hundred microns in size and are not dissimilar to patchy label seen in other sensory areas.

Functional evidence for columnar organization can also be found from optical imaging studies of cortical response to microstimulation (Fig. 6.4). Sawaguchi (1994, 1996) used voltage sensitive dye imaging methods to record prefrontal cortical response to local electrical microstimulation in the anesthetized monkey. Such stimulation produced focal, interdigitated regions of activation and suppression, ranging in size from 200 to 1,000 µms. The enhanced and suppressed activity in these optically detected regions were confirmed by electrophysiological recording. The optical signal at these sites followed the stimulation 3–6 ms after microstimulation, peaked at 70–80 ms, and lasted for 130–150 ms. Hirata and Sawaguchi (2008) also demonstrated the presence of columnar activation in prefrontal brain slices with voltage sensitive dye techniques. These studies demonstrating focal activation support the presence of functional clustering within prefrontal cortex. Although such clustered activation was not observed in another study of macaque prefrontal cortex using voltage sensitive dye recording methods (Seidemann et al. 2002), there were significant differences in the preparation (awake monkey), the stimulation parameters, and locations of stimulation (FEF) and imaging (FEF and 8Ar).

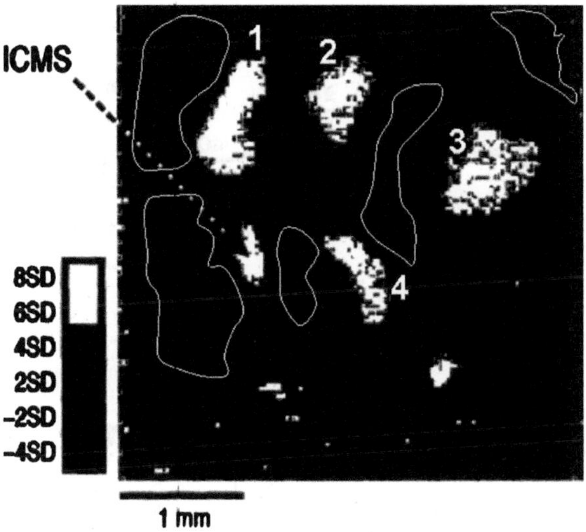

Fig. 6.4 Local activation of clustered activity in prefrontal cortex in response to microstimulation. Intracortical microstimulation (ICMS) at one site leads to focal activation at some nearby sites (1–4) and suppression at other sites (*white outlines*). These sites measure a few hundred microns in size. (Adapted from Sawaguchi (1994, 1996).)

6.4 Topographic Organization of Prefrontal Cortex

Prefrontal cortex contains cells with different response functions. In dlPFC, spike firing has been associated with sensory stimulation, mnemonic response, premotor response, and motor response. Do these response functions have any topographic organization and, if so, do these organizations have any predictable relationship to each other? With respect to visual response, the dlPFC is organized such that the central visual locations are represented ventral and posterior on the arcuate convexity and eccentric locations more dorsal and anterior (Fig. 6.5a, Suzuki and Azuma 1983). Although different studies differ in detail, most studies find a ventral to dorsal topographic map in the PFC. Saccadic responses in the frontal eye fields have an organization such that microstimulation of ventral sites lead to small eye saccades and that of dorsal sites lead to larger saccades (Fig. 6.5b, Bruce and Goldberg 1985).

Since both visual and motor-related responses have a ventral (central, small saccades) to dorsal (peripheral, large saccades) organization in PFC, it is possible that a spatial map for delay period activity may bear a similar organization. Alternatively, as working memory is a dynamic process in which information must be constantly updated, manipulated, and integrated, it is possible that such spatial information changes by the moment or that no spatial organization for mnemonic information exists in prefrontal cortex.

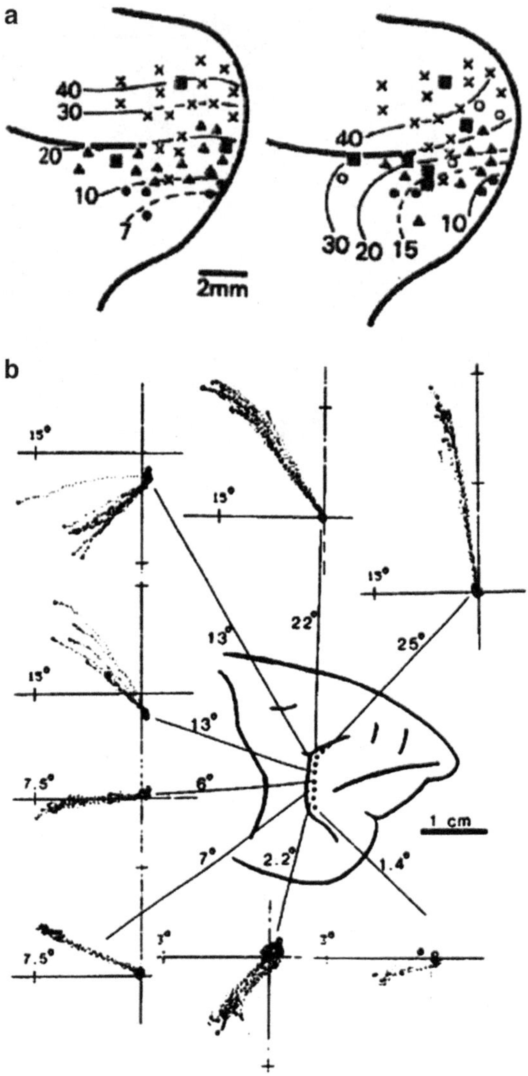

Fig. 6.5 (**a**) Topography of visual response in prefrontal cortex (Suzuki and Azuma 1984). (**b**) Topography of saccadic response to microstimulation at different locations along the arcuate sulcus (frontal eye fields) (Bruce and Goldberg 1985)

6.5 Is There Spatial Organization for Memory Location?

In a series of studies on the functional organization of prefrontal cortex, Sawaguchi and his colleagues (Sawaguchi et al. 1988, 1989; Sawaguchi and Goldman-Rakic 1991, 1994) injected pharmacological agents (such as bicuculline or dopamine antagonists) into restricted regions of prefrontal cortex near the principal sulcus. While monkeys retained their ability to saccade to cued spatial locations, they lost their ability to accurately saccade to remembered locations. These effects were spatially specific, as injection at a site a few millimeters away resulted in a change in the visual location of the deficit. Upper fields were more affected by injections into caudal regions of the dlPFC and amd lower fields by injections into more rostral regions. These studies provided evidence that some spatial map for mnemonic function exists in dlPFC.

One way to directly examine the possible existence of a spatial mnemonic map is to use the intrinsic optical imaging approach. Intrinsic cortical signals are activity-dependent reflectance changes of cortical tissue due to changes in oxygenation of the blood in the microvasculature (Grinvald et al. 1986; Ts'o et al. 1990; Vanzetta et al. 2004) and have both spiking and subthreshhold contributions (Roe and Ts'o 1995, 1999; Issa et al. 2000; Ramsden et al. 2001; Dragoi and Sur 2000; Schwartz and Bonhoeffer 2001; Devor et al. 2003; Thompson et al. 2003). This hemodynamic signal consists of a decrease in reflectance (so-called "initial dip") due to an initial deoxygenation of the tissue (caused by neuronal activity induced increase in oxygen consumption) followed by an increase in reflectance presumably due to the inrush of freshly oxygenated blood (BOLD signal) (see Fig. 6.6a). In sensory cortices, the typical timecourse under 600–630 nm illumination exhibits a time to peak of 2–3 s. The magnitude of reflectance change is typically in the 0.1–1.0% range. Similar signals have been recorded in awake and anesthetized animals (Grinvald et al. 1991; Vnek et al. 1999; Chen et al. 2005). By presenting appropriate stimuli during optical imaging, the functional organizations of the sensory cortices can be mapped at high spatial (50–100 μm) resolution.

The intrinsic signal imaging method was used to map mnemonic prefrontal activity in macaque monkeys trained to perform a delay match to sample task (Roe et al. 2004). During this task (cf. Fig. 6.2), monkeys maintained fixation throughout the imaging period. After presentation of a visual cue (0.5 s), monkeys were required to remember the location of the cue during the delay period (2 s, fixation maintained). Upon offset of the fixation spot, they performed an eye saccade to the previously cued location, indicating they had remembered the correct location. During this task, single condition images were obtained by summing imaged frames acquired during the delay period (see Fig. 6.7, frames 6–15) and subtracting a sum of prestim frames (frames 1–2, first-frame subtraction). During each block of trials, the set of cued locations were presented in an interleaved fashion. Blank trials, during which no cue was presented, were also interleaved.

To explore the presence of a spatial map, imaged delay period activity for eight different locations of identical eccentricity were compared (Fig. 6.6b, inset above). In this task, while the monkey maintained fixation, a cue (Cue) was flashed at one of the eight locations. Memory for this location was maintained through the delay period (Delay) and was indicated by saccade to the correct remembered location (Resp).

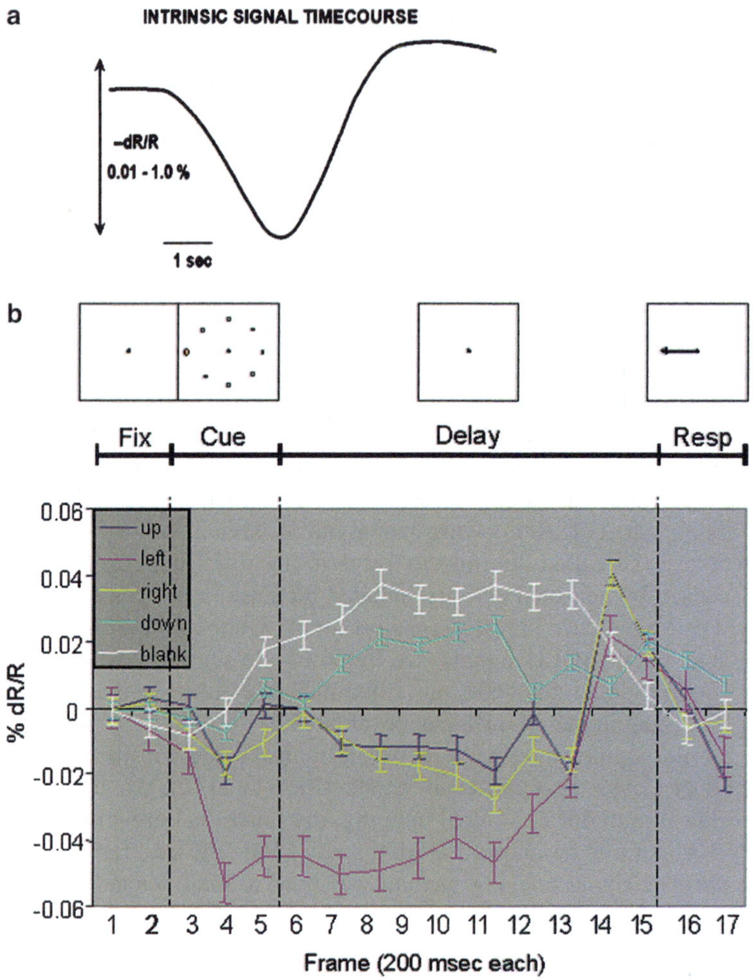

Fig. 6.6 Differential timecourses in dlPFC for remembered location. (**a**) Standard intrinsic signal timecourse. Neural activation is accompanied by a decrease in cortical reflectance (dR/R). (**b**) The intrinsic signal recorded from dlPFC location with strong delay activity for *left* (*pink*), weaker response for *up* (*dark blue*) and *right* (*yellow*), and even weaker (or perhaps suppressive) for *down* (*light blue*) and *blank* (*white*)

Examination of intrinsic signal timecourses suggested spatial specificity. As shown in Fig. 6.6b, cortical response timecourses were sampled from a region responsive to memory for the left location. Delay period activity during memory for up, left, right, down, and blank locations produced different magnitude responses. In this location, response was greatest for the left (pink curve) location, weaker for the up (blue curve) and the right (yellow curve) locations, and weakest (perhaps suppressed?) for the down (light blue curve) and blank

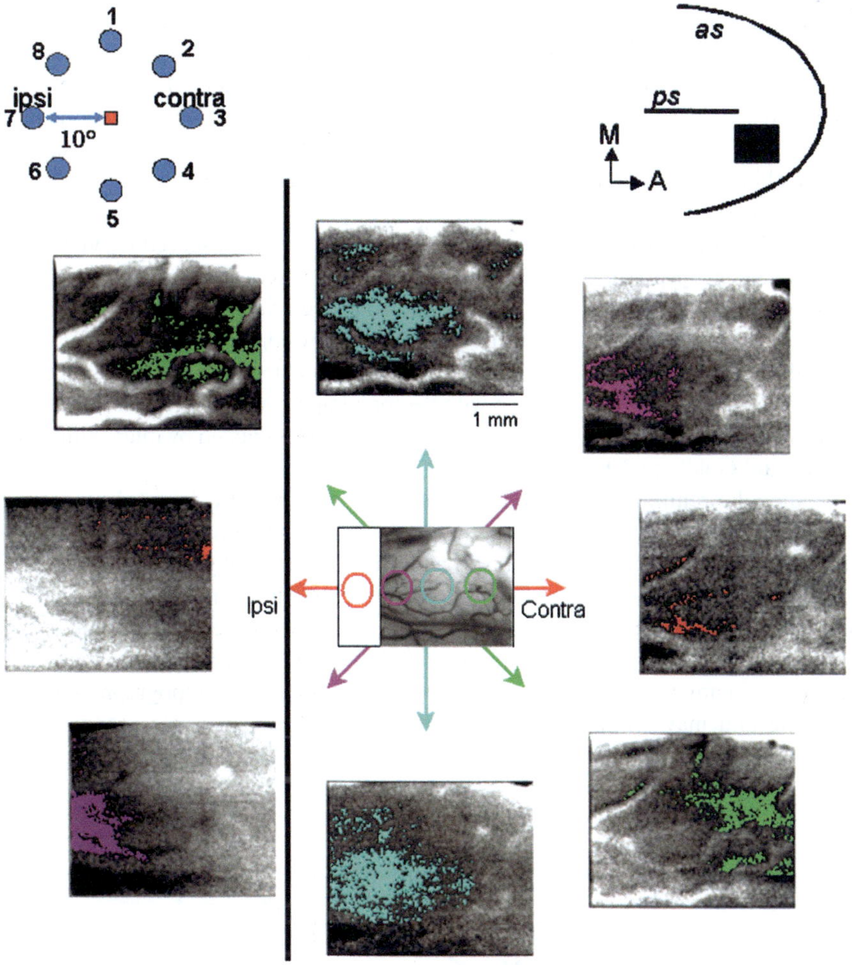

Fig. 6.7 Imaged maps in dlPFC in a delay match to sample task. Each image was obtained by summing the reflectance response during delay period (frames 6–15) and subtracing precue activity (frames 1–2) and represents the delay activity associated with memory for one of the eight locations (see *upper left inset*). *Upper right inset* indicates location imaged in dlPFC. *Ps* principal sulcus; *As* arcuate sulcus; *A* anterior; *M* medial. Central schematic summarizes the activations: from posterior to anterior, mapping progresses from horizontal axis (*red*), to 45° clockwise axis (*magenta*), to vertical (*light blue*), to 45° counterclockwise axis (*green*)

(white curve) conditions. This suggests that each location of dlPFC exhibits maximal delay period activity for one particular remembered location and not for others.

This impression was further supported by examining images of delay activity. Images of prefrontal delay activity revealed that memory for different locations led to different prefrontal activations. Within the region representing a single eccentricity (Fig. 6.7, upper right inset), memory for locations on the vertical axis (locations 1 and 5) appeared central most in the image (light blue pixels). Those on axis 45° clockwise (locations 2 and 6) fell more caudal (magenta pixels), and those on axis tilted 45° counterclockwise (locations 4 and 8) fell more anteriorly (green pixels). Activations to locations on the horizontal axis (locations 3 and 7) produced the least activation (red pixels). Thus, although the topography is crude, this data does suggest that, within a local region in dlPFC, there is a differential mapping with respect to spatial mnemonic location.

How does this activation relate to the dorsal–ventral map observed for visual and eye movement response? To address this question, a second paradigm compared delay period response for more peripheral (10° eccentricity) vs. more central (5° eccentricity) locations (Fig. 6.8, left). Results showed that activation for more central locations (Fig. 6.8d, 5°) were located ventral to those of more eccentric (Fig. 6.8c, 10°) locations, consistent with the topographic maps for visual stimulation (Suzuki and Azuma 1983) and for location of eye saccades elicited by microstimuation (Bruce and Goldberg 1985).

Although preliminary, these data together suggest that PFC in monkeys trained for spatial oculomotor delay response tasks contain some global spatial topography (Fig. 6.9, dashed isoeccentricity lines) and a local topography within each spatial location which is based on different axes of spatial memory (Fig. 6.9, colored arrows). Combining previous studies on topography with these results suggests a possible organization as depicted in Fig. 6.9. However, one must bear in mind that, given the dynamic nature of working memory demands, such a topography may not be static and may simply be one instantiation of many possible organizations

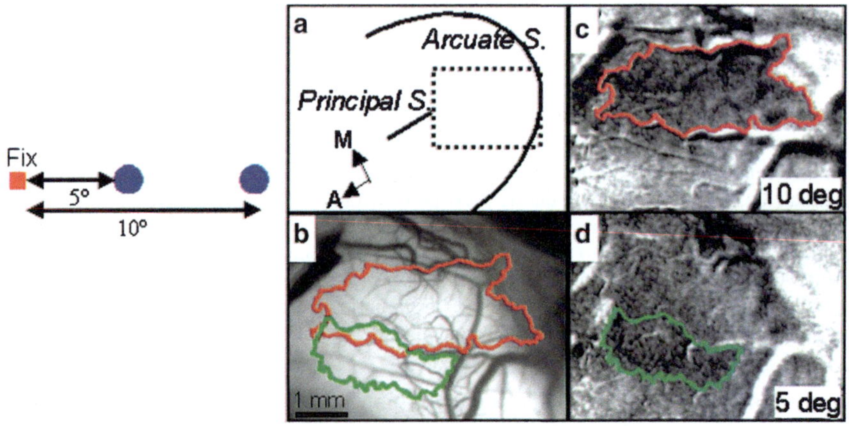

Fig. 6.8 Imaging global organization of central and peripheral mnemonic representation. *Left*: Schematic of mnemonic task for small vs. large eccentricities. (**a**) Diagram of imaged location in dlPFC (*dotted rectangle*). (**b**) Summary overlay of activations seen in (**c**) and (**d**). (**c**) and (**d**) Activation for more central locations ((**d**), 5°) were located ventral to those of more eccentric ((**c**), 10°) locations, consistent with the topographic maps for visual stimulation (Suzuki and Azuma 1984) and for location of eye saccades elicited by microstimuation (Bruce and Goldberg 1985)

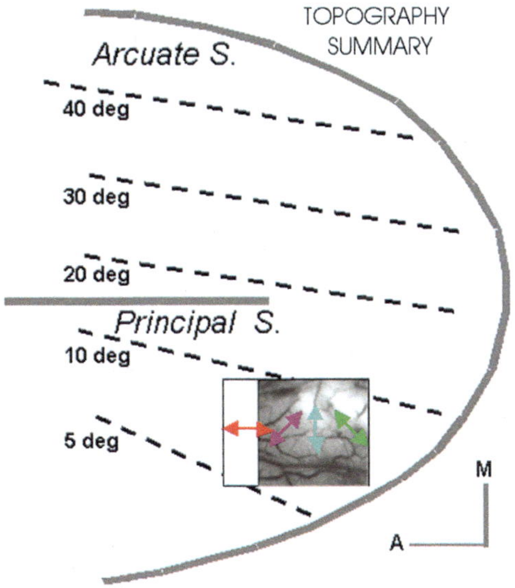

Fig. 6.9 Summary for spatial topography of delay activity in PFC

established in different tasks. Although speculative, we would like to forward the possibility that prefrontal cortex may have a native spatial organization that is then assumed by current task demands.

6.6 Is There a Signal for Suppression in Prefrontal Cortex?

One of the hallmarks for prefrontal function is the ability to appropriately suppress actions. Diseases involving prefrontal dysfunction are often accompanied by inability to appropriately cease or adapt to changing task demand.

To test whether such dysfunction might be detected with imaging methods, during delay match to sample tasks, we presented blank trials interleaved with cued trials. During the blank trials, the monkey visually acquires the fixation spot but is presented no cue. He thus awaits the cue and expects it at a certain time. However, when none is presented he must suppress his characteristic behavior and NOT perform an eye saccade (see Fig. 6.10, schematic at left). During these blank trials, since there is no visual cue and thus nothing to remember, one might expect the lack of any detectable response (no reflectance change). However, a consistent and robust signal was observed during these blank trials. Surprisingly, the observed optical signal was opposite in sign (i.e., increase in reflection) to the normal decrease in reflectance (Fig. 6.10, right; see also Fig. 6.6 white curve). The magnitude of this upward deflection was quite large, in fact equal in magnitude to the downward mnemonic response (Fig. 6.6). This

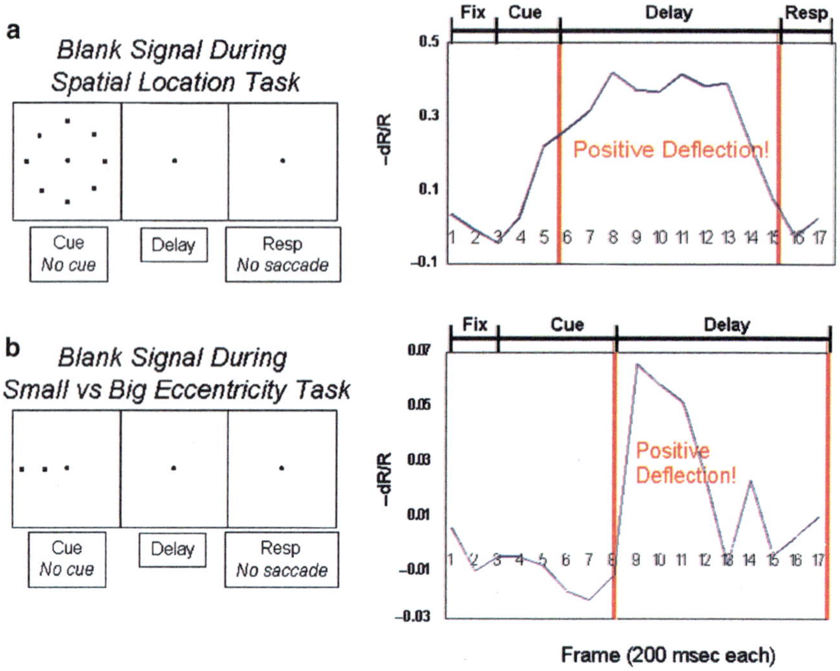

Fig. 6.10 Delay period activity during blank trials. During blank trials, no cue appeared and the monkey was required to maintain fixation at the central fixation spot and not to perform any saccade. (**a**) Eight location task; (**b**) small vs. big eccentricity task. For both tasks, reflectance signal exhibits upwards deflection during the period of task corresponding to the delay period in cued trials

signal, furthermore, was not a general response of the task, but rather followed the timing of the delay period. As seen in the graphs in Fig. 6.10, the upward deflection begins at onset of delay period (left red line: at frame 6 in Fig. 6.10a, at frame 8 in Fig. 6.10b) and falls by the end of the delay period (second red line: at frame 15 in Fig. 6.10a and at frame 17 in Fig. 6.10b).

How should this increase in reflectance be interpreted? It is unlikely to be cue related since it is seen only during delay periods. It is also unlikely to be a mnemonic response since there was no cue to remember. One possibility is that this positive signal deflection could be interpreted as a suppressive signal, one which is needed to suppress the saccadic response. Electrophysiological recording would be required to evaluate the neuronal response (e.g., decrease in neuronal firing) underlying the increase in optical reflectance. We suggest one possible corollary of this finding. If this signal arises due to an active suppression of behavioral response, it is possible that the interference of this suppressive response would result in inappropriate saccade behavior. In other words, lack of such a suppressive signal could underly perseveration behavior characteristic of prefrontal dysfunction (cf. Gusnard et al. 2003). This is an exciting possibility that could have significant clinical implications.

6.7 Summary

The experiments described here summarize some of the initial attempts at elucidating the functional organization of dorsolateral prefrontal cortex in the macaque monkey. While such studies are still in their early stages, they are important for guiding future studies on prefrontal organization. The key observation that delay period activity in dlPFC is tuned was highly instrumental in driving investigations into spatial organization of mnemonic representation. In principal, such tuning could be viewed as analogous to orientation tuning in visual cortex. As demonstrated in visual cortex, such tunings can self-organize to form systemic maps (cf. Swindale 2004). Similarly, prefrontal areas may share such structural framework. The studies described here have introduced the idea that such tuning can, at least at the population level, form crude spatial organizations in prefrontal cortex. Some important directions that remain for future studies include: (1) What are the dynamics of such maps, especially as they relate to task demand? (2) Is there differential localization for different types of mnemonic activity, such as spatial vs. nonspatial memory (McCarthy et al. 1996; Rainer et al. 1998; Kojima et al. 2007)? and (3) How does the organization of mnemonic function relate to other prefrontal functions such as active suppression of persistent behavior? Advances in neuroimaging and electrophysiological approaches, both at the cellular, population, and behavioral levels, will undoubtedly elucidate these questions in the near future.

Acknowledgment This chapter is written in memory of Dr. Patricia Goldman-Rakic. Much of the optical imaging work described here was conducted in collaboration with Dr. Goldman-Rakic at Department of Neurobiology at Yale University School of Medicine, New Haven CT. Dr. Goldman-Rakic was a pioneer in prefrontal function and encouraged me to explore and extend ideas regarding functional organization to prefrontal areas. I am grateful to have had her support and mentorship. Others who contributed to this work were E Sybirska and Douglas Walled. Supported by Packard Foundation, NIMH P50MH068789, NEI EY11744, NIDA DA023002.

References

Bruce CJ, Goldberg ME (1985) Primate frontal eye fields. II. Physiological and anatomical correlates of electrically evoked eye movements. J Neurophysiol 54:714–734

Cavada C, Goldman-Rakic PS (1989) Posterior parietal cortex in rhesus monkey: II Evidence for segregated corticocortical networks linking sensory and limbic areas with the frontal lobe. J Comp Neurol 287(4):422–445

Chen LM, Friedman RM, Roe AW (2005) Optical imaging of SI topography in anesthetized and awake squirrel monkey. J Neurosci 25:7648–7659

Devor A, Dunn AK, Andermann ML, Ulbert I, Boas DA, Dale AM (2003) Coupling of total hemoglobin concentration, oxygenation, and neural activity in rat somatosensory cortex. Neuron 39:353–359

Dragoi V, Sur M (2000) Dynamic properties of recurrent inhibition in primary visual cortex: contrast and orientation dependence of contextual effects. J Neurophysiol 83:1019–1030

Friedman HR, Goldman-Rakic PS (1994) Coactivation of prefrontal cortex and inferior parietal cortex in working memory tasks revealed by 2DG functional mapping in the rhesus monkey. J Neurosci 14:2775–2788

Funihashi S, Bruce CJ, Goldman-Rakic PS (1989) Mnemonic coding of visual space in the monkey's dorsolateral prefrontal cortex. J Neurophysiol 61:331–349

Fuster JM (1997) The prefrontal cortex. Lippincott-Raven, Philadelphia, PA

Georgopoulos AP, Kettner RE, Schwartz AB (1988) Primate motor cortex and free arm movements to visual targets in three-dimensional space. II. Coding of the direction of movement by a neuronal population. J Neurosci 8(8):2928–2937

Ghose GM, Ts'o DY (1997) Form processing modules in primate area V4. J Neurophysiol 77(4):2191–2196

Goldman-Rakic PS (1987) Circuitry of primate prefrontal cortex and regulation of behavior by representational memory. In: Mountcastle VB (ed) Higher functions of the brain. Part 1, Handbook of physiology. Section I: The nervous system, vol 5. American Physiological Society, Bethesda, MD, pp 374–417

Goldman-Rakic PS (1999) The physiological approach: functional architecture of working memory and disordered cognition in schizophrenia. Biol Psychiatry 46(5):650–661

Goldman-Rakic PS, Schwartz ML (1982) Interdigitation of contralateral and ipsilateral columnar projections to frontal association cortex in primates. Science 216(4547):755–757

Grinvald A, Lieke E, Frostig RD, Gilbert CD, Wiesel TN (1986) Functional architecture of cortex revealed by optical imaging of intrinsic signals. Nature 324:361–364

Grinvald A, Frostig RD, Siegel RM, Bartfeld E (1991) High-resolution optical imaging of functional brain architecture in the awake monkey. Proc Natl Acad Sci U S A 88(24):11559–11563

Gusnard DA, Ollinger JM, Shulman GL, Cloninger CR, Price JL, Van Essen DC, Raichle ME (2003) Persistence and brain circuitry. Proc Natl Acad Sci U S A 100(6):3479–3484

Hirata Y, Sawaguchi T (2008) Functional columns in the primate prefrontal cortex revealed by optical imaging in vitro. Neurosci Res 61:1–10

Inoue M, Mikami A, Ando I, Tsukada H (2004) Functional brain mapping of the macaque related to spatial working memory as revealed by PET. Cereb Cortex 14:106–119

Issa NP, Trepel C, Stryker MP (2000) Spatial frequency maps in cat visual cortex. J Neurosci 20:8504–8514

Kojima T, Onoe H, Hikosaka K, Tsutsui K, Tsukada H, Watanabe M (2007) Domain-related differentiation of working memory in the Japanese macaque (Macaca fuscata) frontal cortex: a positron emission tomography study. Eur J NeuroSci 25:2523–2535

Kritzer MF, Goldman-Rakic PS (1995) Intrinsic circuit organization of the major layers and sublayers of the dorsolateral prefrontal cortex in the rhesus monkey. J Comp Neurol 359:131–143

Lund JS, Yoshioka T, Levitt JB (1993) Comparison of intrinsic connectivity in different areas of macaque monkey cerebral cortex. Cereb Cortex 3(2):148–162

McCarthy G, Puce A, Constable RT, Krystal HJ, Gore JC, Goldman-Rakic P (1996) Activation of human prefrontal cortex during spatial and nonspatial working memory tasks measured by functional MRI. Cereb Cortex 6:600–611

Petrides M (1994) Frontal lobes and working memory: evidence from invesitgation of the effects of cortical excisions in nonhuman primates. In: Boller F, Spinnier H, Hendler JA (eds) Handbook of neuropsychology, vol 9. Amsterdam, Elsevier, pp 59–82

Petrides M, Pandya DN (1999) Dorsolateral prefrontal cortex: comparative cytoarchitectonic analysis in the human and the macaque brain and corticocortical connection patterns. Eur J NeuroSci 11(3):1011–1036

Raffi M, Siegel RM (2005) Functional architecture of spatial attention in the parietal cortex of the behaving monkey. J Neurosci 25(21):5171–5186

Rainer G, Asaad WF, Miller EK (1998) Memory fields of neurons in the primate prefrontal cortex. Proc Natl Acad Sci U S A 95(25):15008–15013

Ramsden BM, Hung CP, Roe AW (2001) Real and illusory contour processing in Area V1 of the primate – a cortical balancing act. Cereb Cortex 11:648–665

Roe AW (2003) Modular complexity of Area V2 in the Macaque monkey. In: Collins C, Kaas J (eds) The primate visual system. CRC Press, New York, pp 109–138

Roe AW (2008) Optical imaging of visual feature representation in the awake, fixating monkey. In: Wang R, Gu F, Shen E (eds) Advances in cognitive neurodynamics ICCN 2007: proceedings of the International Conference on Cognitive Neurodynamics 2007. Springer, New York. ISBN: 978-1-4020-8386-0

Roe AW, Ts'o DY (1995) Visual topography in primate V2: multiple representation across functional stripes. J Neurosci 15:3689–3715

Roe AW, Ts'o DY (1999) Specificity of color connectivity between primate V1 and V2. J Neurophysiol 82:2719–2731

Roe AW, Walled D, Sybirska E, Goldman-Rakic PS (2004) Optical imaging of prefrontal cortex during oculomotor delay response task in Macaque monkey. Soc Neurosci Abstract, San Diego, CA

Sawaguchi T (1994) Modular activation and suppression of neocortical activity in the monkey revealed by optical imaging. NeuroReport 6:185–189

Sawaguchi T (1996) Functional modular organization of the primate prefrontal cortex for representing working memory process. Brain Res Cogn Brain Res 5(1–2):157–163

Sawaguchi T, Goldman-Rakic PS (1991) D1 dopamine receptors in prefrontal cortex: involvement in working memory. Science 251(4996):947–950

Sawaguchi T, Goldman-Rakic PS (1994) The role of D1-dopamine receptor in working memory: local injections of dopamine antagonists into the prefrontal cortex of rhesus monkeys performing an oculomotor delayed-response task. J Neurophysiol 71(2):515–528

Sawaguchi T, Matsumura M, Kubota K (1988) Delayed response deficit in monkeys by locally disturbed prefrontal neuronal activity by bicuculline. Behav Brain Res 31(2):193–198

Sawaguchi T, Matsumura M, Kubota K (1989) Delayed response deficits produced by local injection of bicuculline into the dorsolateral prefrontal cortex in Japanese macaque monkeys. Exp Brain Res 75(3):457–469

Schwartz TH, Bonhoeffer T (2001) In vivo optical mapping of epileptic foci and surround inhibition in ferret cerebral cortex. Nat Med 9:1063–1067

Seidemann E, Arieli A, Grinvald A, Slovin H (2002) Dynamics of depolarization and hyperpolarization in the frontal cortex and saccade goal. Science 295(5556):862–865

Siegel RM, Raffi M, Phinney RE, Turner JA, Jando G (2003) Functional architecture of eye position gain fields in visual association cortex of behaving monkey. J Neurophysiol 90:1279–1294

Suzuki and Azuma (1983) Topographic studies on visual neurons in the dorsolateral prefrontal cortex of the monkey. Exp Brain Res 53:47–58

Swindale NV (2004) How different feature spaces may be represented in cortical maps. Network 15(4):217–242

Takeda K, Funihashi S (2004) Population vector analysis of primate prefrontal activity during spatial working memory. Cereb Cortex 14(12):1328–1339

Tanigawa H, Lu HD, Chen G, Roe AW (2008) Functional subdivisions in macaque V4 revealed by optical imaging in the behaving Macaque monkey. Vision Sciences Society, Naples, FL

Thompson JK, Peterson MR, Freeman RD (2003) Single-neuron activity and tissue oxygenation in the cerebral cortex. Science 299:1070–1072

Ts'o DY, Frostig RD, Lieke EE, Grinvald A (1990) Functional organization of primate visual cortex revealed by high resolution optical imaging. Science 249:417–420

Tsunoda K, Yamane Y, Nishizaki M, Tanifuji M (2001) Complex objects are represented in macaque inferotemporal cortex by the combination of feature columns. Nat Neurosci 4:832–838

Vanzetta I, Slovin H, Omer DB, Grinvald A (2004) Columnar resolution of blood volume and oximetry functional maps in the behaving monkey; implications for FMRI. Neuron 42:843–854

Vnek N, Ramsden B, Hung C, Goldman-Rakic PS, Roe AW (1999) Optical imaging of functional domains in the cortex of the awake and behaving primate. Proc Natl Acad Sci U S A 96:4057–4060

Walker AE (1940) A cytoarchitectural study of the prefrontal area of the macaque monkey. J Comp Neurol 73:59–86

Chapter 7
Intraoperative Optical Imaging of Human Cortex

Minah Suh, Hongtao Ma, Mingrui Zhao, Andrew Geneslaw, and Theodore H. Schwartz

Abstract Optical imaging of the human brain can be performed during neurosurgical operations. In spite of large sources of noise it is possible to obtain maps of somatosensory, motor and language cortices. These changes in intrinsic light reflectance tend to overestimate the area of cortex essential for these functions, representing hemodynamic alterations associated with neuronal activity. Imaging of abnormal cortical activity such as afterdischarges or seizures can also be performed intraoperatively and the larger magnitude of the signal improves the singal-to-noise ratio. The utility of intrinsic signal imaging during neurosurgical operations and as a tool for human brain mapping is still evolving but certainly holds great promise based on the success of the preliminary experiments to date.

While optical imaging techniques have had a profound and powerful impact on laboratory investigations into cortical organization in animals, human studies have been limited. Several factors account for this paucity of optical data on human cortical organization. The thickness of the human skull and dura severely limit the spatial resolution of transcranial optical imaging in subjects beyond the first or second year of life. Thus, the only access to exposed human brain must occur during the course of therapeutic neurosurgical interventions. In the operating room, for ethical reasons, any research investigation using imaging must be a secondary priority to the therapeutic goal of the surgery and the well-being of the patient. Hence, the location of the exposed cortex is dictated by the pathology in the brain. Exposed cortex, if not itself pathological, is usually adjacent to pathology and thus may not be entirely "normal." In addition, the time allotted for each experiment is limited. There is little time for "troubleshooting" any technical difficulties. Operating room time is also quite expensive and at a busy medical center, surgeries are expected to be completed in a certain timeframe after which another surgery is usually scheduled. In addition, the sources of noise in the human cortex are much larger than encountered in the animal cortex. Cranial windows are on the order of several centimeters and the curvature

T.H. Schwartz (✉)
Department of Neurological Surgery, Weill Medical College of Cornell University, New York, NY 10021, USA
e-mail: schwarh@med.cornell.edu

A.W. Roe (ed.), *Imaging the Brain with Optical Methods*,
DOI 10.1007/978-1-4419-0452-2_7, © Springer Science+Business Media, LLC 2009

of the surface of the cortex is significant. Movement artifact from heartbeat and respiration are enormous without mechanical dampening and post hoc processing. Controlling ambient light is also challenging since nurses and anesthesiologists must enter and leave the operating room, answer the phone, and be able to use computer screens. As can be imagined, the priority of these aforementioned essential OR staff members is not the quality of the optical data, but rather the stability of the patient and the smooth operation of the operating room. Finally, few neurosurgeons are willing to donate their operating room time to a PhD neurobiologist when they have patients waiting in their offices and other cases following in the operating room. For all of these reasons, and many others, the hurdles in acquiring high quality optical imaging data in adult human cortex are considerable.

Nevertheless, several groups have pursued intraoperative optical imaging of human brain with varying results. Overall, the technical hurdles and uncontrolled environment have, to date, limited the impact of these studies. However, what is clear is that optical imaging techniques have enormous potential for contributing to human cortical investigation and as the technical hurdles are overcome, more and more clinical and basic applications will emerge. In this chapter, we will summarize this work to date and then discuss methods to eliminate noise and future directions for study.

7.1 The Intrinsic Optical Signal

Since voltage sensitive and calcium dyes are not approved for use in humans, most work has relied on intrinsic signals. The intrinsic optical signal (IOS) is based on the work of Hill and Keynes (1949), who showed that the light absorption (and hence reflection) properties of neuronal tissue change with activation. These signals were then used to map cortical architecture on a macroscopic level in anesthetized animals (Grinvald et al. 1986; Bonhoeffer and Grinvald 1991; Frostig et al. 1990). The sources of the intrinsic signal are wavelength dependent (Bonhoeffer and Grinvald 1996). At isosbestic wavelengths of hemoglobin where oxygenated hemoglobin (HbO_2) and deoxygenated hemoglobin (Hbr) reflect light equally (e.g., 525, 545, 570.5 and 583 nm) (Sheth et al. 2003), the IOS measures total hemoglobin (Hbt), which is directly proportional to cerebral blood volume (CBV) and cerebral blood flow (CBF), if the hematocrit is constant(Nemoto et al. 2004; Mayhew et al. 2000). At higher wavelengths (600–650 nm), the majority of the signal arises from the oxygenation state of hemoglobin since Hbr absorbs light with three times the absorption coefficient of HbO_2 (Malonek and Grinvald 1996; Sato et al. 2002a). With multiwavelength optical spectroscopic imaging, it is possible to directly calculate and quantify Hbr, $HbO_{2,}$ and Hbt using a modified Beer–Lambert law (Sheth et al. 2003). At higher wavelengths (>650), scattering of light from cell (neuron and glia) swelling and changes in the extracellular volume become increasingly important, although hemoglobin oxygenation still comprises a significant proportion of the signal at these wavelengths when recorded in vivo.

7.1.1 Neurovascular Coupling

To understand the correlation between the IOS and neuronal activity, we must quickly review our current understanding of neurovascular coupling. As neurons become active, they increase their metabolic demand for oxygen ($CMRO_2$) causing a rapid, focal decrease in tissue partial pressure of oxygen (pO_2) and an increase in the oxygen extraction ratio leading to a increase in the concentration of Hbr relative to the concentration of oxygenated hemoglobin HbO_2(Malonek and Grinvald 1996; Vanzetta and Grinvald 1999; Pouratian et al. 2003a). This "initial dip" in local oxygenation is spatially highly colocalized with neuronal activity and begins within 100 ms of neuronal activation (Frostig et al. 1990). Approximately 300–500 ms later, the arterial microvasculature begins to dilate, which causes a subsequent increase in CBF and CBV. The pressure gradient between the arterial, capillary, and venous compartments drives the flow of blood. As the pressure increases against the pressure of the surrounding brain the capacitance of the vascular compartment decreases leading to a saturation in the nonlinear increase in volume expansion (Huppert et al. 2007). Initially, this increase in perfusion appears to colocalize with the population of active neurons, but soon (>2 s) spreads to a larger area of cortex (Malonek and Grinvald 1996). A few seconds later the diffuse increase in CBF delivers an oversupply of HbO_2 to a large region of surrounding brain, which causes a relative decrease in the concentration of Hbr (Fox and Raichle 1986). This latter component is one of the main contributors to the blood oxygen level dependent (BOLD) signal imaged with fMRI, which has its onset ~ 2s after stimulus presentation, although other factors are also involved (Vanzetta and Grinvald 1999; Pouratian et al. 2003a; Kwong et al. 1992). Colocalization of the increase in the BOLD signal with neuronal activity is less precise and the signal is often seen in the draining veins (Pouratian et al. 2003a).

7.2 The History of Human IOS

The first report of IOS imaging in humans was an abstract presented at the Society of Neuroscience by MacVicar et al. (1990). The authors performed cortical stimulation on the surface of the human brain during surgery for epilepsy and noted a focal decrease in the reflectance of scattered light in the stimulated gyrus. Soon after, Haglund et al. (1992) recorded the IOS at 695 nm in a series of patients undergoing epilepsy surgery during cortical stimulation and language tasks and again noted focal changes in light absorption. Soon after, several other groups began reporting human IOS data (Shoham and Grinvald 2001; Schwartz et al. 2004; Sato et al. 2002b; Toga et al. 1995).

7.3 Imaging Normal Cortical Architecture

7.3.1 Somatosensory Cortex

The first studies of human somatosensory cortex with intraoperative IOS were performed by Haglund et al. (1992) at 690 nm, acquired through the operating microscope. A glass footplate was used for cortical stabilization. An awake patient moved his or her tongue and reflectance changes were recorded from the palate sensory strip. Monophasic reflectance changes of ~20% which lasted for tens of seconds were reported. At the wavelengths recorded, given the length of the IOS response, the signal likely represented a combination of decreased Hbr and light scatter.

Further work on somatosensory processing was then reported by Toga et al. (1995) at UCLA. In these and subsequent studies (Cannestra et al. 1996; Cannestra et al. 1998; Pouratian et al. 2000) his group recorded the IOS though the optics of a Zeiss operating microscope. The cortex was illuminated with white light and recorded through 610 nm transmission filters using a 200 ms frame duration and 500 ms frame resolution. Somatosensory cortex was excited with 2s transcutaneous stimulation of the median or ulnar nerves, digit stimulation or tongue movement. Noise was reduced by time-locking image acquisition with heart beat and respiration, and using post hoc motion correction algorithms and principal component analysis (PCA). The recorded IOS was monophasic and peaked at 2–3 s after stimulation and lasted for ~6 s. Responses were seen at the crest of the gyrus, corresponding to Area 1. These authors found a good correlation between the IOS and other methods of mapping somatosensory cortex such as somatosensory evoked potentials (SSEPs), electrical stimulation mapping and fMRI (Pouratian et al. 2003a; Pouratian et al. 2003b). While stimulation mapping provided the most focal map of either sensory or motor cortex, the IOS spread to a region 25% larger. Compared with the fMRI signal, the IOS demonstrated a region of ~500 mm^3 while late positive BOLD fMRI signal spread to an area of ~8,000 mm^3 (Cannestra et al. 2001) (Fig. 7.1). The authors argue that the initial dip recorded with the IOS is more tightly linked to neuronal metabolism than the late positive BOLD signal, which is related to the influx of HBO_2 and an increase in CBV. Using this technique, the authors were able to image sensory representation of individual digits from the peak of the IOS (Cannestra et al. 1998). However, there was significant spread of the signal and overlap between digits over time. While the Toga group performed the most comprehensive comparison between the human IOS signal and other mapping modalities, it is unclear from their data whether the IOS they report in humans is actually the initial dip (increase Hbr) or the later decrease in Hbr. Certainly, the timecourse and monophasic signal reported by the authors is more consistent with the later decrease in Hbr than the initial dip. Likewise, the authors recorded only at one wavelength in their human data and so their 610 nm data is not a pure measure of Hbr.

Shoham and Grinvald (2001) also used the IOS at either 605nm or 630 nm to record from human somatosensory cortex during median nerve stimulation and confirmed the location of the decrease in reflection with electrical SSEPs (Fig. 7.2).

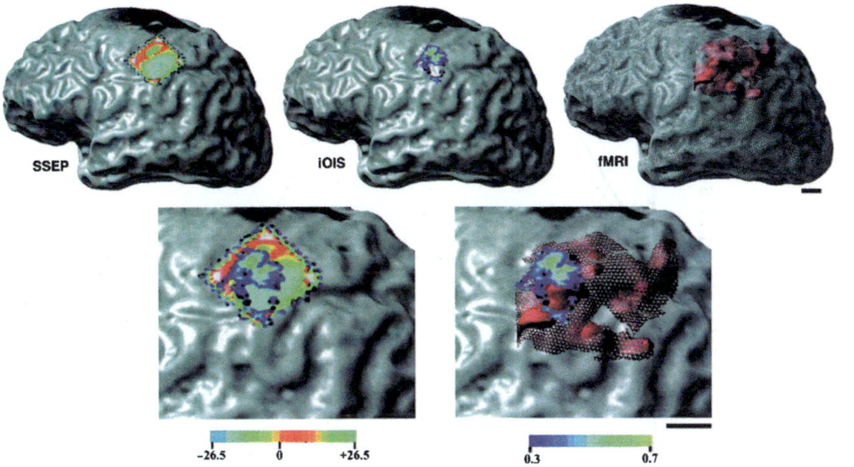

Fig. 7.1 Differences in spatial localization of finger stimulation in cortical area with SSEP, iOIS, and fMRI (monophasic correlation). Stimulations were provided by a 110 Hz pneumatically driven finger vibrator. The scale bar is 1 cm. (*Source:* Adapted frpm Cannestra et al., 2001, Fig. 1)

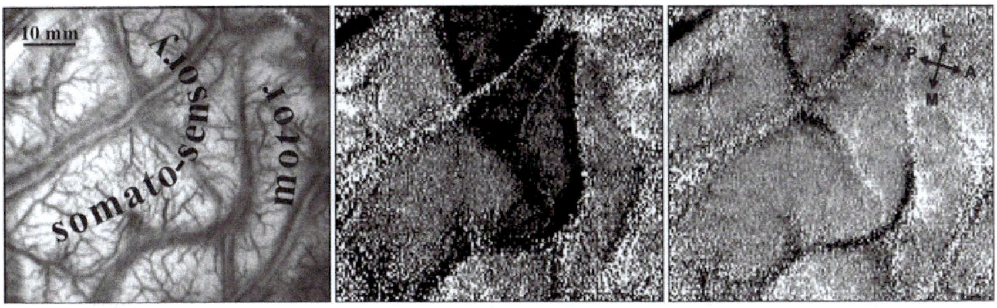

Fig. 7.2 Intraoperative optical imaging of the activated somatosensory hand area following median nerve stimulation. (*Source:* Adapted from Shoham and Grinvald, 2001, Fig. 14)

However, the authors did not report on the timecourse and magnitude of the recorded signal and were not able to image the IOS following stimulation of individual digits due to large vasomotor fluctuations. These authors emphasize the difficulty in recording the intraoperative IOS in human cortex.

Sato et al. (Sato et al. 2002b) performed intraoperative IOS of median nerve, individual digit and face stimulation at 605 nm, and stabilized the cortex with a glass footplate. Optical responses were biphasic, indicating that the initial dip (increased Hbr) was present followed by a decrease in Hbr and maps were made from the early, more localized dip. The authors report that the IOS was sufficiently sensitive to record nonoverlapping signals from individual digits and the locations of activation correlated with magnetoencephalographic maps of sensory cortex. In some patients, two separate areas were identified for each digit, one of which was unique to that digit while the other was common to both digits (Fig. 7.3).

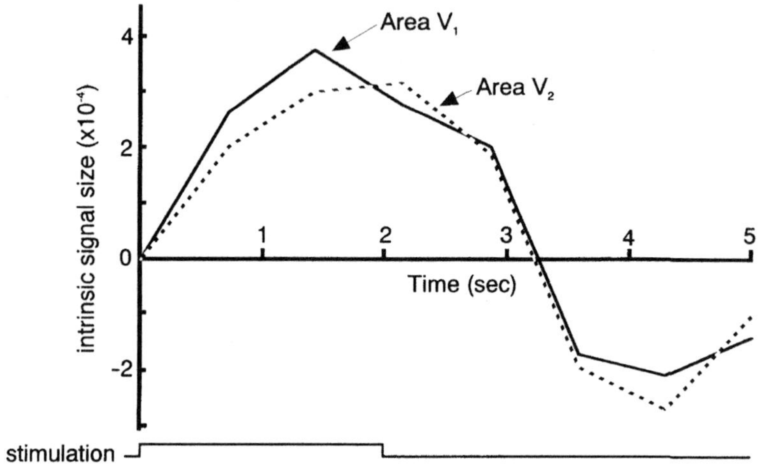

Fig. 7.3 Time courses of changes in the intrinsic optical signal size in area V1 (a solid line) and area V2 (a dotted line). (*Source:* Adapted from Sato et al. 2002b, Fig. 4)

This "common patch" was found in only a minority of humans, whereas it is more commonly present in the macaque (Shoham and Grinvald 2001). Sato et al. (Sato et al. 2002b) also reported that the first digit had a larger representation than other digits and that the IOS overlapped slightly between digits, indicative either of a common processing module or overspill of the IOS.

In a single case study, Schwartz et al. (2004) also used intraoperative IOS to map the representation of upper and lower face during epilepsy surgery at 650 nm (Fig. 7.4). In order to minimize motion artifacts and vascular noise, a glass footplate was used to dampen cortical pulsations. The authors report separate areas of activation for upper and lower face but with significant overlap. Based on the locations of stimulation and the resulting changes in the IOS, the authors calculated the cortical magnification factor to be 0.36 mm of cortex or every cm of skin.

Whether IOS mapping of somatosensory cortex will ever become routine during neurosurgical operations is unclear. The current methods of SSEPs and direct cortical stimulation can be performed rapidly with high reproducibility and adequate spatial resolution. For the purpose of most neurosurgical interventions, gross identification of the central sulcus is generally sufficient to guide any intraoperative decisions. Higher resolution data may not be necessary. For the purpose of research into human somatosensory processing, IOS certainly holds great promise. One possible future application would be the development of a small strip of light emitting diodes and optical sensors that can be placed on the cortex during somatosensory stimulation to quickly map the central sulcus. This avoids the tedium of setting up a camera, fiberoptic light guides, and the need for cortical stabilization since the sensors are placed directly on the cortex.

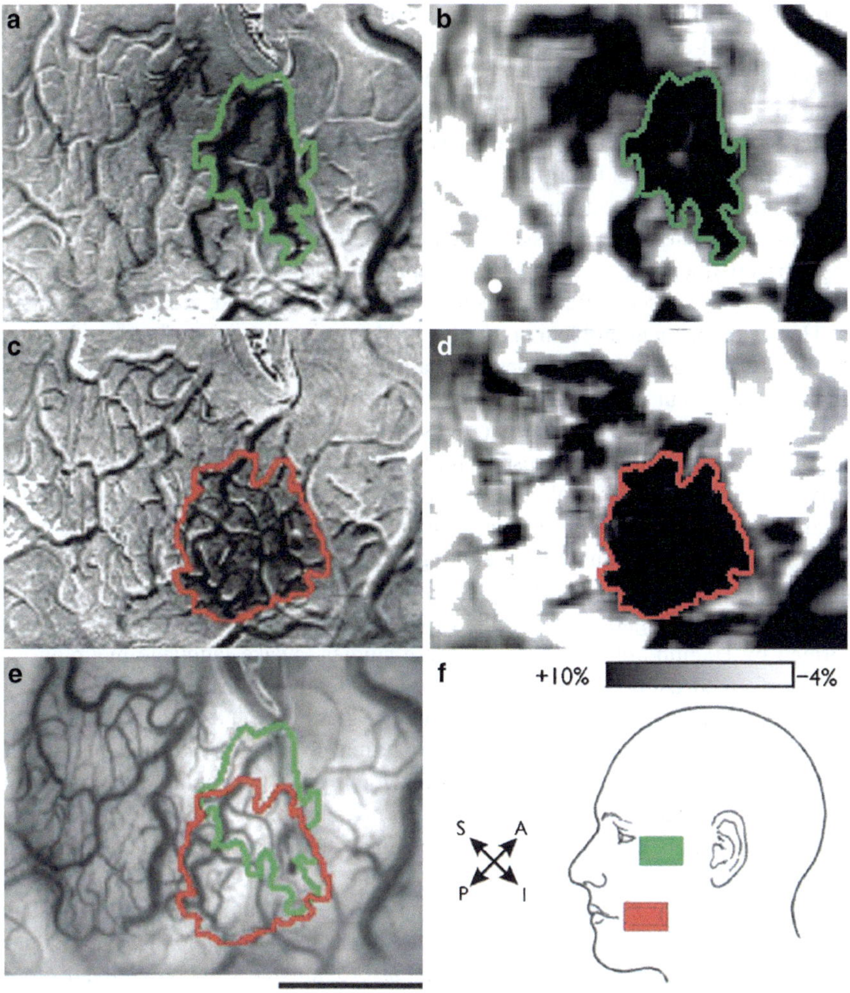

Fig. 7.4 Topography of face representation in S1. (*Source:* Adapted from Schwartz et al. 2004, Fig. 3)

7.3.2 Language Cortex

Intraoperative IOS of language processing in awake patients during neurosurgery has the potential to make significant contributions to our understanding of the cortical organization of high-level cognitive processes. After all, language processing is a unique feature of human cerebral function that cannot be investigated in laboratory animals since the complexity of their symbolic methods of communication does not approach that of a human. In addition, unlike current methods for identifying somatosensory cortex intraoperatively in humans such as cortical stimulation or SSEPs, which are well-established and rapid techniques, current intraoperative identification of language cortex is a time-consuming process. The surgeon must apply an electrical current to a localized area of the brain while the patient is performing one of a variety of language tasks. Essential areas are identified when the performance of the task is interrupted. It is important not only to sample multiple 1 cm^2 areas of cortex several times to ensure adequate sampling and reproducibility but also to test different aspects of language processing such as reading, picture naming, auditory naming, speech fluency etc. IOS offers the promise of simultaneously sampling the entire area of exposed cortex and identifying multiple discrete areas of activation.

However, the main problem in using hemodynamic surrogates of neuronal activity to identify an ongoing higher order association task is that perfusion and oxygenation changes may occur in a larger region of the brain than is truly necessary for that particular function. In other words, not only is the area of brain participating in a function larger than the area of brain necessary for that function, but perfusion and metabolism may overestimate the area of active cortex involved in higher order cognitive tasks such as language processing, as opposed to hierarchically simpler tasks such as movement or sensation. For this reason, although there have been a few studies using IOS to map language cortex in humans, the significance of these data are unclear.

The first investigation into IOS responses to language processing were performed by Haglund et al. (1992). In these studies, awake language mapping was first performed using stimulation mapping to identify both frontal (Broca's) and temporal (Wenicke's) areas. At 690 nm with a glass footplate for stabilization, the authors reported a change in reflection maximal in premotor cortex that did not correlate with the essential naming sites identified with stimulation mapping. Rather an inverted signal was recorded from these areas. Although the authors interpret these data in light of the positron emission tomography studies, which also show changes in premotor areas during naming, another interpretation is conceivable. The negative signal recorded from the essential language areas may have represented the "initial dip" or increase in Hbr, which would look dark at 690 nm. The increased signal from the premotor areas probably correspond with a decrease in Hbr similar to a less specific BOLD signal that correlates more with an increase in CBV. Haglund et al. (1992) also recorded monophasic optical changes from posterior (Wernicke) sites at 690 nm, some of which correlated with essential naming areas disrupted with cortical electrical stimulation. However, the majority of the optical signal arose from areas where stimulation mapping did not reliably disrupt language, indicating that the decrease in Hbr (BOLD-like) spreads to a larger area of cortex than is critical for language activation (Fig. 7.5).

Cannestra et al. (2000) at UCLA also used IOS imaging at 610 nm to study language processing in 10 awake patients undergoing brain surgery. Monophasic IOS changes during visual object naming, auditory discrimination, and auditory responsive naming were compared with cortical stimulation mapping of visual object naming and counting (Fig. 7.6). As in Haglund's study (Haglund et al. 1992) the regions of IOS activation were far larger than those defined as essential with stimulation mapping. As in the somatosensory processing trials, given the duration

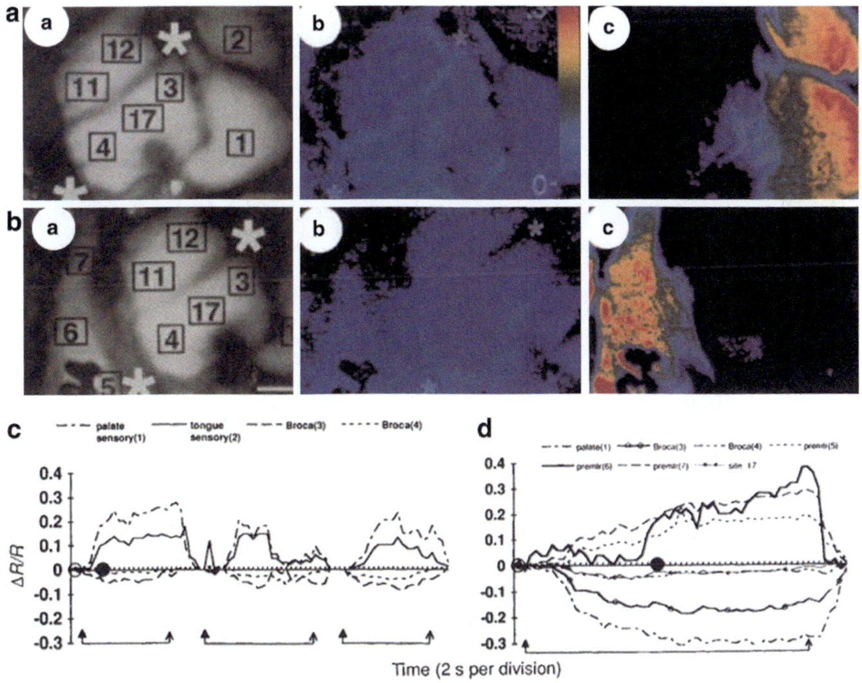

Fig. 7.5 Images of cortical surface following stimulation (*Source:* Adapted from Haglund et al., Fig. 2)

of the responses, it is unlikely that the authors were imaging the initial dip, or increase in Hbr, but rather the BOLD-like decrease in Hbr, which is spatially less restricted to the population of active neurons. In addition, the authors claim that that the distribution IOS activation differed according to the linguistic task, indicating subregions within Broca's and Wernicke's areas specific to semantic vs. phonologic processing. However, it is not clear why the authors did not test the same tasks during stimulation mapping as they did during IOS imaging for comparison.

Pouratian et al. (2000) from UCLA also used IOS at 610 nm to investigate the functional organization of language areas in the brain of a bilingual patient. The authors identified not only single language-specific areas but also common cortical areas, activated by both languages. These results are similar to those reported with stimulation mapping (Ojemann 1978), although again the monophasic responses were more widespread. Correlations between stimulation mapping sites, IOS sites, and fMRI sites were not particularly strong.

Overall, although IOS responses can be recorded during language tasks, few studies have been undertaken and the signal most likely arises from CBV-related decreases in Hbr rather than the highly localized initial dip. The significance and spatial resolution of these widespread changes in reflectance are unclear and require further investigation.

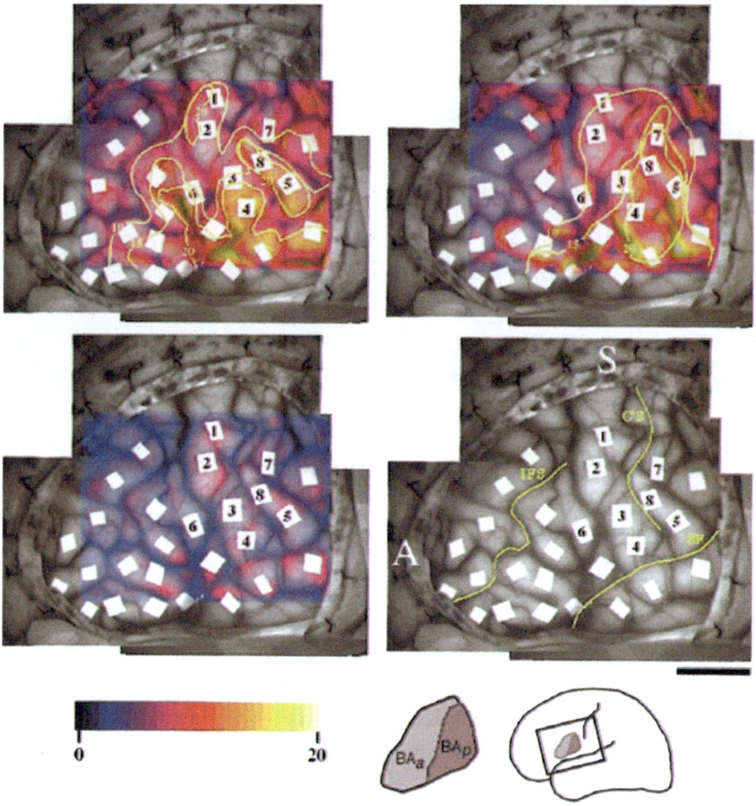

Fig. 7.6 Intrinsic signal response over Broca's area demonstrates topographical specificity dependent on task. (*Source:* Adapted from Cannestra et al., 2000, Fig. 2)

7.4 Imaging Pathologic Cortical Activity

7.4.1 Cortical Stimulation

The simplest, robust, and reproducible method for activating the human cortex to elicit IOS changes involves direct bipolar cortical stimulation. Cortical stimulation is frequently used by neurosurgeons to map the human brain intraoperatively by activating and deactivating small populations of neurons. MacVicar et al. (1990, 1993) first recorded the IOS after stimulating the human brain at a single wavelength and reported a monophasic 3% increase in light absorption that lasted for 25 s after the termination of stimulation. More extensive studies involving IOS spectroscopy were reported by Suh et al. (2006) (Fig. 7.7). In these studies, the authors recorded the IOS at 546 nm and 605 nm at a temporal resolution of 100 ms to characterize the Hbt and Hbr responses associated with varying amplitudes of

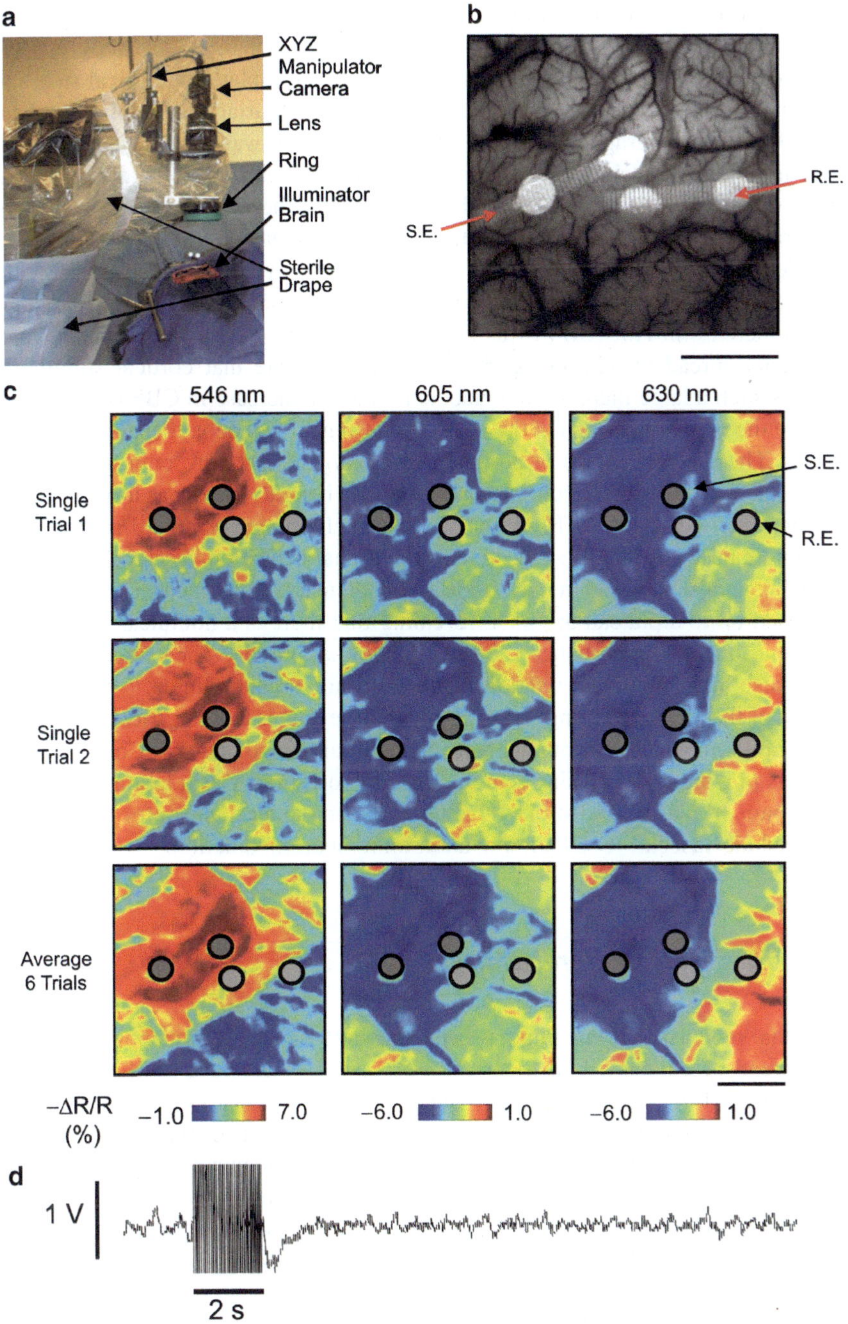

Fig. 7.7 Experimental set-up and reproducibility of the stimulus-induced optical signal. (*Source:* Adapted from Suh et al., 2006, Fig. 1)

human cortical stimulation. Stimulation was delivered at 50 Hz and varying stimulation amplitudes under 4 mA to avoid eliciting afterdicharges. The IOS response occurred rapidly, within 100–200 ms. Using a glass footplate for cortical stabilization, the IOS response was so robust that even a single trial was highly correlated ($r>0.9$) with an average of 6 trials. The Hbr response was biphasic, indicating a large initial dip (4% change in reflectance) that was localized to the gyrus beneath the electrodes within the first 2 s. At later timepoints, the signal inverted and spread to surrounding gyri (8% change in reflectance). These data indicate that the BOLD-like decrease in Hbr may not be a good mapping signal based on poor spatial resolution. The Hbt signal, on the other hand, was monophasic. Again this larger (10%) increase in Hbt (CBV) was only well-localized or the first 2 s after which the signal spread to adjacent gyri. These data indicate that cortical stimulation induces such a large increase in metabolism that the increase in CBF is inadequate to meet the metabolic needs of the neurons causing a focal deoxygenation of hemoglobin. These optical maps of perfusion and oxygenation were only localized for the first 2 s, indicating that clinical tools that use hemodynamics as a surrogate for neuronal activity such as fMRI must have a high enough temporal resolution to generate data within the first second after neuronal activation. The changes in Hbt and Hbr signals increased in magnitude as the intensity of direct cortical stimulation increased up to a plateau at 3 mA, revealing that there is a nonlinear relationship between optical signal changes and stimulation amplitude. The spread of optical signal also had a nonlinear relationship with stimulation amplitude.

One of the most intriguing ideas raised by these data is that electrical stimulation of the brain may have profound effects on cerebral hemodynamics. Cortical stimulation is playing an increasing role in clinical neurosurgery as a treatment for epilepsy, pain, headache, and stroke rehabilitation and the mechanism for these effects may include vascular modulation in addition to neuromodulation.

7.4.2 Triggered Afterdischarges

In animal models, IOS imaging has been shown to be useful in mapping the onset and spread of epilepsy, both interictal and ictal (seizure) events (Schwartz and Bonhoeffer 2001; Suh et al. 2005; Bahar et al. 2006; Zhao et al. 2009). Since spontaneous seizures rarely occur in the operating room, triggered epileptiform events, or “afterdischarges” provide a reproducible mechanism to elicit and image seizure-like activity. Haglund et al. (1992) first reported monophasic changes in the IOS recorded at 695 nm following triggered afterdischarges. The intensity and duration of afterdischarges correlated with the amplitude and horizontal spread of the IOS.

Ma et al. (2009a) performed IOS spectroscopy at 570 and 610 nm to quantify the amplitude and timecourse of perfusion and oxygenation changes associated with afterdischarges using suprathreshold cortical stimulation of human cortex (Fig. 7.8). These studies reveal that during seizure-like events, there is an increase in

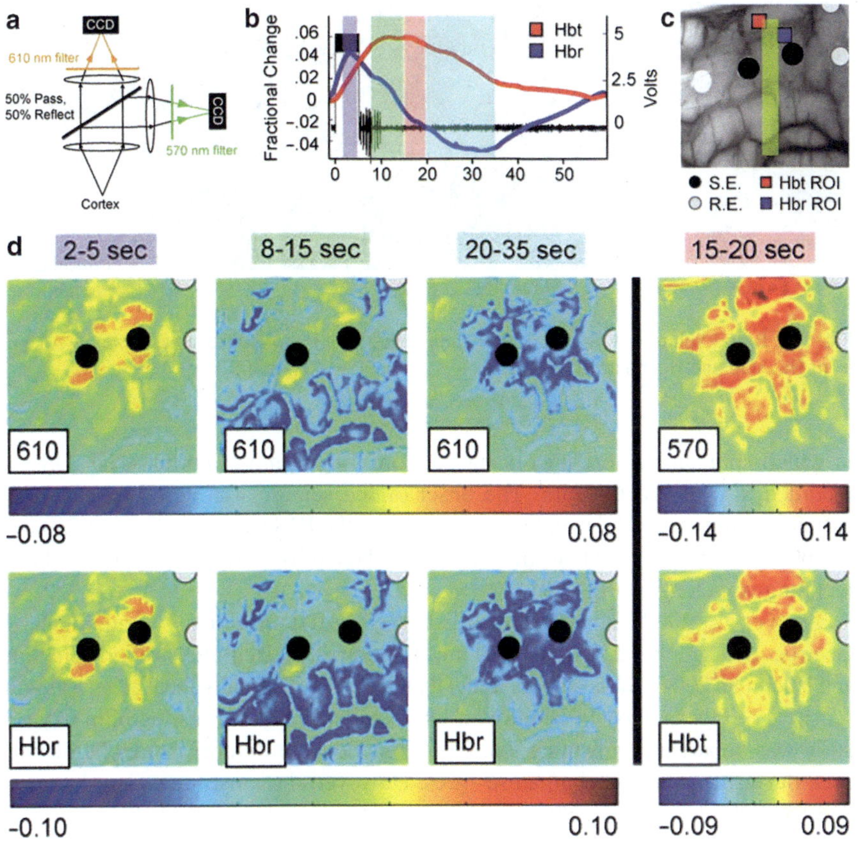

Fig. 7.8 Experimental set up and optical imaging of intrinsic signal responses following direct cortical stimulation in human cortex. (*Source:* Adapted from Ma et al. 2009a, Fig. 1)

Hbr or "initial dip" that lasts for several seconds and in some cases for the duration of the afterdischarge. CBF is again, in this situation, inadequate to meet metabolic demand. These Hbr responses are biphasic compared with the Hbt (CBV) responses which are monophasic. As was described with the subthreshold stimulation, the late increases in Hbt and decreases in Hbr are less well localized than the earlier (<2 s) changes. However, the later BOLD-like decrease in Hbr, although lasting longer than the afterdischarge, had a more linear relationship with the duration of the afterdicharge than the early increase in Hbr.

7.4.3 *Spontaneous Seizures*

Although rare, there have been two reports of spontaneous seizures recorded using IOS imaging in the operating room. Haglund and Hochman (2004) reported a case

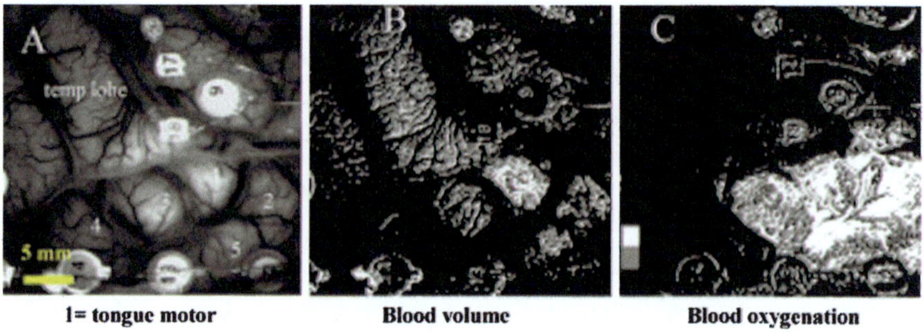

Fig. 7.9 Blood volume and blood oxygenation changes in motor cortex area. (*Source:* Adapted from Haglund and Hochman 2004, Fig. 2)

in which spontaneous seizures occurred from tongue motor cortex. The IOS was recorded at 530 and 660 nm. The onset of the seizure was noted based on the patients symptoms and the authors reported that the IOS was more focal at 530 than at 660 nm indicating that in their hands the Hbt (CBV) signal provided a better mapping wavelength (Fig. 7.9).

Zhao et al. (2007), however, recorded several seizures along with simultaneous electrocorticography at 570 and 610 nm. In contrast to Haglund's report (Haglund and Hochman 2004), these authors found that the IOS was equally focal at wavelengths sensitive to Hbr as Hbt. In addition, a long "epilepsy dip" in hemoglobin oxygenation was apparent, confirming that CBF is inadequate to meet metabolic demand even during spontaneous human seizures. However, the most intriguing data were that the IOS changes actually preceded the electrographic changes by ~20 s. The authors postulated that the IOS may demonstrate hemodynamic responses associated with the movement of extracellular potassium, astrocyte metabolism, and swelling or infragranular electrographic activity that my not be recorded with surface electrodes. These signals may be useful in predicting not only the location by the time of onset of seizures and may hold promise in triggering abortive therapies. Both CBV and Hbr signals changes persisted for a long time after the offset of the seizures (Fig. 7.10).

The persistent dip in oxygenation during epileptiform events may explain the tissue damage following epilepticus or cognitive decline in chronic epileptic patients. These findings are critically important to the interpretation of the perfusion-based imaging studies, such as PET, SPECT, and fMRI.

7.4.4 Spontaneous Interictal Spikes

There have been no reports of human IOS imaging of spontaneous interictal spikes. Whether the signals are too small to record or the noise is too high is unclear but such data would be a valuable addition to the literature.

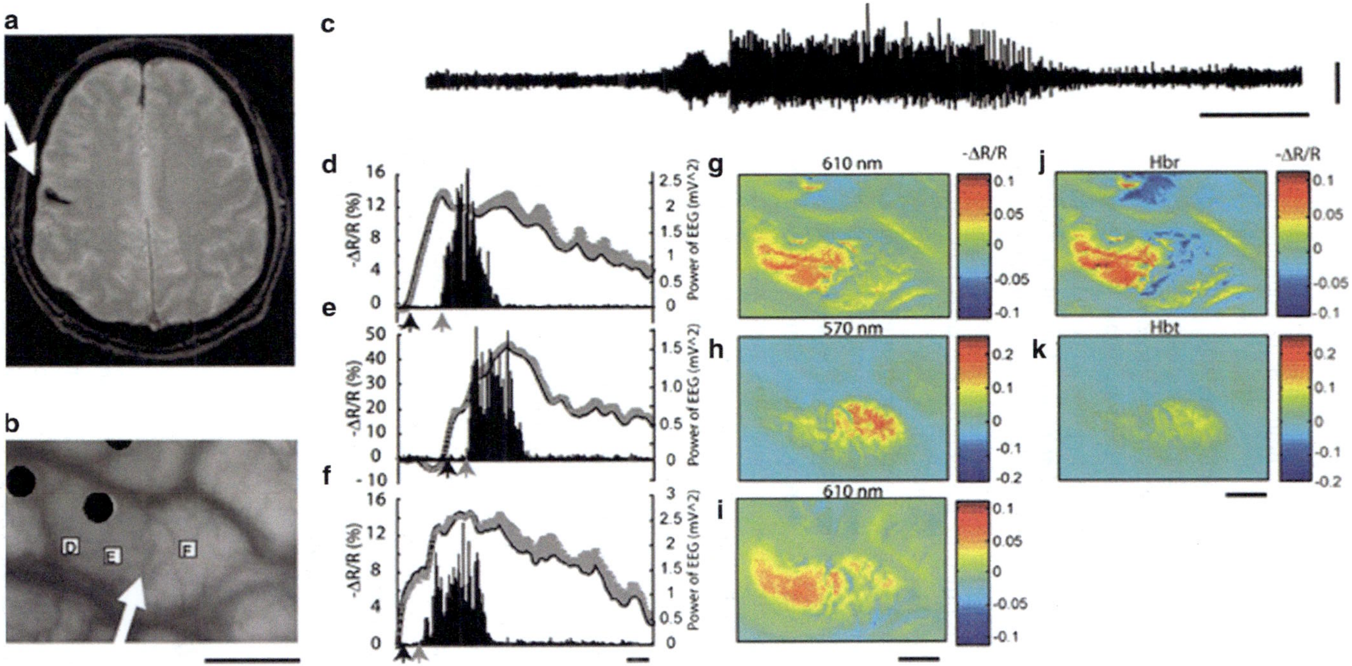

Fig. 7.10 (**a**) Gradient echo axial MRI scan. (**b**) Surface of the brain under glass footplate. (**c**) ECoG recording of a typical seizure. Scale bars: 20 seconds and 1 millivolt. ORIS maps for each of the three seizures in (**d**), (**e**) and (**f**) at the time of the maximum change in the intrinsic signal. (**j** and **k**) Hbr (**j**) and Hbt (**k**) maps calculated using modified Beer-Lambert law on seizures shown in (**d**) and (**e**). Scale bar: 1 cm. (*Source:* Adapted from Zhao et al., 2007, Fig. 1)

7.5 Noise Reduction

Perhaps the most crucial factor when imaging the intrinsic optical signal (IOS) is image stability. In IOS studies, the brain is being compared at each moment to some reference state. If the position of the brain at any moment no longer correlates to the position of the brain in that reference state – due to head motion or motion of any other sort – then any data taken subsequent to that motion is invalid because a comparison between the two states is now meaningless.

In the case of IOS imaging, the reference state is an image or image average of the cortex at some time when it is assumed to be in a baseline state of activity. Such an image is referred to as the *baseline image*. This image is digital; while it may seem elementary, it is important to grasp the meaning of this term with respect to the IOS. A digital image is an array of pixels, the numerical magnitude of which encodes the intensity of light recorded by the camera. The scale runs such that small values are dark and large values are bright. When processing and analyzing the IOS, each image of the cortex is compared – whether through division, subtraction, or a number of more complex processes – to the baseline image. So, as the cortex moves in the camera's field of view, the recorded and processed signal will undergo a series of concomitant increases and decreases, depending on the brightness of each pixel compared to the baseline pixel to which it is incorrectly referenced. Since the amplitude of legitimate changes in the IOS is generally much smaller than the differences in the magnitudes of values representing light and dark regions of the image, any movement of that image ensures that those motion-related changes in signal amplitude will mask, or in the worst cases obscure completely, valid IOS.

Particularly for intraoperative optical imaging of human cortex, it is important to understand the nature of baseline vascular activity because it can affect stability and reproducibility of optical imaging maps. Shoham and Grinvald (2001) explored the microvasculature's activity in anesthetized human cortex and found that artifacts, such as vasomotor fluctuations (Mayhew et al. 1996) can be confused with the functional mapping signals (Fig. 7.11). Therefore, special attention is needed for interpreting human intraoperative IOS data. In this chapter, we discuss problems and solutions for human optical imaging in relation to noise, biological responses, and movement artifacts (Suh et al. 2006; Zhao et al. 2007; Ma et al. 2009a; Zhao et al. 2009; Ma et al. 2009b).

7.5.1 *Periodic Motion*

This is perhaps the most difficult yet common type of motion to prevent during recording of the IOS; luckily, it is also one of the simplest to correct. Periodic motion most frequently results from heartbeat, breathing, or other essential, repetitive biological processes. This type of noise is readily observable in the initial monochromatic recordings, in which the cortex can be seen to pulse clearly and

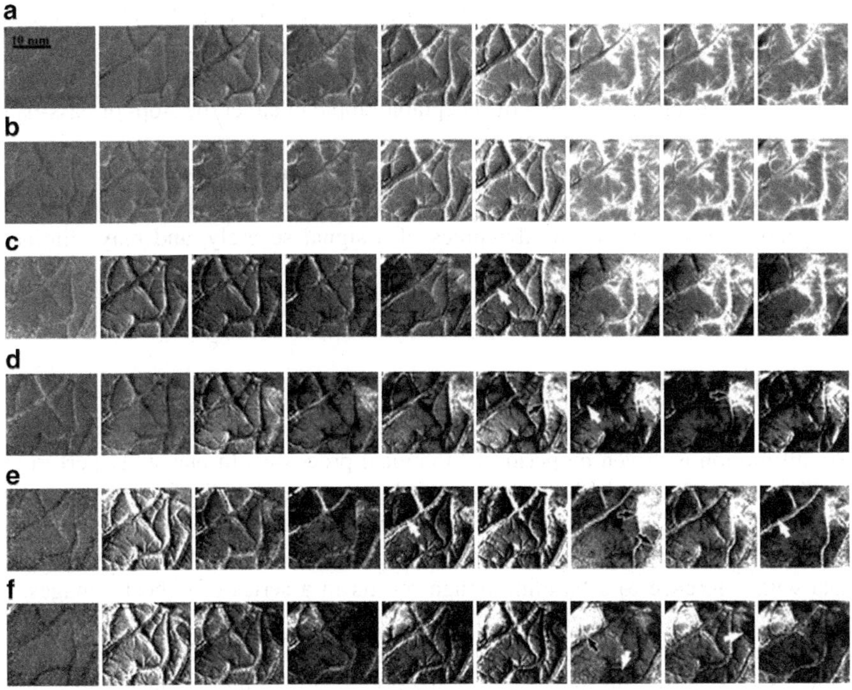

Fig. 7.11 Visualization of spontaneous vascular activity in the human cortex. Scale bar, 10 mm. (*Source:* Adapted from Shoham and Grinvald, 2001, Fig. 13)

regularly. This pulsation causes a slight deformation with respect to the baseline image, and as such causes the light to reflect differently at any given point in time (Ma et al. 2009a).

In order to remove periodic noise from the signal trace, it is first necessary to determine the frequency of motion. One of the simplest and quickest methods is through Fourier analysis. Recall that each image is a two dimensional array of pixels, recorded throughout time. A recording of the IOS, then, is technically a three-dimensional matrix of size length (L) by width (W) by time (t). Performing a one-dimensional fast-Fourier transformation on each pixel's value through time provides another three-dimensional matrix of size L by W by frequency (f). This three-dimensional matrix is rearranged into a two-dimensional matrix of size ($L \times W$) by f, where rows correspond to individual pixels and columns correspond to frequencies. Since each pixel displays noise of the same periodicity, this noise can be intensified and readily observed by summing up the columns and plotting the resultant vector. This plot can be characterized as frequency vs. "frequency intensity" across all pixels.

Meaningful signal shifts tend to be limited to specific cortical regions, whereas all regions display periodic noise, and so the clearest peaks should be located at the frequencies of the noise. By designing a digital band-stop filter (or several filters) spanning these frequency peaks, the periodic noise can be dampened. Choosing the proper filter is often a matter of preference and of the specific nature of the IOS recorded: the proximity of signal frequencies to noise frequencies, the slope of the roll–off needed, and the amount of acceptable ripple in either the stop- or pass-band must each be taken into account.

Caution must be exercised when using filters, and it is of utmost importance to understand the general shape of the signal. Without this knowledge an improperly designed filter can affect the dynamics of a signal severely, and may eliminate crucial information.

7.6 Aperiodic Motion

Not all motion is driven by periodic biological processes. In one study performed by Zhao et al. (2007), the IOS was recorded at two wavelengths from the cortex of an awake patient undergoing recurrent focal seizures. During the seizures, the patient's head underwent continuous and rapid positional shifts. Processing such data with reference to a baseline image results in a series of "ghost" images, in which the high contrast edges of cortical surface features (blood vessels, electrodes, etc.) stand out due to mismatches between cortical pixels. In addition, as the patient's head moves in the x- and y-directions, cortical regions meant to remain outside of the camera's field of view were brought into focus.

A periodic motion does not have an observable frequency, so filtering the data digitally to remove specific frequencies is not a viable option. Moreover, a baseline image does not exist in the normal sense because there is no static frame of reference. Instead, a particular image frame (often the first one) must be chosen as the reference image, and all other images can then be repositioned to best align with this baseline.

The first step in this process is designed to correct for any exterior cortical regions brought into the camera's field of view. Note that these methods should be applied before processing the data with respect to any baseline. The data are first run through a threshold function in order to generate an equivalent number of binary images. This assures that only the cortical surface features, and not the IOS, are taken into account when determining image position. Next, a reference image is chosen and cropped by a few pixels in both dimensions, generating a "centered" version of the original image. Each sequential binary image is then automatically and successively cropped by the same number of pixels as this reference image, incrementally and in both spatial dimensions. The effect is to cut portions of each image of the same size as the reference image, but from a number of different shifted positions. These images are then compared with the reference image via two-dimensional correlation. The highest correlation coefficient for each image is determined, and the particular image position associated with this coefficient is recorded.

Next, each image in the original (nonbinary) dataset is cropped according to the image positions just recorded. This has the effect of observing each image through a window, positioned such that the window allows for the highest possible correlation between the two images.

A further method to account for aperiodic head motion involves taking a weighted average across each image using the formula:

$$\frac{\Sigma_i^n = 1 d_i p_i}{\Sigma_i^n = 1 p_i}$$

where d_i is the amplitude of a given pixel and p_i is a pixel's horizontal or vertical position. This equation offers a "center of mass" for each image in the x- and y-dimensions, which can then be compared to the same values taken from a reference image. Each image can then be offset by the difference between the two center of mass coordinates. Processing the data with a threshold function is useful in order to diminish the effect of the intrinsic optical signal, instead relying upon cortical surface features.

Both of these methods can quickly and automatically correct the largest portion of translational shifts in the cortex. Due to the digital nature of the images, repositioning any pixel is the computerized equivalent of picking it up and moving it using the correct number of steps in either direction. An offset of 0.5 in the x-direction, however, is more complicated – the phrase "half a pixel" bears little meaning, as pixels cannot be divided into halves and combined with their neighbors. One such method to accurately reposition images by fractional pixel increments utilizes the Fourier Shift theorem, which states that:

$$f(x-\alpha, y-\beta) = F(\mu, v)\, e^{-i(\alpha\mu+\beta v)}$$

where α and β are constants displacing function f, F is the two-dimensional Fourier-transform of f, and μ and v are the spatial frequency vectors of F. In other words, any constant displacement of a function can be accomplished by modifying the complex exponentials of that function's Fourier transform, and then multiplying the Fourier transform by these new complex exponentials.

So, in order to best align an entire dataset with a reference image, find the increments by which each image should be offset using the methods discussed above. Once those increments have been determined, the displacement can be accomplished in three steps: first, simply realign each image by the whole-number part of the offset value, pixel-by-pixel; second, find the two-dimensional Fourier-transform of each image in the dataset and multiply its spatial frequency vectors by the fractional part of the offset values; and third, multiply the Fourier transform by these new complex exponentials. This new function can then be returned to the spatial-temporal domain via an inverse Fourier transformation, accomplishing the necessary displacement to the image. A number of functions have been written for various programming environments, intended for free use, to simplify this process.

Theoretically, these steps should be sufficient to account for any aperiodic motion of sufficiently small amplitude to leave the IOS intact. However, images can be poorly focused, grainy, or otherwise flawed, and so these methods will sometimes fail to accurately reposition the images to account for motion. It is important to detect when this is the case and to attempt a more rigorous check of image position. One detection method is to calculate the mean pixel difference of each image from the baseline reference image, according to the equation:

$$\frac{\sum_{j=1}^{m}\sum_{k=1}^{n} | p_{j,k} - r_{j,k} |}{mn}$$

where m and n are the number of image rows and columns, $p_{j,k}$ refers to a pixel of row j and column k, and $r_{j,k}$ refers to that same pixel in the baseline reference image. Any image whose mean pixel difference is greater than some user-defined threshold value can then be subjected to further motion correction. This threshold value determines the precision with which images should match the baseline image. If the remaining number of incorrectly aligned images is small, then slower, more exhaustive methods can be utilized.

It is possible to combine the "window" method, in which the image is successively cropped to find the best fit, with the Fourier shift algorithm. Offset values can be chosen incrementally in both dimensions. Each image is successively shifted by these increments. The images can then be correlated with the baseline reference image, and the shifted image displaying the best correlation is chosen. For example, each image can be Fourier shifted horizontally between –5% and 5% of its width, and vertically between –5% and 5% of its height, at 21 intervals in either dimension. This creates 441 possible displacements for each image in the dataset, all of which must be correlated to the baseline before selecting the best fit. In a dataset consisting of 1,000 images, this entails 441,000 Fourier shifts and comparisons; hence this method is best used to correct a limited sample of poorly positioned images, rather than all available data.

These aperiodic motion correction methods rely upon automated comparison of image characteristics. Since the intrinsic optical signal is recorded as a series of images, any correlation between images entails legitimate signal as well as false signal from incorrectly matched cortical pixels. As mentioned, this can be somewhat surmounted by threshold functions, but care must be taken to avoid overusing these methods.

7.7 Transient Linear Motion

It is also possible for the patient's head to move briefly across the camera's field of view, due to a number of reasons: a poorly mounted camera, human error in the operating room, a motor response to stimulation. Such motion can obviously occur at any point during the recording, and leaves a very clear artifact in the timecourse of all cortical pixels. Motion can clearly be observed by viewing monochromatic, nonprocessed IOS recordings. In individual plots of pixel intensity vs. time, this motion artifact takes the form of large, transient increases or decreases in intensity.

If the head or camera soon returns to the baseline position, however, it is sometimes possible to correct for this type of motion artifact. In particular, the possibility of removing this artifact is dependent upon the timing of the head motion; if the motion occurs during a period of rapid signal change, or during the baseline period, motion correction is difficult.

If these changes are brief and occur during a time in which the IOS is changing at a low frequency, the transient signal increases can be removed by interpolation. Many methods of interpolation are available, depending on the characteristics of the data – cubic spline, piecewise cubic Hermite polynomial (pchip), and linear are a few commonly applied options, all of which are built into Matlab and other analysis software packages. Interpolation relies upon marking the precise times at which motion begins and ends. The signal amplitude immediately pre and postmovement can be sampled from each cortical pixel. Using the sampled signal amplitudes, the temporal regions affected by the motion artifact can be replaced by interpolated data. An additional tactic is to apply a highpass filter to each pixel before interpolation, and then to add the resultant high-frequency limited signal to the interpolated signal region. This ensures that the interpolated signal shape will closely match the surrounding unaffected regions.

Inherent in this method is the fact that the IOS is replaced at least in part by artificially generated data. Because of this, it is important to limit the application of this interpolation method to times during which the signal undergoes well-understood, gradual change. If the transient motion corrupts images in the baseline period, or temporal regions that are poorly understood, than replacing the IOS with interpolated data may mislead at best and outright deceive at worst.

7.8 Future Directions

The future of IOS as an application for human intraoperative mapping will be based on our ability to overcome the technical hurdles we have outlined in these chapters. We have tried to provide some possible solutions based on intraoperative stabilization methods (glass footplate) and post hoc data analysis. In addition, the technique will have to become more rapid both in the acquisition of the data and its analysis and interpretation since operating rooms are based on principals of efficiency and rapid turn-over. The preferred wavelength or focal mapping is still unresolved. Clearly the wavelengths which are sensitive for perfusion and oxygenation provide data about these hemodynamic phenomena associated with neuronal activity. However, recordings at higher wavelengths more sensitive to light scatter may be faster and more focal thereby providing better maps of neuronal activity.

Another possible direction for the IOS application will be to develop a method for chronically implanting a thin array of LEDs and photodiodes (or other illuminating and recording technology) in the sudural space for chronic optical monitoring of epileptic activity and physiologic brain mapping. Movement artifact may be minimal if the array lies directly adjacent to the cortex. Empiric data is required to evaluate this hypothesis. Likewise, diffuse optical tomography applied from the subdural space in the humans may provide data on light reflection in the depths of the brain rather than just the surface monitoring provided with current methods of IOS. Finally, IOS endoscopes may also play a role recording from deep structures to record deep epileptic foci or monitor blood flow or light scatter to provide feedback for neuromodulatory devices.

7.9 Summary

The IOS holds great promise as a method for high resolution mapping of perfusion, oxygenation and light scatter signals associated with neuronal activity, not only improved localization of functional architecture but also more effective ways to map abnormal physiologic activity, such as seizures. To date, limited studies have been performed due to the technical hurdles outlined in this chapter.

In addition to applications in brain mapping, IOS studies in humans will lead to a better understanding of the origin and characteristics of the connection between neuronal activity and hemodynamic changes, which will aid the interpretation of fMRI experiments concerned with much higher levels of cortical function such as cognition. Furthermore, the results of this research will serve as a stepping stone for developing better diagnostic and treatment tools for brain disorders associated with hemodynamic abnormalities.

References

Hill DK, Keynes RD (1949) Opacity changes in stimulated nerve. J Physiol 108:278–281

Grinvald A et al (1986) Functional architecture of cortex revealed by optical imaging of intrinsic signals. Nature 324:361–364

Bonhoeffer T, Grinvald A (1991) Iso-orientation domains in cat visual cortex are arranged in pinwheel-like patterns. Nature 353:429–431

Frostig RD et al (1990) Cortical functional architecture and local coupling between neuronal activity and the microcirculation revealed by in vivo high-resolution optical imaging of intrinsic signals. Proc Natl Acad Sci 87:6082–6086

Bonhoeffer T, Grinvald A (1996) In: Toga AW, Mazziota JC (eds) Brain mapping. The methods. Academic Press, San Diego, pp. 55–99.

Sheth SA et al (2003) Evaluation of coupling between optical intrinsic signals and neuronal activity in rat somatosensory cortex. NeuroImage 19:884–894

Nemoto M et al (2004) Functional signal- and paradigm-dependent linear relationships between synaptic activity and hemodynamic responses in rat somatosensory cortex. J Neurosci 24:3850–3861

Mayhew JEW et al (2000) Spectroscopic analysis of neural activity in brain: increased oxygen consumption following activation of barrel cortex. NeuroImage 12:664–675

Malonek D, Grinvald A (1996) Interactions between electrical activity and cortical microcirculation revealed by imaging spectroscopy: implications for functional brain mapping. Science 272:551–554

Sato C et al (2002a) Reassessment of activity-related optical signals in somatosensory cortex by an algorith with wavelength-dependent path length. Jpn J Physiol 52:301–312

Vanzetta I, Grinvald A (1999) Increased cortical oxidative metabolism due to sensory stimulation: implications for functional brain imaging. Science 286:1555–1558

Pouratian N et al (2003a) Spatial/temporal corelation of BOLD and optical intrinsic signals in human. Magn Res Med 47:766–776
Huppert TJ et al (2007) A multicompartment vascular model for inferring baseline and functional changes in cerebral oxygen metabolism and arterial dilation. J Cereb Blood Flow Metab 27:1262–1279
Fox PT, Raichle ME (1986) Focal physiological uncoupling of cerebral blood flow and oxidative metabolism during somatosensory stimulation in human subjects. Proc Natl Acad Sci USA 83:1140–1144
Kwong KK et al (1992) Dynamic magnetic resonance imaging of human brain activity during primary sensory stimulation. Proc Natl Acad Sci USA 89:5675–5679
MacVicar BA et al (1990) Stimulation evoked changes in intrinsic optical signals the human brain. Soc Neurosci Abstr 16:309
Haglund MM et al (1992) Optical imaging of epileptiform and functional activity in human cerebral cortex. Nature 358:668–671
Shoham D, Grinvald A (2001) The cortical representation of the hand in macaque and human area S-1: high resolution optical imaging. J Neurosci 21:6820–6835
Schwartz TH et al (2004) High resolution intraoperative optical imaging of human face cortical topography: a case study. Neuroreport 15:1527–1531
Sato K et al (2002b) Intraoperative intrinsic signal imaging of neuronal activity from subdivisions of the human primary somatosensory cortex. Cereb Cortex 12:269–280
Toga AW et al (1995) The temporal/spatial evolution of optical signals in human cortex. Cereb Cortex 5:561–565
Cannestra AF et al (1996) The evolution of optical signals in human and rodent cortex. Neuroimage 3:202–208
Cannestra AF et al (1998) Topographical and temporal specificity of human intraoperative optical intrinsic signals. Neuroreport 9:2557–2563
Pouratian N et al (2000) Optical imaging of bilingual cortical representations. Case report. J Neurosurg 93:676–681
Pouratian N et al (2003b) Shedding light on brain mapping: advances in human optical imaging. Trends Neurosci 26:277–282
Cannestra AF et al (2001) Temporal spatial differences observed by functional MRI and human intraoperative optical imaging. Cereb Cortex 11:773–782
Cannestra AF et al (2000) Temporal and topographical characterization of langauge cortices utilizing intraoperative optical intrinsic signals. NeuroImage 12:41–54
Ojemann GA (1978) Organization of short-term verbal memory in language areas of human cortex: evidence from electrical stimulation. Brain Lang 5:331–340
Suh M et al (2006) Blood volume and hemoglobin oxygenation response following electrical stimulation of human cortex. Neuroimage 31:66–75
Haglund MM, Hochman DW (2004) Optical imaging of epileptiform activity in human neocortex. Epilepsia 45:43–47
Zhao M et al (2007) Focal increases in perfusion and decreases in hemoglobin oxygenation precede seizure onset in spontaneous human epilepsy. Epilepsia 48(11):2059–67
Mayhew JEW et al (1996) Cerebral vasomotion: a 0.1-Hz oscillation in reflected light imaging of neural activity. Neuroimage 4:183–193
Schwartz TH and Bonhoeffer T (2001) In vivo optical mapping of epileptic foci and surround inhibition in ferret cerebral cortex. Nat Med 7(9):1063–1067
Bahar S et al (2006) Intrinsic optical signal imaging of neocortical seizures: the 'epileptic dip'. Neuroreport 3;17(5):499–503
Ma H et al (2009a) The importance of latency in the focality of perfusion and oxygenation changes associated with triggered afterdischarges in human cortex. J Cereb Blood Flow Metab 29(5):1003–1014
Ma H et al (2009b) Hemodynamic surrogates for excitatory membrane potential change during interictal epileptiform events in rat neocortex. J Neurophysiol 101(5):2550–2562. Epub 2009 Feb 25
Zhao M et al (2009) Spatiotemporal dynamics of perfusion and oximetry during ictal discharges in the rat neocortex. J Neurosci 4;29(9):2814–2823

Chapter 8
Using Optical Imaging to Investigate Functional Cortical Activity in Human Infants

Susan J. Hespos, Alissa L. Ferry, Christopher J. Cannistraci, John Gore, and Sohee Park

Abstract Imaging methods have caused a revolution in cognitive science for research on adult brain function. It is clear that many of these neuroscience methods can be applied to younger populations to investigate the relationship between cognition and brain development. The goal of this chapter is to describe the feasibility of using optical imaging on human infants. The motivation for using neuroimaging techniques on young populations is that they allow us to address critical issues of continuity and change over development. When a young infant, an older child, and an adult all exhibit a behavioral discrimination, how can we tell whether the underlying mechanisms producing the discrimination are the same (a case of developmental continuity) or different (a case of qualitative developmental change)? Imaging studies of infants, combined with and guided by the findings of imaging studies with adults and by behavioral studies, should let us approach this question. If we see convergent findings in infants and adults at both the behavioral level and the neural level, we can infer continuity across development. Conversely, non-convergence between infants and adults may signal points of significant developmental change, as well as providing a window for examining such change.

Imaging tasks allow us to test young infants in a state of full attention. These same tasks have been used with great success in fMRI with adults using freeviewing, providing opportunities of comparisons across tasks. However, fMRI techniques with infant participants are challenging because infants lack the ability to remain still while awake. With near-infrared spectroscopy (NIRS), the infant is allowed to move with some degree of freedom and they are tested while sitting upright on their caretakers lap. The vast majority of neuroscience research on infants has been done with ERP (Gliga and Dehaene-Lambertz 2007; Kuhl 2004; Nelson and Monk 2001; Neville 2005), and so comparisons to fMRI research are indirect and spatial localization for neural activity is minimal. Using optical imaging on infants provides a window to the characteristics of the hemodynamic response associated with functional activity. These data will allow for more direct comparisons with fMRI studies on adults and children. For this reason optical

S.J. Hespos (✉)
Department of Psychology, Northwestern University, Evanston, IL, 60208, USA
e-mail: hespos@northwestern.edu

A.W. Roe (ed.), *Imaging the Brain with Optical Methods*,
DOI 10.1007/978-1-4419-0452-2_8,

imaging holds promise as a technique for elucidating the links between behavioral and neural development. In this chapter, we will discuss the feasibility of one type of optical imaging called NIRS as a tool for cognitive scientists who study development. We will describe the technology, review existing studies using NIRS on infants, and speculate about how NIRS technology will impact the emerging field of developmental cognitive neuroscience.

8.1 How Does NIRS on Infants Work?

In a typical experiment, an array of fiber optic probes is placed on the infant's head. Half of the probes emit light, the other half of the probes serve as detectors that absorb reflected light (see Fig. 8.1). The emission probes emit light at two specific frequencies; one frequency, 830 nm, is optimally absorbed by oxygenated blood while the other frequency, 780 nm is optimally absorbed by deoxygenated blood.[1] The dependent measure is the amount of light at each wavelength that is reflected from the emission probe to detection probes. Increases in oxygenated and deoxygenated hemoglobin in functionally activated cortex will produce an increase

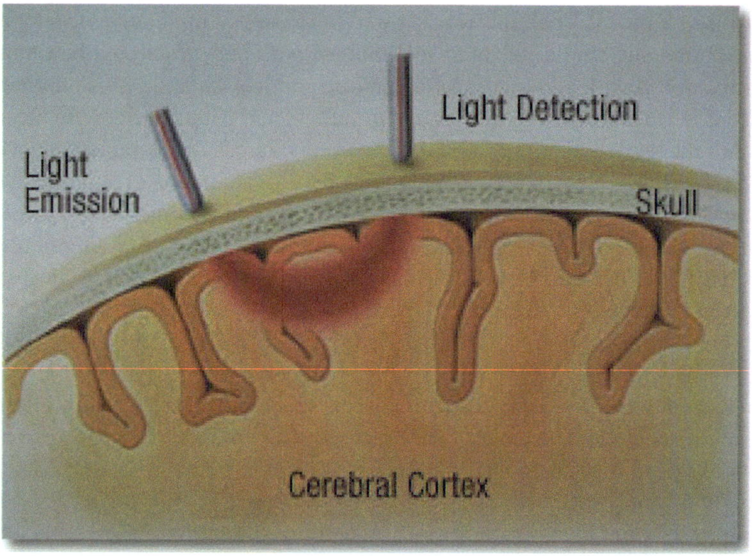

Fig. 8.1 This figure demonstrates the principle of NIRS technology. Light enters the skull from the emission probe placed on the surface of the head. Some of the light passes through the path depicted by the *red banana-shaped curve* and is absorbed by the detection probe

[1]Data collected at the 780 nm frequency has been noisy and newer machines use frequencies in the 690 nm range.

in the amount of light that is absorbed, and therefore a decrease in the amount of light that is reflected back to the detection probe. By gathering data on the changes in blood flow over time, we can make inferences about the cortical activity occurring in the underlying brain regions.

There are numerous reasons why the NIRS technique is the best choice for studying cortical functioning in infants. First, while the technology is different from fMRI, both methods use changes in blood flow to study neural activity. The parallel use of the hemodynamic response allows us to compare NIRS and fMRI data and examine changes in the same variable from birth through adulthood, making NIRS ideal for longitudinal and cross-sectional designs. The fiber optic probes are built into a fleece cap that is slipped over the infant's head allowing their head to move with some freedom and permitting the infant to be seated on their caretaker's lap throughout the experiment. Freedom of movement makes NIRS have a distinct advantage over other neuroimaging techniques used on young participants because with ERP and fMRI data are compromised due to motion artifacts. Moreover, NIRS allows the infants to move around on their caretaker's lap where they are more comfortable and consequently more likely to complete the study. Unlike fMRI experiments, the NIRS instrument is silent, making the presentation auditory stimuli easy. Because infants tend to have fine hair and their skulls are thin and small, the proportion of signal loss due to scattering is less than that for adult participants. Finally, while fMRI only measures deoxygenated blood flow, the NIRS instrument measures temporal changes in both oxyhemoglobin and deoxyhemoglobin with a high signal-to-noise ratio. The ability to measure changes in deoxyhemoglobin is especially important for infant research because fMRI studies suggest that the infant visual cortex does not display the characteristic activity-dependent increase in blood oxygenated level-dependent (BOLD) response: (Anderson et al. 2001; Dehaene-Lambertz et al. 2002; Martin et al. 1999). Further research needs to be done to verify these initial reports. Meek (2002) reviewed a few examples in the NIRS literature on clinical populations that did not display the characteristic BOLD response as well. NIRS currently has two limitations as a research tool: it is restricted to measurements near the cortical surface and it is not yet used widely with infants (or adults).

8.2 Review of the Existing Studies Using NIRS on Infants

One of the exciting aspects of using NIRS on infants is that the technology improves dramatically and quickly. There are two broad categories of machines that have been used in NIRS research on infants. The earlier machines have one to four emission/detection probe pairings (called channels) and more recently developed machines have had 24 channels or more. The advantage of fewer channels is that the machines generate less data and therefore data set size and analyses are more manageable. The disadvantage is that the placement of the probes is critical and displacement of less than 1 cm can result in a substantial loss of NIRS signal

(Kleinschmidt et al. 1996). One solution to this problem is to increase the number of probes so that a larger area is sampled. With more probes it is possible to use marker tasks in the procedure to identify specific regions of interest with functional activation in addition to external anatomic landmarks. Two inherent problems incurred by having more channels and they are: (1) the generation of large data sets requiring an interdisciplinary team of computer scientists, engineers, and physicists to deal with the data in an appropriate manner and (2) getting multiple probes to rest well on the curved surface of a infant's head is a behavioral and engineering challenge.

The research on infants using a small numbers of probes has demonstrated that NIRS is a promising and feasible method for noninvasive tracking of functional activation in diverse cortical regions (for a review see Meek (2002)). Studies have shown measurable responses in neonates to flickering checkerboard (Meek et al. 1998), pictures of faces (Csibra 2004), and various smells (Bartocci et al. 2000). Studies on both young and old infants have revealed the potential of using this method for a broad range of ages. Baird et al. (2002) performed a longitudinal study on infants from 5 to 12 months of age comparing the hemodynamic response in frontal cortex before and after infants succeeded on the Piagetian object permanence/manual search tasks. They found enhanced activity in prefrontal areas on visits when infants succeeded in the task providing converging evidence for the behavioral tasks that each infant performed simultaneously. Wilcox et al. (2005) found increased activation in temporal and visual areas in 6-month-old infants when watching visual displays involving objects compared to baseline. Bortfeld et al. (2007) extended these findings to infants who were 6–9 months of age and showed that during visual presentation alone versus language and visual presentation there was constant activation in visual areas but the activation in temporal areas increased during the conditions with language. In addition, Franceschini et al. (2007) measured regionally specific increases in blood volume and oxygen consumption during resting states in healthy infants over the course of the first year of life. This work provides critical baseline information. For example, they demonstrate that as early as 6 weeks of age infants show predictable patterns of cortical blood flow.

There is a growing set of findings from machines that use 24 channels. A study by Taga et al. (2003) tested 2- to 4-month-old infants and measured from two brain areas –frontal and occipital – while presenting infants with a flickering checkerboard stimulus. They used an event-related design to demonstrate that the visual area showed significantly greater activation than the frontal area. These findings demonstrate that NIRS can be used to differentiate global cortical locations. Peña et al. (2003) tested sleeping neonates and measured auditory cortical activation bilaterally while the infants were presented with infant-directed speech, backward speech, and silence. They found left hemisphere superiority during forward speech but bilateral or no activation during the backward speech and silence. These data demonstrate that even neonates, with their limited language experience, have a specific brain response to normal speech. In our preliminary studies, we made strides in replicating and extending the work by Taga et al. and Peña et al.

8.3 Preliminary Studies on Motor and Visual Responses

The first experiment focused on measuring the cortical response in infants and adults in response to functional activation of the visual and motor areas. Thirtynine people from three different age groups were tested. We tested 25, healthy, full-term infants. Eleven were in the "young" age group from 4 to 6 months of age and 14 were in the "old" age group from 7 to 9 months of age. Fourteen healthy adults ranging from 19 to 35 years old ($M=23$ years) were also recruited. It is worth noting that only 1 infant out of 25 produced unusable data. This is an extremely low drop out rate for infant research and bodes well for the success of the technique.

NIRS was carried out using a 24-channel 780/830 nm spectrometer (NIROT ETG-100; Hitachi Medical). Signals were acquired at a sample rate of 10 Hz from 24 cortical regions. The probes consist of 9 probes in a 3×3 grid with 5 emission probes and 4 detection probes that formed 12 source-detector pairings (i.e., channels). There was a 3 cm distance between the emission and detection probes for the adult probes and a 2 cm distance between the probes for the infants. Probes were positioned to maximize the likelihood of monitoring motor and visual areas. They were placed by using skull landmarks. One set of probes was centered over the inion and the other set was centered over the middle of the left ear with the top edge of the probes touching vertex (see Fig. 8.2). The vertex was determined as the point that a line going from the nasion to the inion intersects a line going from the left to the right preauricular lobule. The infant probes were held in black fleece hats with a chinstrap. On the infant probes there was a 90-degree bend in the fiber optic cable, which allowed us to wrap the infant's head with an ace bandage to insure that the probes stayed in the same place throughout the experiment and minimized motion artifacts in the data.

Absorbance data were processed using a method established by Horovitz and Gore (2004) and Folley and Park (2005). The raw data were exported from the Hitachi machine, and then processed using Matlab (The Math Works). The first stage was to filter the data in the temporal domain to remove artifacts due to respiration and cardiac variations (using a band pass filter with range 0.01–0.5 Hz). The data were down sampled (10–1 Hz), and converted to measurements of hemoglobin levels according to the modified Beer-Lambert Law, arranged into epochs, and an average time course was obtained for each subject at each location. Brain Voyager QX (Brain Innovation, Maastricht, Netherlands) was used for linear drift correction and statistical analyses. The false discovery rate statistic q(FDR) protected contrasts from alpha inflation.

There are a variety of possible analyses to perform as there is little precedence to follow at this time in the literature. While we collected data for oxyhemoglobin, deoxyhemoglobin and total hemoglobin we focus on oxyhemoglobin only. The rational is that the deoxyhemoglobin signal tends to return flat data across all conditions, and the total hemoglobin tends to follow the trend of the oxyhemoglobin with the constant variance added by the deoxyhemoglobin signal. The flat deoxyhemoglobin signal is not unique to our machine and since a larger proportion of the

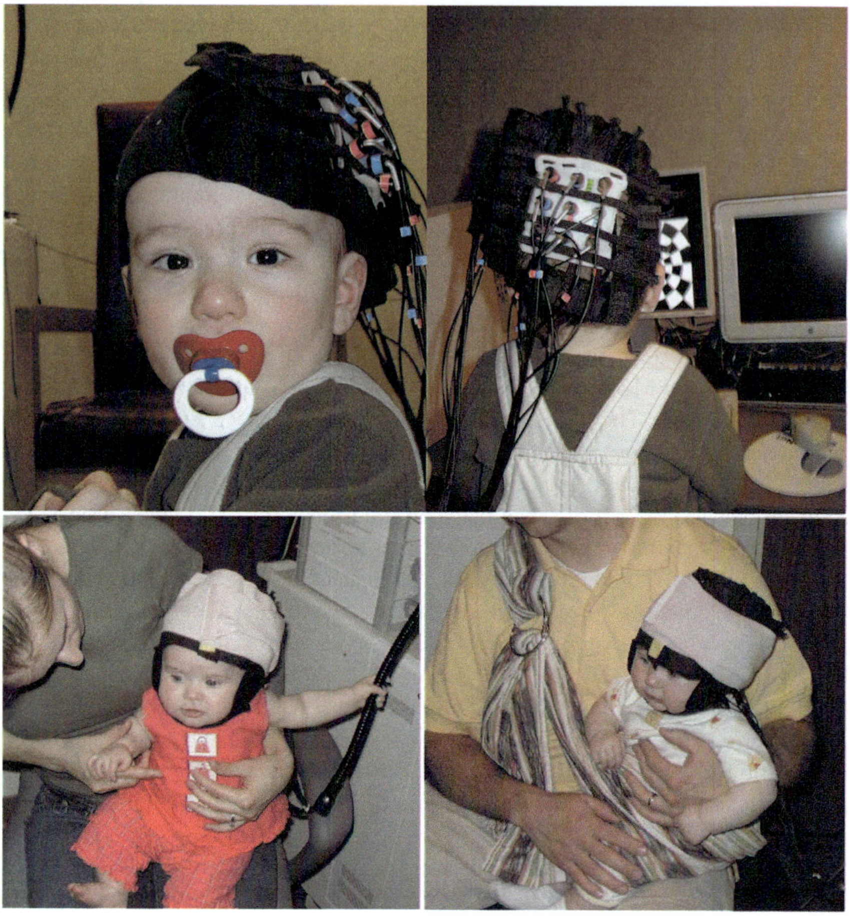

Fig. 8.2 Photos of infants who were in our Experiments. The photos on the left reveal the placement of the probes over motor and visual area, but this was an early model of the hat before the chinstrap was added. After the placement of the probes was checked we wrapped the head with an elastic bandage to insure that the probes did not change position when the infants moved

deoxyhemoglobin signal is absorbed at the 780 nm there has been an adjustment in the frequency of this laser in newer machines to 690 nm.

The experimental design with our predicted response is pictured in Fig. 8.3. We wanted to find out if the NIRS system could measure functional changes in activation over global areas of surface cortical activity. Furthermore we wanted to characterize the shape of a typical BOLD response for motor and visual areas and get insight into the differences across cortical areas, individuals, age, and gender. Since we measured from the motor and visual areas continuously and we presented all participants with motor and visual stimulus blocks we have data from each area during activation and rest and contrast the findings across stimulus conditions.

Fig. 8.3 The design for the experiment on motor and visual activation with the predicted responses

Stimuli and Design

	Motor Cortex	Visual Cortex
Vibrating Toy	Activity	No Activity
Visual Flicker	No Activity	Activity

The motor stimuli consisted of holding a vibrating toy in the subject's right hand (if the infants did not voluntarily grab the object it was held against their hand by the parent). The visual stimuli consisted of watching a colorful *Finding Nemo* video clip. Stimuli were presented for 8 continuous seconds followed by a "rest" period that varied between 15 and 25 s, and an attention-capturing sound was continuous during stimuli and "rest" periods. The variable rest period decreased the chances of entrainment caused by blocked designs. Each testing session consisted of six to eight stimuli presentations. Between sessions the participant was allowed to move or stretch and infants were assessed as to whether they could continue through another 2-min testing period. Each infant contributed at least one block of visual and motor data and not more than six blocks total. On average infants contributed four blocks (two motor, two visual). All adults contributed six blocks (three motor, three visual) each. Whether the initial presentation was a visual or motor stimulus was counterbalanced across subjects. The stimuli presentation was controlled by E-prime software (Schneider et al. 2001) and the display computer and the ETG-100 were connected through a serial port so that stimulus presentation and data collection were synced.

The results confirmed our two predictions. First, there was more blood flow to active areas as compared to rest areas and the response in the motor cortex was distinct from the response in the visual cortex. Figure 8.4 shows the results of the spatial analysis for three individual subjects. There were 12 channels over the motor area and an additional 12 channels over the visual area. In Fig. 8.4 we present a 5×5 grid for each cortical area. The pixels between each of the 12 actual channels were interpolated using a nearest neighbor function yielding a 25-pixel grid for each cortical area. The mean number of pixels that showed significant increases in blood flow is displayed in Fig. 8.5.[2] While the effect is not surprising, this finding confirms that optical imaging can be used to distinguish global cortical areas in infants and adults. The second main finding is that the characteristics of the hemodynamic

[2] We acknowledge that voxel counting is not an optimal data analysis method but it is unlikely to give false positives and we are currently developing new spatial analysis techniques.

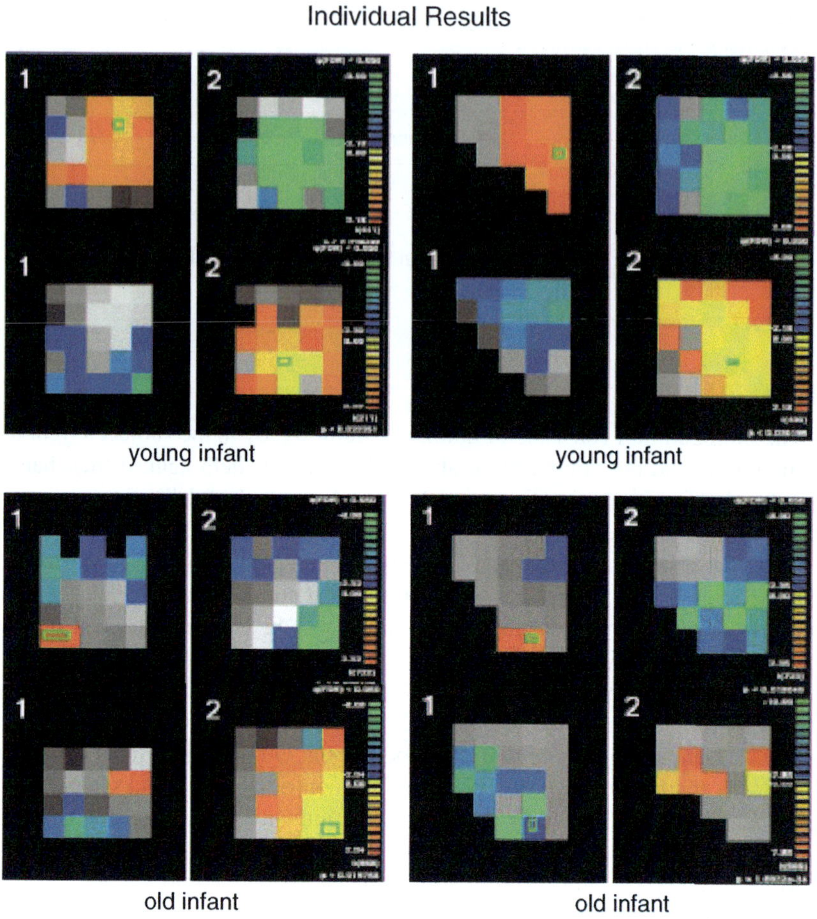

Fig. 8.4 Activation patterns for four individual participants. The design in Fig. 8.3 maps onto the data in this figure. Warm colors (*orange* and *yellow*) indicate significant increase in blood flow and cool colors (*blue* and *green*) indicate significant decrease

response are different across the motor and visual cortex but there are similarities across ages within these areas. In Fig. 8.6, we depict the average blood oxygen level dependency (BOLD) response separated by age. The hemodynamic curve from the motor area peaks earlier than the visual area. This difference needs further investigation. In fMRI research the BOLD response typically has a 6 s delay from the onset of the stimulus. It is not clear why the time to peak for the visual cortex is 13 s after stimulus onset. One variable could be the nature of visual stimulation that we used in this particular experiment. A second variable that could contribute to the differences is that NIRS depends on the reflection of light as it passes through tissue. Consequently it is possible that differences in the anatomy across channels (e.g., veins or arteries near the surface of the skin) will contribute variance in the signal.

Average Number of Voxels Active

	Young infants		Old infants		Adults	
	Motor	Visual	Motor	Visual	Motor	Visual
Toy	5	1	5	0	12	5
Video	2	10	0	12	1	12

Fig. 8.5 The means represent the number of channels that had a significant increase in activation across conditions

For this reason data analysis should always compare difference scores for probe pairs that have not moved over the time between activation and nonactivation.

There are two additional questions that need further investigation: (1) The non-activated area appears to show a decrease. The decrease could be due to depletion from measuring the same area during activation and rest over multiple trials. (2) It is not clear why the percent signal change has a smaller range for the infants than the adults. One possible explanation for the differences is that we used the same analysis on the infants and adults. However in the modified Beer-Lambert equation (used to transform the raw absorption data into hemoglobin levels) there is a variable for path length. In these experiments the path length/distance between the probes was 2 cm for infants and 3 cm for adults but this difference was not included in the analysis. The differences between conditions would not change but the percent signal change for the infant data would be larger.

Our final analysis looked at individual differences between subjects (see Fig. 8.7). These graphs show that each subject demonstrated the predicted pattern of response and that there is a greater signal change in the adult data compared to the infants.

These data provide an exciting step forward in understanding the nature of the hemodynamic response using optical imaging. We find that there are global similarities across infants' and adults' responses to motor and visual stimulation. All participants showed increased blood flow in response to functional activation. There were measurable differences between global cortical areas suggesting that even in early infancy the hemodynamic response is not global. Furthermore, there were no cases of reverse BOLD response which counters previous fMRI studies (Anderson et al. 2001; Dehaene-Lambertz et al. 2002; Martin et al. 1999). Although there were differences across ages there are global similarities across brain areas. In general all ages have similar time to peak response in motor and visual areas. The individual data suggest that there is a certain amount of variance that is most likely due to anatomical differences in thickness of hair, scalp, and skull.

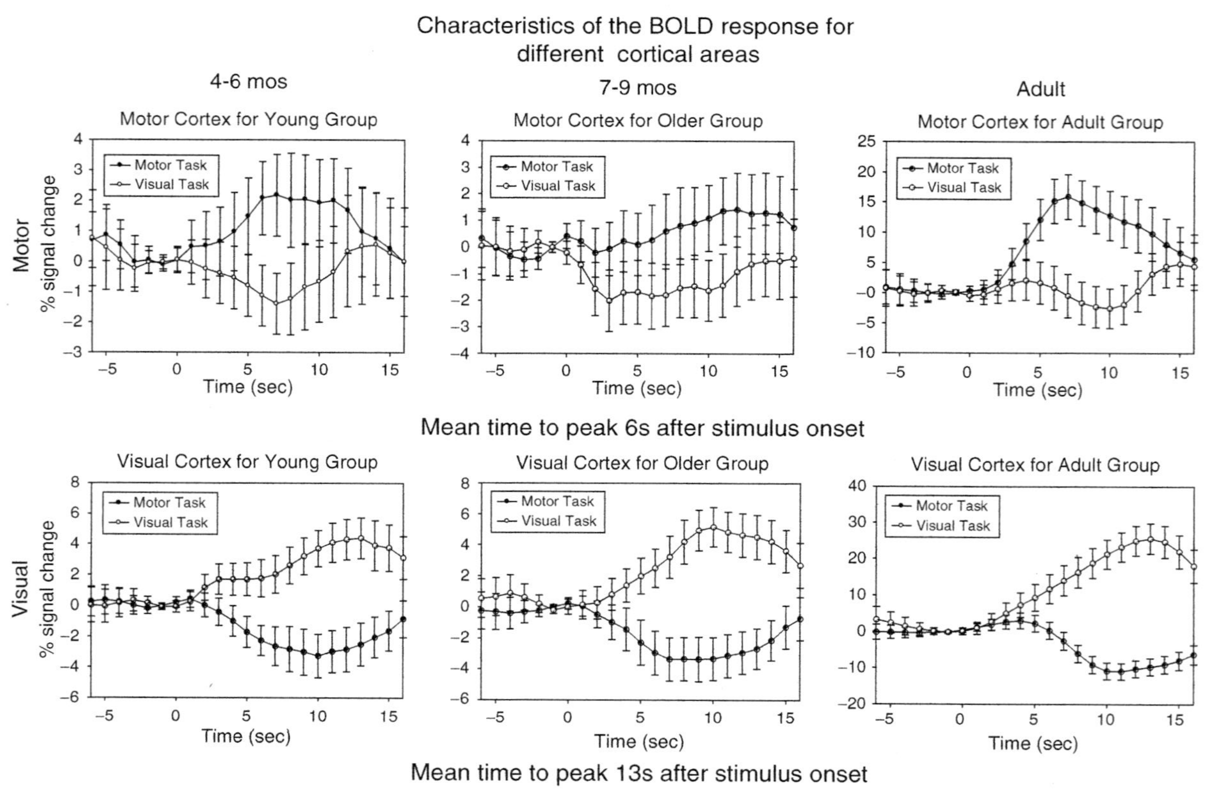

Fig. 8.6 For these graphs, the *x*-axis represents time and zero is stimulus onset. The *y*-axis is percent signal change. The *top row* shows measurements from the motor cortex and the *bottom row* shows measurements from the visual cortex. Within each graph, the line on top is the BOLD response during the active epochs and the bottom line is the same voxel during the nonactive epochs

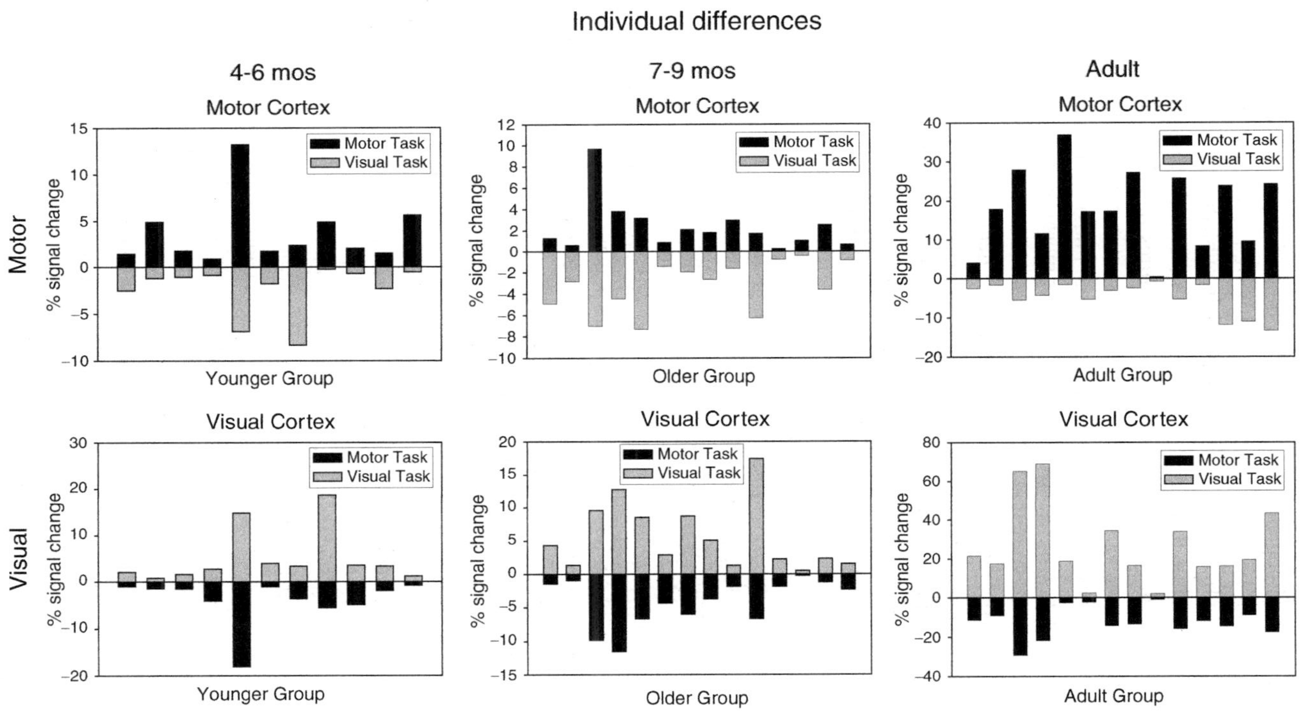

Fig. 8.7 Each bar represents the average signal change for the most active voxel per subject over motor (*black*) and visual (*gray*) areas separated by age (young infants, old infants, and adults from left to right). The *top row* represents measurements from the motor cortex and *the bottom row* represents measurements from the visual cortex

8.4 Methodological Advances

There are *four* methodological advances in this preliminary experiment that bode well for the success of future research. *First*, in this study we tested approximately three times as many subjects as any of the previously published papers using optical imaging on infants. While there is a good signal-to-noise ratio for optical imaging, the variation that is inherent in infant behavior requires that there be at least ten babies in each age group or condition. Furthermore this experiment tested adult control subjects. To draw comparisons between optical imaging on infants and fMRI on adults it is necessary to create two bridges: first we must compare adult fMRI and adult optical data; next we will compare infants and adults in optical.

The *second* improvement was making the experiment more infant-friendly to enhance the likelihood of getting a lower dropout rate, more samples per subject, and higher quality data that was not riddled with motion artifacts. These achievements were accomplished by making a comfortable fleece hat with a chin strap that was slipped on the infant's head while an assistant blew bubbles to entertain the infant. The elastic bandage wrapped on top of the fleece cap made a vast improvement in the reduction of motion artifact and since the infants were able to make modest movements throughout the experiment they were happy to engage in the task for longer. Also, the parent held the infant on their lap throughout the experiment which made the infants more relaxed and comfortable. A common initial study in adult fMRI and optical imaging is to use finger-tapping to stimulate the motor cortex. This is a fine motor skill that is beyond the capacity of most young infants. In pilot studies we found that there was no significant difference in adults' activation levels when we use finger tapping versus holding a vibrating toy. Encouraging an infant to hold a vibrating toy (e.g., electric toothbrush or hand-held back massager) was an engaging task for the infants and kept them from falling asleep during the task. We made similar adjustments to the visual stimuli. A flickering checkerboard is used in many preliminary studies using adult fMRI or optical imaging. We found that the infants (and adults!) had an adverse response to watching the flickering checkerboard for extended periods, but the infants would watch a *Finding Nemo* video for extended periods of time. Again, in pilot studies we found no significant difference in the adults' activation levels to the flickering checker board compared to the *Finding Nemo* video. Finally, we were able to reduce motion artifacts in our infant data because we wrapped their head (lightly) with elastic bandage. This ensured that the probes did not change position during the course of the experiment and allowed infants to move with some degree of freedom.

One of the biggest differences between paradigms that work well for infants and adults is that adults have the capacity to sit still and rest, and infants do not. The *third* improvement in the experimental design was devising a stimulus presentation that involved constant stimulation. The 8 s period of active stimulation engaged the infants, but in the subsequent 15–25 s rest period we presented the alternate stimulus. For example in one motor stimulus block there were six presentations of the vibrating toy that lasted 8 s and the 15–25 s period that followed each stimuli consisted of watching a *Finding Nemo* video. Since the focus of the analysis was activity level in the motor cortex, there was no reason to expect that visual stimulation during rest would influence the motor response. The benefit was that the infants "rested" in that they remained still and did not fuss. Within the adult literature for fMRI and optical there has been discussion that even for adults there should be better control of activity during rest periods because the individual differences are likely to contribute unwanted variances (Birn et al. 2004). By tailoring the paradigm to infants, we got a low dropout rate, a higher quality of data, and better spatial resolution.

The *fourth* improvement in the experiment was the design. Like Taga et al. (2003) we felt it was important to measure from two different cortical areas simultaneously so that we could differentiate a local from a global response. In Taga et al.'s experiment they measured occipital and frontal areas but only presented visual displays. In our experiment, we set up a double dissociation. We measured from visual and motor areas and presented stimuli that would activate each at different times. During a motor stimulus, the visual area probes served as a control and vice versa during the visual stimuli.

While the advances in methodology are notable, there is still plenty of room for further advances. For example, in this experiment it is clear that the motor response showed less activation overall and had more noise compared to the visual response for infants. It is possible that this is the nature of the hemodynamic signal in infancy but it is also possible that young infants do not sit still during stimulus or rest and this activity causes an increased baseline. This later hypothesis is bolstered by the fact that there was less variability in the young compared to the old infant group. Since the average onset of crawling in the US is 9 months of age, a larger proportion of the older infants were mobile compared to the younger group and it is likely that increased mobility of the older infants contributed to the variance in the data.

8.5 Preliminary Studies on Auditory Activation

In the next experiment we investigated a third surface cortical area. These studies were inspired by the Peña et al. (2003) article described above. Our goal was to replicate and extend their initial findings with older, awake infants. There are interesting developmental changes in how infants process language sounds over the course of the first year of life. Infants are able to discriminate the phonemes of native and nonnative language sounds early in the first year, but between 6 and 12 months of age infants' tend to discriminate only speech sounds available in their ambient language. In this experiment, we investigated whether these developmental changes in processing language sounds would be evident through measuring cortical activity in brain areas associated with language processing. Peña et al. (2003) used NIRS to test neonates' responses to linguistic stimuli. They found that infants in the first week of life showed increased activation in their left hemisphere (LH) when presented with their native language compared to backward speech, or silence. Our questions were: Did LH superiority to language stimuli continue to be measurable over the course of the first year? Were there nonlanguage stimuli that showed LH superiority? Were the responses similar across development (e.g., young infants, old infants, and adults)? We presented infants with native and nonnative speech, scrambled versions of both languages, and tone sounds. We tested infants in two age groups to see if there were developmental differences to nonnative speech sounds over the course of the first year. In addition, we tested monolingual English-speaking adults.

We tested 80, healthy, full-term infants. All participants were from monolingual English-speaking families. Forty were in the young age group from 3 to 6.5 months of age and 40 were in the older age group from 6.5 to 10.5 months of age. In addition, we tested 16 monolingual English-speaking adults as control subjects. The apparatus and data analysis were identical to the motor visual experiment except that the probes were positioned to maximize the likelihood of monitoring auditory areas. We used the same skull landmarks described above and centered each probe set over the preauricular lobule. There were five different stimulus conditions: English, Scrambled English, Korean, Scrambled Korean, and Tone (continuous sine wave). The stimuli were recorded segments of infant-directed speech from an English/Korean children's story book read by a bilingual female. The stimulus presentation was similar to Experiment 1. To keep infants engaged in the task a *Finding Nemo* video with no sound was playing continuously throughout the experiment. Stimuli were presented for 8 continuous seconds followed by a "rest" period that varied between 15 and 25 s. Each block consisted of six to eight stimuli presentations. We obtained three blocks (6+6+8=20 stimuli presentations) for each condition from each adult participant. Infants contributed two blocks on average. The presentation order of conditions was counterbalanced across subjects. The scrambled conditions were created by splitting each language condition into 50 ms components and randomizing the order of each segment. In one sense these stimuli were not language like at all because the randomized pattern violated the continuity and prosody of much of the stimuli. However, the scrambled conditions could also be described as very language like because they preserved the onset of segmental consonants and vowels.

We replicated and extended the findings of Peña et al. (2003). More specifically, we found significantly more activation in LH compared to right hemisphere (RH) for the native language at all ages. Fig. 8.8 shows the results of the spatial analysis for two individual subjects separated by the oxy-, deoxy-, and total hemoglobin data. Comparison across the rows reveals that the deoxyhemoglobin data tends to return flat and variable data so many studies focus on the oxyhemoglobin data alone. Using the oxyhemoglobin data we found that optical imaging is capable of detecting differences across conditions, between LH and RH, and between upper and lower channels within each hemisphere. Similar to Peña et al. (2003) we found that there was significantly more activation in the lower probes compared to probes near the top of the head. We were able to capture developmental differences as well. The younger infants showed the most LH activation to the English and the

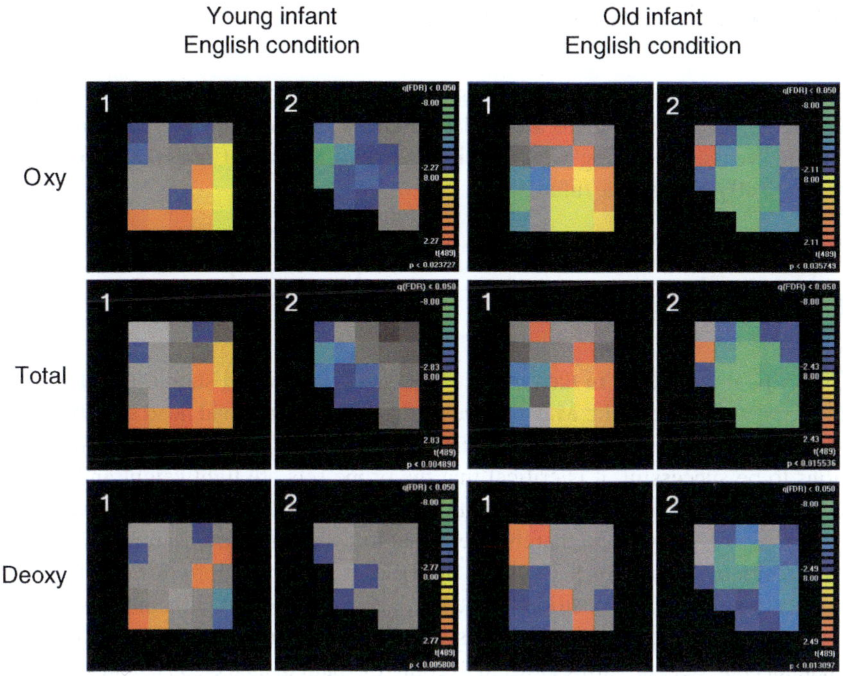

Fig. 8.8 Activation patterns for two individual participants. The probes were placed bilaterally over auditory areas (LH is on the left, RH is on the right). Warm colors (*orange* and *yellow*) indicate significant increase in blood flow and cool colors (*blue* and *green*) indicate significant decrease. The *top row* is absorption data for oxyhemoglobin only, the *bottom row* is deoxyhemoglobin only and the middle is the total or oxy- and deoxyhemoglobin combined

Scrambled English conditions. The older infants showed more LH activation to the straight compared to the scrambled language conditions for both English and Korean. The adults showed LH superiority to English and Korean conditions and RH superiority to scrambled conditions. The tone condition revealed bilateral and significantly less activation for all age groups. There was a significant effect of age comparing the young infants and adults on the language conditions but there was no significant difference between the older infants and adults. An analysis of the individual differences across infants was performed to capture the variability across subjects. There were 62 infants that were tested in the English condition. Of those infants 71% showed LH superiority, 11% showed RH superiority, 2% showed equal activation, and 16% showed no activation. These findings replicated and extend the work of Peña et al. (2003) by showing the LH superiority to native language with older infants who were awake instead of sleeping. Furthermore, these data demonstrate that optical imaging can capture developmental differences between infants of different ages.

8.6 Speculations About the Future

The future of using optical imaging to measure functional activation in human infants is promising. There are still challenges that need to be overcome in the coming years. In closing we will highlight two areas that are likely to improve dramatically in the near future.

8.7 Probe Design

The critical ingredient in data collection is the comfort and stability of the probes that emit and detect light. Many of the prototype machines had head sets that weigh several pounds (some even have counterweights). For research on infant populations to be successful, it is critical that the probe sets are lightweight and durable. One cost of lightweight probes is that the fiber optic cable is smaller. These smaller cables carry less light and thereby the signal to noise ratio gets diminished. Second, as the number of probes per system increase, it becomes more difficult to get the probes to have a stable position on the head. The most recent Hitachi probes are flat on the end and reside in a slightly sticky silicon sheath that works well with 24 channels but there are difficulties in keeping the path distance consistent on a round head with more than 24 probes.

8.8 Analysis Methods

Currently there is no agreed upon analysis method for NIRS research on infants. In the current literature there is a wide variety of techniques. NIRS can borrow some of the lessons learned from statistical analysis for fMRI data. However there are issues that are unique to NIRS research alone. For example, there is variability in wavelength frequency used (this is particular to the deoxy signal that ranges from 690 to 780 in recent machines). Because many have found little or no signal in the deoxy range many papers report only oxyhemoglobin changes. Early papers reported total change (oxy + deoxy) but since these signals tend to go in opposite directions it might be better to report them separately. In addition, there seems to be different peak responses based on different tissue areas (recall that in our preliminary data the time to peak in the motor data was 6 s and the visual data was 13 s). In analysis done with Brain Voyager the data are fit to a line called the hemodynamic response function (HRF). This line has a peak at 6 s based on the BOLD response from fMRI. It is likely that the HRF for infants is different and may change over development these nuances would be missed in the current analysis techniques. Currently we are developing analysis techniques with Analysis of Functional NeuroImaging software (Afni) and fitting individual HRF functions to individual participants and cortical areas. Since the light reflectance can vary as a function of the different anatomy under the probes, we predict that this analysis technique will give better spatial resolution and clearer data.

In collaboration with Paul Reber we will apply new spatial analyses to these data. First, we will identify in each subject, the channel that reveals the best activation in the respective motor and visual areas. Then we will align the group data with respect to the best channel and evaluate the intensity and range of activation. These analyses will lend insight to the individual differences in response, differences in activation levels, and the shape of the hemodynamic curve in terms of time to peak and period before return to baseline. The preliminary analyses described above demonstrate global differences but the new spatial analysis should reveal more subtle aspects of the data and potentially uncover developmental differences that were not revealed in the first analysis. Second, we will do a channel-by-channel analysis. We will accomplish this in two ways: first with a smoothing function and second without smoothing, which will allow us to explore and discover the best approach. There are two important factors to consider about smoothing: smoothing improves the signal-to-noise ratio, which increases the likelihood of finding existing changes in activation, but at the cost of losing spatial localization.

To speculate on the impact of NIRS on the field of cognitive development one can look at the insights gained in recent decades from applying neuroimaging methods to the study of mature brains. As Spelke (2002) has pointed out this progress has been mixed. There have been discoveries of neural signatures for phenomena that were well understood at the behavioral level. Although occasions of neurological findings that overturn previously held conclusions or offered radically new conclusions about the underlying mechanisms have been rare. Spelke (2002) also points out that what distinguishes behavioral and imaging data in the realm of infant work is that imaging data are rich, and structured. In contrast infant behavioral data are often a limited set of data points indicating a successful discrimination of change versus a lack thereof (e.g., negative finding). It is likely that cognition, even at the immature state of the infant, is more nuanced and complex than dichotomous data reveal and neuroimaging may be in window into a new level of analysis.

Acknowledgments We are indebted to Anna Lane for her help with data collection and processing. This research was supported by a McDonnell Junior Scientist Research Award, Vanderbilt Discovery Grant, and NSF (0718513) all awarded to the first author. We thank the parents of the children who participated in this research and the undergraduate research assistants in the Infant Cognition Lab at Vanderbilt.

References

Anderson AW, Marois R, Colson ER, Peterson BS, Duncan CC, Ehrenkranz RA et al (2001) Neonatal auditory activation detected by functional magnetic resonance imaging. Magn Reson Imaging 19(1):1–5

Baird AA, Kagan J, Gaudette T, Walz KA, Hershlag N, Boas DA (2002) Frontal lobe activation during object permanence: data from near-infrared spectroscopy. Neuroimage 16(4):1120–1126
Bartocci M, Winberg JAN, Ruggiero C, Bergqvist LL, Serra G, Lagercrantz H (2000) Activation of olfactory cortex in newborn infants after odor stimulation: a functional near-infrared spectroscopy study. Articles. Feedback. Pediatr Res 48:1
Birn RM, Cox RW, Bandettini PA (2004) Experimental designs and processing strategies for fMRI studies involving overt verbal responses. Neuroimage 23(3):1046–1058
Bortfeld H, Wruck E, Boas DA (2007) Assessing infants' cortical response to speech using near-infrared spectroscopy. Neuroimage 34(1):407–415
Csibra G (2004) Near infrared spectroscopy reveals neural activation during face perception in infants and adults. J Pediatr Neurol 2(2):85–89
Dehaene-Lambertz G, Dehaene S, Hertz-Pannier L (2002) Functional neuroimaging of speech perception in infants. Science 298(5600):2013–2015
Folley BS, Park S (2005) Verbal creativity and schizotypal personality in relation to prefrontal hemispheric laterality: a behavioral and near-infrared optical imaging study. Schizophr Res 80(2–3):271–282
Franceschini MA, Thaker S, Themelis G, Krishnamoorthy KK, Bortfeld H, Diamond SG et al (2007) Assessment of infant brain development with frequency-domain near-infrared spectroscopy. Pediatr Res 61(5 Pt 1):546–51
Gliga T, Dehaene-Lambertz G (2007) Development of a view-invariant representation of the human head. Cognition 102(2):261–288
Horovitz SG, Gore JC (2004) Simultaneous event-related potential and near-infrared spectroscopic studies of semantic processing. Hum Brain Mapp 22(2):110–115
Kleinschmidt A, Obrig H, Requardt M, Merboldt KD, Dirnagl U, Villringer A et al (1996) Simultaneous recording of cerebral blood oxygenation changes during human brain activation by magnetic resonance imaging and near-infrared spectroscopy. J Cereb Blood Flow Metab 16(5):817–826
Kuhl PK (2004) Early language acquisition: Cracking the speech code. Nat Rev Neurosci 5(11):831–841
Martin E, Joeri P, Loenneker T, Ekatodramis D, Vitacco D, Hennig J et al (1999) Visual processing in infants and children studied using functional MRI. Pediatr Res 46(2):135–140
Meek JH (2002) Basic principles of optical imaging and application to the study of infant development. Dev Sci 5(3):371–380
Meek JH, Firbank M, Elwell CE, Atkinson J, Braddick OJ, Wyatt JS (1998) Regional hemodynamic responses to visual stimulation in awake infants. Pediatr Res 43(6):840–843
Nelson CA, Monk CS (2001) The use of event-related potentials in the study of cognitive development. In: Nelson CA, Luciana M (eds) Handbook of developmental cognitive neuroscience. MIT, Cambridge, MA, pp 125–136
Neville HJ (2005) Development and plasticity of human cognition. In: Mayr U, Awh E, Keele SW (eds) Developing individuality in the human brain: a tribute to Michael I. Posner. American Psychological Association, Washington, DC, pp 209–235
Peña M, Maki A, Kovacic D, Dehaene-Lambertz G, Koizumi H, Bouquet F et al (2003) Sounds and silence: an optical topography study of language recognition at birth. Proc Natl Acad Sci USA 100(20):11702–11705
Schneider W, Eschman A, Zuccolotto A (2001) E-Prime user's guide. Psychology Software Tools, Inc., Pittsburgh, PA
Spelke ES (2002) Developmental neuroimaging: a developmental psychologist looks ahead. Dev Sci 5(3):392–396
Taga G, Asakawa K, Hirasawa K, Konishi Y (2003) Hemodynamic responses to visual stimulation in occipital and frontal cortex of newborn infants: a near-infrared optical topography study. Early Hum Dev 75(Suppl):S203–S210
Wilcox T, Bortfeld H, Woods R, Wruck E, Boas DA (2005) Using near-infrared spectroscopy to assess neural activation during object processing in infants. J Biomed Opt 10(1):11010

Chapter 9
In Vivo Dynamics of the Visual Cortex Measured with Voltage Sensitive Dyes

Per E. Roland, Xiaoying Huang, and Jian-Young Wu

Abstract Voltage sensitive dye (VSD) imaging is an optical method for measuring trans-membrane potentials. This method has developed into a powerful tool for studying brain dynamics. VSD molecules bind to the membranes of excitable cells and report changes in membrane potential by shifting their absorbance or florescence spectra within a microsecond. In this chapter we provide the details of the method, the various sources of noise and how to avoid them, and the advantages of the new blue dyes. We further describe technical issues such as staining, imaging, and reduction of pulsation artifact and noise control for achieving a high sensitivity in the measurements. Voltage sensitive dye recording is now superior to local field potential recording from the supergranular layers of the cortex. In in vivo applications, the voltage sensitive dye signals stem from thousands of cells in the upper layers of cortex. Pulsation artifacts can be nearly completely removed by analytical or ad hoc subtraction methods. The resulting $\Delta F/F$ then shows the changes in the membrane potentials for thousands of neurons in the upper layers. Net excitation or inhibition of the neuronal population can be measured when the $\Delta F/F$ changes its polarity. The dynamics of the stimulus evoked changes in population membrane potentials as measured with voltage sensitive dye seriously challenge the view that cortical computations of sensory stimuli are confined to columns or patches processing particular stimulus properties. Rather, the cortical dynamics of the evoked changes in the membrane potentials indicate that the whole sensory scene is computed in each of the early sensory areas. Furthermore, the dynamics observed depend on the physical nature of the sensory stimuli. All focal stimuli evoke a feed-forward depolarization and lateral spreading depolarization in the area(s) reacting to the stimuli. The further computations and the engagement of different visual areas depend on the initial computations and involve within area dynamics as well as dynamics of the population membrane potentials driven from other visual areas. This implies that the computations in the cortex are not confined to particular cortical subregions, but highly dependent on intra-area as well as inter-area communications.

P.E. Roland (✉)
The Laboratory of Brain Research, Department of Neuroscience, Karolinska Institutet, Retzius vaeg 8, S171 77, Solna, Sweden
e-mail: Per.roland@neuro.ki.se

A.W. Roe (ed.), *Imaging the Brain with Optical Methods*,
DOI 10.1007/978-1-4419-0452-2_9,

9.1 The Method

Voltage sensitive dye (VSD) imaging, an optical method of measuring trans-membrane potential, has developed into a powerful tool for studying brain activity ever since a pioneering work was published about 40 years ago (Cohen et al. 1968; Tasaki et al. 1968). VSD molecules bind to the membranes of excitable cells and report changes in membrane potential by shifting their absorbance or fluorescence spectra. While the optical signal of VSD has excellent linearity with the membrane potential (within a range of approximately ±300 mV, Ross et al. 1977) and very fast response time (<1 μs) (Ross et al. 1977), the VSD signal is small. Fractional changes in absorption or fluorescence are only about 10^{-2}–10^{-5} of the resting absorption/fluorescence intensity per 100 mV of membrane potential change (Fig. 9.1). In biological experiments, such small signals are quite vulnerable to noise and artifacts. For imaging in vitro preparations such as brain slices and cultured cells, noise and artifact can be controlled well and the sensitivity of VSD imaging rivals that of local field potential

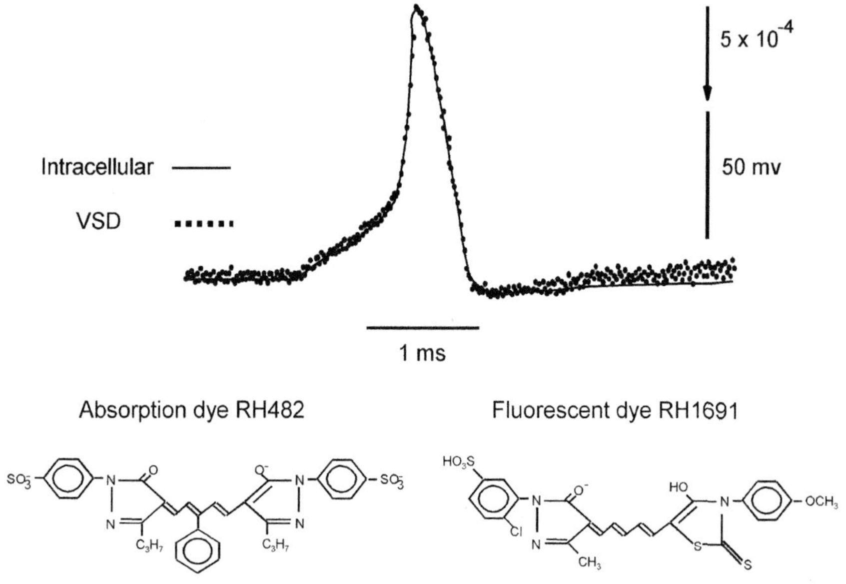

Fig. 9.1 Voltage sensitive dyes and dye signal. *Top*, Simultaneous VSD and intracellular recordings from squid giant axon. The squid giant axon was stained by an absorption dye XVII. The optical signal is the change in light absorption at 705 nm (*dotted trace*). The actual membrane potential was simultaneously recorded by an intracellular electrode (*smooth trace*). During an action potential, the time course of the VSD signal matched accurately to that of the membrane potential change. Note that the amplitude of the VSD signal is very small (modified, with permission, from Ross et al. 1977). *Bottom*, Chemical structures of two commonly used dyes. The absorption dye, RH482, (also known as NK3630, JPW1132) is commonly used for brain slice recordings because of its excellent signal and ignorable phototoxicity (see Fig. 9.2). The fluorescent dye, RH1691, is commonly used for in vivo recordings

recordings (Jin et al. 2002, Fig. 9.2). Imaging the mammalian cortex in vivo is more difficult because hemoglobin absorbance causes a large pulsation artifact. This artifact can sometimes exceed evoked cortical signals by an order of magnitude (London et al. 1989; Shoham et al. 1999; Grinvald and Hildesheim 2004; Ma et al. 2004).

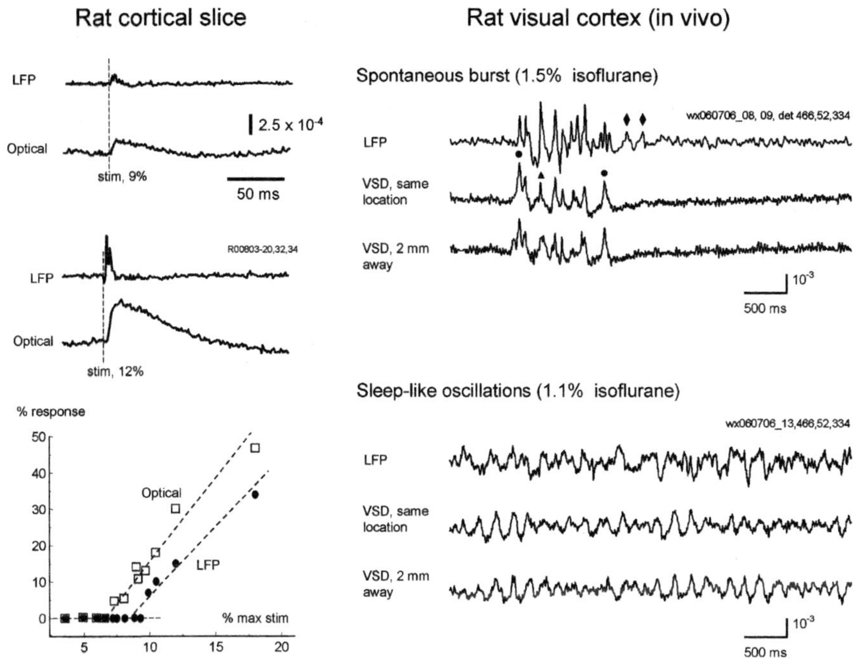

Fig. 9.2 Sensitivity of voltage sensitive dye imaging. *Left*, Sensitivity of VSD recordings in brain slice. A local field potential microelectrode is placed in cortical layer II–III and the same location is measured by optical imaging. A stimulation electrode is placed in the white matter, 2 mm away from the measuring point. The evoked activity can be seen in single trials of both VSD and local field potential (LFP) recordings (*top* and *middle panel*). The relationship between stimulus intensity and response amplitude is plotted in the *bottom panel*, showing that the VSD signal has a similar (or better) sensitivity to that of local field potential in detecting the activity evoked by the stimulus. The signals have been low pass filtered at 50 Hz (Modified, with permission, from Jin et al. 2002.) *Right*, Dye signals in vivo. The top two traces in each panel are local field potential and VSD recordings from the same location in the visual cortex. Another VSD recording (bottom trace in each panel) was 2 mm away from this location. (**a**) Under 1.5% isoflurane, spontaneous bursts seen in the local field potential recordings can also be seen in the VSD recordings. However, the peaks in the two recordings do not match exactly, probably, because the field potential comes from deep layers or subcortical sources while VSD signal is mainly from the superficial layers. *Dots* mark events in which VSD has higher amplitude than local field potential. *Triangles* mark events in which local field potential has higher amplitude than VSD. *Diamonds* mark events seen in local field potential but not in VSD. (**b**) Recordings from the same animal as in **a**, about 5 min later, with isoflurane anesthesia lowered to 1.1%. Most of the peaks in local field potential are also seen in VSD, but the correlation between the two signals is lower. The baseline fluctuations cannot be attributed to noise, since the analogous recordings had much less fluctuation when anesthesia was increased (*top left*) (Modified, with permission, from Lippert et al. 2007.)

New "blue" dyes, developed by Amiram Grinvald's group (Shoham et al. 1999), have brought a revolutionary potential for in vivo imaging of mammalian cortex. The excitation wavelength of blue dyes has minimal overlap with the absorption of hemoglobin, and hence blue dye recordings have minimal pulsation artifact (Shoham et al. 1999). Authors using blue dyes have shown excellent results (Slovin et al. 2002; Petersen et al. 2003a, b; Ferezou et al. 2006; Xu et al. 2007; Eriksson et al. 2008). With careful control of the sources of noise, the detecting sensitivity of in vivo VSD imaging can also be comparable to that of local field potential recordings (Fig. 9.2) (Lippert et al. 2007). In this section, we will discuss technical issues such as staining, imaging, reduction of pulsation artifact and noise control for achieving a high sensitivity in in vivo optical imaging from mammalian cortex. With a sensitivity that is comparable to that of local field potential recording one can record spontaneous and evoked activity in single trials without spatial or temporal averaging. We will also discuss dye bleaching, washout, phototoxicity, and the total recording time for optical experiments, and limitations of current method.

9.1.1 Noises and Noise Reduction

Usually more than one source of noise interferes with the VSD recording. The total noise of the measurement is a Pythagorean summation of all sources:

$$N_{\text{total}} = \sqrt{\left[N_1^2 + N_2^2 + \cdots + N_n{}^2\right]}$$

From the equation one can see that if noise from one source is ten times larger than other noises, it would dominate the signal-to-noise ratio and other noises in the system can be ignored. The sources of noise can be divided into dark noise and light noise. Dark noise is defined as the noise when there is no light, and thus is often refered to as the intrinsic noise of the imaging device. The light noise is the noise related to the illumination light. For in vivo imaging, the light noise mainly comes from three sources: the shot noise or random fluctuation of photo flux, the fluctuation in the hemoglobin due to blood circulation and mechanical movement of the preparation pulsation, respiration, and floor vibration. In the section below we will discuss these noises and how to control them.

9.1.1.1 Dark Noise

An imaging device consists of an array of photo sensors and a readout circuit (Fig. 9.3). After the readout circuit, the signals are digitized and stored as a data file in a computer. Noise is generated at each of these stages from the sensor to the digitizer, and a detailed analysis can be complicated. However, usually at one stage one type of noise dominates, and so the noises at other stages can be ignored. Therefore,

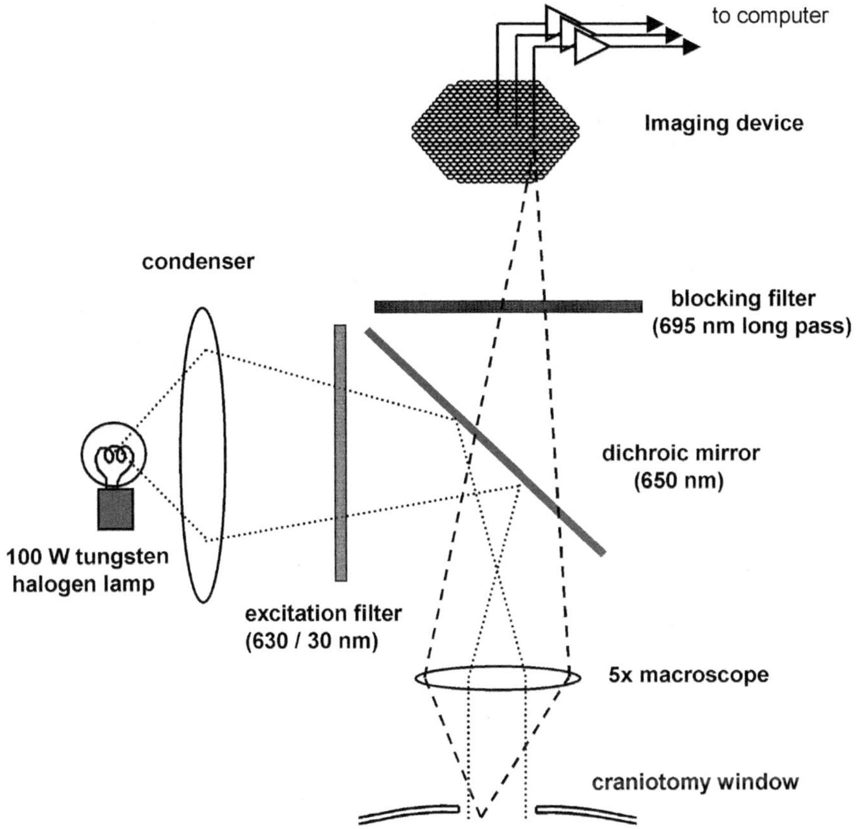

Fig. 9.3 High throughput recording apparatus. This 5× macroscope permits light to pass through about 100 times brighter than does an ordinary microscope with 5× objective lens. The main lens is a commercial video lens (Navitar 25 mm 1/F 0.9), providing a numerical aperture of 0.4. Since the macroscope has a high light throughput, an ordinary microscope halogen tungsten filament lamp (Zeiss) provides adequate light intensity for the recording. The filter assembly is constructed by an Olympus filter cube and customer ordered filters (Modified, with permission, from Lippert et al. 2007.)

it is important for a user to know what noise dominates his/her imaging system. Silicon photodiodes are most common sensors used in diode array, CCD, and CMOS cameras (a CMOS camera is a camera that has photodiodes in an integrated circuit, a so-called active pixel sensor). At very low light, such as when one records from a dendrite of a neuron, the dark noise of the photodiode can be larger than the shot noise. Cooling down the sensor, such as is done in a cooled CCD camera, would thus be necessary. However a cooled CCD is not necessary in most other circumstances (e.g., imaging brain slices or cortex in vivo) because the shot noise is ten times larger than the thermal noise of the photodiodes.

Since VSD signal is only 10^{-5}–10^{-2} of the resting light intensity, the digitizing noise may become significant. For example, with a 12 bit camera, the resting light intensity is represented by 12 bits. If the signal is 1/1,000th of resting light intensity, then there are only 2 bits left to represent the signal. The digitizing noise would be about ¼ of the signal. Photo-diode arrays avoid the digitizing noise by using a two-stage amplifier. The resting light intensity is subtracted after the first stage, and then the signal is amplified ~100 times. This extends the effective dynamic range to ~19 bits (Cohen and Lesher 1986; Wu and Cohen 1993). In the examples shown in this section, the resting light intensity was about 1 V at the output of the first stage. The signal, at approximately 10^{-3} of the resting light intensity, therefore produced a change of ~1 mV riding atop this resting 1 V. A second stage amplifier was used to remove the DC component (resting light intensity) and differentially amplify the signal. After this second stage amplification, the 1 mV signal is amplified to 500 mV. This signal was digitized with an A/D converter of 12 bits at a range of ±10 V, resulting in the 500 mV signal being represented by ~9 bits. Since the original DC component was approximately 10^3 of our signal amplitude (i.e., ~10 bits), our effective dynamic range in this system is ~19 bits. Such high dynamic range is essential for the overall sensitivity of optical recordings.

9.1.1.2 Shot Noise

Shot noise is the random fluctuation of the photon flux in the light. Shot noise is proportional to the square root of the light intensity. In optical recordings the optimum situation is that shot noise dominates the total noise, i.e., shot noise is at least twice as large as all other noises. In such an optimized situation, the sensitivity of the optical recording increases with the illumination intensity and the signal-to-noise ratio is proportional to the squire root of the illumination intensity. However, in practice the light intensity cannot be increased without a limitation. Some imaging devices, such as CCD cameras, have limited well size and will become saturated at a high light. Dye bleaching and photo toxicity are other major factors limiting the total recording time under high illumination intensity. Movement artifacts are also proportional to the light intensity and if this noise source were not controlled, increasing the illumination intensity would not increase the signal-to-noise ratio.

9.1.2 Signal-to-Noise Ratio and Spatial–Temporal Resolution

Increasing the spatial or the temporal resolution will reduce the signal-to-noise ratio. Higher spatial resolution leads to a smaller area of stained tissue under each detector, and therefore to a smaller photon flux. The smaller photo flux will have larger relative fluctuations in the flux or larger shot noise, resulting in a decreased signal-to-noise ratio. Likewise, higher temporal resolution leads to shorter sampling duration of the photon flux, less photons sampled, and a decreased signal-to-noise ratio. Therefore, the spatial and temporal resolution of the VSD imaging is usually limited by the signal-to-noise ratio, not by the spatial resolution or the sampling rate of the imaging device.

In actual in vivo VSD recordings from rodent cortex, spontaneous and sensory evoked signals are about 10^{-3} of the resting fluorescent intensity. Without averaging, a signal-to-noise ratio of about 5 can be achieved at a pixel size of approximately 100 μm in diameter and a temporal resolution of 1.6 kHz. If the spatial resolution is increased fourfolds, or the pixel size reduced to 50 μm in diameter, the noise would double and the signal-to-noise ratio should be reduced to 5/2.

9.1.2.1 Pulsation Artifact

The pulsation artifact is a combination of hemoglobin absorption and physical movement of the tissue being imaged (Figs. 4, 5 of Lippert et al. 2007). When people used conventional dyes (e.g., RH795), the pulsation artifact was the limiting factor for signal-to-noise ratio of imaging in vivo. This is because the dye's excitation spectrum overlaps with that of the hemoglobin absorption (~530 nm, Shoham et al. 1999). The pulsation artifact of RH795 had an amplitude of 0.5–1% of the resting light intensity. This is similar to amplitude of the dye signal during epileptiform spikes in the EEG induced by bicuculline, or about five to ten times larger than the sensory evoked cortical activity (Lippert et al. 2007; London et al. 1989). Without blue dyes, only large signals (e.g., epileptic spikes) can be imaged without averaging under in vivo conditions in mammalian cortex (however, see, Arieli et al. 1995; Civillico and Contreras 2005). Therefore, with conventional dyes, single trial sensory-evoked waves and other small signals could most easily be imaged in the brain of turtle (Prechtl et al. 1997; Senseman and Robbins 1999, 2002; Lam et al. 2000, 2003), salamander (Cinelli et al. 1995), and in invertebrates (Delaney et al. 1994; Kleinfeld et al. 1994), where heartbeat is not present or can be paused with cooling.

"Blue" dyes (Shoham et al. 1999) brought a great improvement for imaging mammalian cortex, because they show virtually no pulsation artifact (Ferezou et al. 2006; Lippert et al. 2007). However, pulsating movements can still cause pulsation artifact even when RH1691 is used. The pulsation artifact can be removed by spectral based single value decomposition (Pesaran et al. 2005) or an ad hoc ECG-triggered subtraction procedure (Lippert et al. 2007).

9.1.3 Methods for Achieving High Signal-to-Noise Ratio

9.1.3.1 High Numerical Aperture Optics

Collecting more light from the preparation is essential for a high signal-to-noise ratio. High numerical aperture optics will not only effectively illuminate the cortex, but also collect a large fraction of the fluorescent emission into the imaging detector. A macroscope with numerical aperture (NA) of 0.45 can give ~100 times brighter light than an ordinary 4x, 0.10 NA microscope objective. The macroscope can be assembled from low-cost commercial optics for video cameras (e.g., Navita 25 mm 1/F 0.9), following the designs of Kleinfeld (Kleinfeld and Delaney 1996) and Cohen (Prechtl et al. 1997). Figure 9.3 shows a schematic drawing of the optical design for the macroscope with illuminator and filter cube. Since this home made 0.45 NA macroscope is bright, a standard microscope illuminator with a 100 W halogen tungsten filament lamp provides adequate illumination intensity.

9.1.3.2 Staining

Good staining is a key to obtain high signal-to-noise ratio recording because voltage-sensitive dye signal can only come from neuronal membranes properly stained with the dye. Under in vivo conditions, staining is usually achieved by applying the dye to the surface of the cortex. Then it is always a concern, how deep will the dye stain the cortex? Figure 9.4 shows cortical sections of rat and monkey cortex stained with fluorescent dye from the surface. In both cases, the dye fluorescence is seen ~500 μm deep into the cortex, adequate for imaging from the top of the cortex. For rodent preparations, the dura mater is thin and can be made permeable to the dye. Thus the dura can be kept intact to provide good protection to the cortex (London et al. 1989; Lippert et al. 2007). In order to increase the dye permeability, one should, before staining, dry the dura for 3–4 min until the dura becomes transparent ("glassy"). This glassy dura would allow for good staining (Lippert et al. 2007). Careful and nontraumatic craniotomy is important for better staining. An irritated brain can appear reddish (due to vessel dilatation), or raise the intra-cranial pressure, which leads to poor staining. Dexamethasone sulfate (1 mg/kg IP) may be given 2–24 h prior to the experiment, to reduce the inflammatory response of the dura and an eventual edema of the brain in experiments with durotomy.

In experiments with dura left intact, one can use approximately 200 μl of dye solution (approx. 2 mg/ml of RH 1691 or RH 1838) to stain an area 5 mm in diameter. In experiments with durotomy, about 0.8 mg/ml will work. A thin layer of silicone grease can be applied to the edge of the craniotomy window so that the dye solution can be held inside the window. During staining, a small perfusion pump continuously circulates the dye solution (London et al. 1989; Lippert et al. 2007). The pump is built with a gear motor, which gently presses the rubber nipple of a Pasteur pipette once every few seconds. The tip of the pipette is placed in the staining solution and performed a gentle, back-and-forth circulation of a small amount of dye (~100 μl). Such circulation is necessary because the cerebrospinal fluid (CSF), slowly exudes from the dura and dilutes the dye concentration locally at the dye–dura interface. The staining process is about 90 min, followed by washing with dye-free artificial CSF for >15 min.

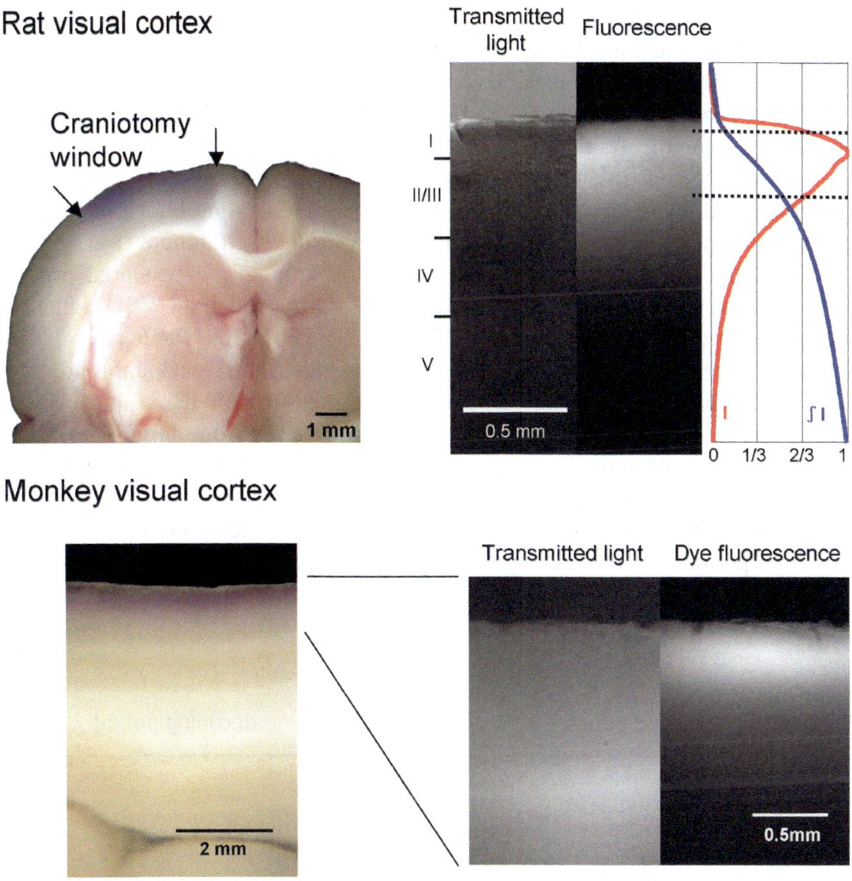

Fig. 9.4 Profile of voltage sensitive dye staining. Rat (*top*) and monkey (*bottom*) cortex stained with RH1691. In both *top* and *bottom panels*, photographs on the *left* are sections of the stained cortex. Staining by RH1691 can be seen by eye as a *light blue* hue. On the *right*, there are transmitted light and fluorescent images from the same cortical section stained with RH1691. The fluorescence image is excited at 630 nm, and the emission is >690 nm. The graph at the *top right* shows the fluorescence intensity (*red*) and integrated fluorescence intensity (*blue*) over cortical depth. *Broken lines* mark the tissue depth with fluorescent intensity higher than 2/3 of the maximum intensity (*Top panel*, modified, with permission, from Lippert et al. 2007.)

9.1.3.3 Dye Bleaching, Washout, and Phototoxicity

In brain slices, the recording time of VSD imaging is mainly limited by dye bleaching (Momose-Sato et al. 1999; Jin et al. 2002). This is partially because the absorption dyes have less phototoxicity. The total optical recording time is about ~1,800 s before the VSD signal declines to half of the initial amplitude (Jin et al. 2002). Under in vivo conditions, the dye will eventually dissolve into and be washed out by the cerebrospinal fluid and blood circulation. This may also reduce the total recording time in addition to that due to bleaching and phototoxicity. Therefore, the total recording time must be determined empirically.

To determine the total feasible recording time, the decrease of VSD signal (RH1691) can be measured under an illumination intensity that allows for a high sensitivity imaging. (e.g., using a 100 W halogen tungsten lamp, ~1.5×10^{11} photo-electrons/mm^2 ms, Fig. 9.5, Lippert et al. 2007). We measured the decay of the amplitude of epileptiform spikes induced by bicuculline under intermittent illumination of the cortex, with 20 s of light exposure each minute (Fig. 9.5).

The decay of the VSD signal did not follow a simple exponential process. Instead, there was a period of some 300–800 ms during which the VSD signal was stable and thereafter the signal declined (Fig. 9.5). During the stable period, there was no apparent decay in the signals, and in some areas the signal amplitude actually increased. Following the flat period, the amplitude of the signal declined steadily (Fig. 9.5). The length of the flat period varies from animal to animal. In one (out of four) animals, the staining was less optimal and the flat period was much shorter (stars in Fig. 9.5), suggesting that the stable period is related to the wash-out of the dye from the stained tissue. Since the boundary between stable and declining periods was not clearly distinguishable, we use the "half-time," i.e., the

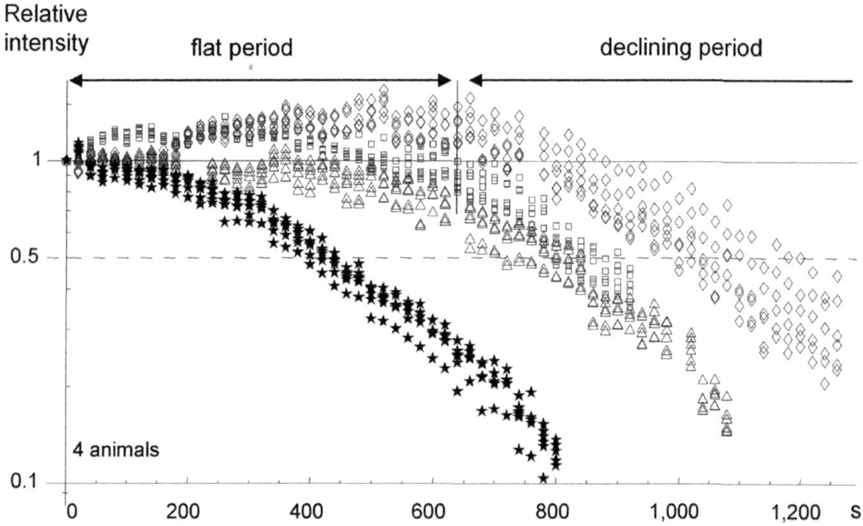

Fig. 9.5 Dye bleaching and total recording time. The VSD signal decline is plotted against recording light exposure time. The signal is bicuculline induced spikes in cortex which are relatively stable over time. Data from four animals were normalized to the amplitude of the first recording trial. In three animals (*diamonds*, *triangles*, *squares*), the signal amplitude remained stable (or slightly increased) for a period (flat period). After the flat period, the signal had a higher rate of decline with light exposure (declining period). Note that one animal did not have optimum staining, and the stable time was much shorter (*stars*) (Modified, with permission, from Lippert et al. 2007.)

duration of light exposure for the signal to reduce to 50% of its beginning amplitude, as the total recording time. In three animals with good staining, the total recording time was 800–1,000 s.

To distinguish dye wash-out from light-related amplitude reduction, part of the cortex was shielded from light during a light exposure experiment. After the signal in the exposed area transitioned to the declining period, the shield was removed. The signal from the previously shielded area had similar amplitude as that in the flat period. However, the signal in the shielded area did not show a flat period, instead, the signal declined at a rate similar to the unshielded regions. This suggests that the flat period was related to the dye wash-out, which only occurred within a certain period after the staining.

The bleaching curve (Fig. 9.5) shows that it took about 800–1,000 s of exposure to illumination for the signal to reach 50% of the initial amplitude. This time would allow for an intermittent recording of 80–100 trials of 10 s each, enough for most types of experiments. Increasing illumination intensity can increase the signal-to-noise ratio for small signals, by compromising the total recording time.

Phototoxicity might become a concern when increasing the illumination intensity. While the literature reports that RH1691 has no obvious toxicity (Slovin et al. 2002; Petersen et al. 2003a), our illumination intensity was higher than previous reports, and we did observe rapid signal decline (Fig. 9.5). To address the concern of phototoxicity, results from later recording trials should be compared with those from the beginning of the experiment for verification.

9.1.4 A Concrete Example from the Barrel Cortex

Signals in barrel cortex evoked by deflecting a single whisker can be reliably detected without averaging. Figure 9.6 shows signals recorded from individual detectors (160 μm in diameter) in single trials, demonstrating a combined high spatial and temporal resolution. In the barrel cortex, the evoked wave started from the correspondent whisker barrel in all trials and propagated to the entire barrel field. The evoked activity had large trial-to-trial variations, usually lower amplitude when evoked at certain phases of spontaneous activity. Trial-to-trial variations in propagation pattern manifest as anisotropy of propagating pattern and velocity. In Fig. 9.6, the propagation pattern of individual trials was compared with that of the multiple trial average. While all trials show a similar overall spatiotemporal pattern, significant trial-to-trial variations can be seen in individual trials: The onset times for trials 2, 3, and 10 are shorter than the average, while in trials 1 and 8, the onset times are longer. In the bottom row of Fig. 9.6, the contour lines of the propagation patterns for trials 5, 6, 8, 10 are superimposed, in order to compare them with the contours of the average pattern. The contour lines are highly variable, suggesting that the propagation of an evoked wave is highly dynamic. Since spontaneous waves occurred frequently, they may interact with the evoked response and contribute to anisotropic propagation patterns. In trial 5, the evoked activity appeared to merge

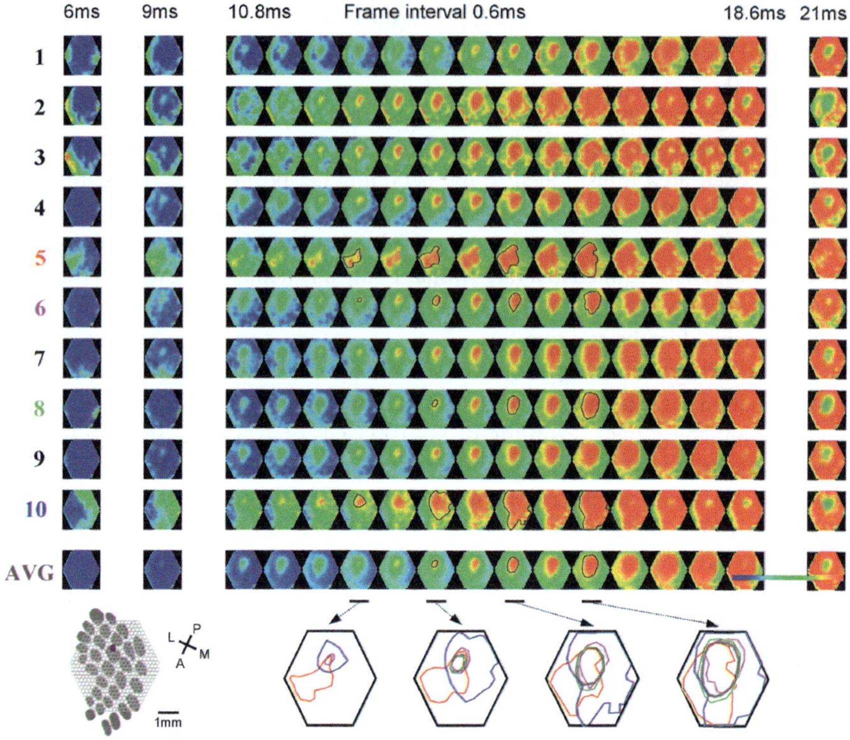

Fig. 9.6 Trial-to-trial variations in barrel cortex. Image rows 1–10: Ten consecutive trials with identical whisker deflection. The bottom images (AVG) are averaged from 105 trials from the same animal. *Bottom row left:* Barrel pattern and the imaging field. *Bottom row right:* The contour lines show the iso-levels of the relative amplitude of the population membrane potentials. The different colors indicate different trials. Each of the four maps then consists of contour lines superimposed from trials 5 (*red*), 6 (*light purple*), 8 (*green*) and 10 (*blue*), along with contour lines from the averaged data (*gray*) (Reprinted, with permission, from Lippert et al. 2007.)

with a spontaneous event and propagated faster in the direction toward the spontaneous event (Fig. 9.6).

9.1.5 Limitations of the Current Method

The main limitation of VSD imaging methods is that the signals are from the superficial layers of the cortex (Fig. 9.4). While layers II–III contain apical dendrites of deep layer neurons, the majority of stained membranes are from local neurons. Further improvements in this regard will await newer staining methods, or bioengineered VSDs. The spatial resolution is also limited to about 100 μm due to light scattering of the cortical tissue, thus cameras with large number of small pixels do not provide a better spatial resolution. A further limitation to the present method is the use of anesthetized and firmly fixed animals. Elimination of motion is necessary for high sensitivity recordings. This could perhaps be alleviated by novel and humane awake fixation methods, or by refinement of a revolutionary optical-fiber based tether method reported by Petersen's group (Ferezou et al. 2006).

In conclusion, the methods described in this chapter offer high signal-to-noise ratio recordings for a number of applications, such as the investigation of trial-to-trial variability of sensory evoked waves, interaction between evoked and spontaneous activity, event-related activity in higher cortical areas, the interactions between individual neurons and local population, and the complex dynamics of epileptic or other pathological activity.

9.1.6 Interpretation of the Voltage Sensitive Dye Signal

In vitro the dye signal is a linear function of the membrane potential (Cohen et al. 1974; Grinvald and Hildesheim 2004). However, as the absolute dye signal depends on the amount of staining one divides the raw signal by $F_{0\,xy}$. Further, the dye signal must be calibrated by intracellular recordings/patch clamping. This is not possible in vivo where large populations of neurons and glia cells are stained. Furthermore, in vivo, the photons from deeper layers of cortex are attenuated and those from the upper layers are scattered. In addition the in vivo signals, $F(t)_{xy,\mathrm{stim}}$ have a pulse artifact. The pulse artifact can in practice be removed (see above) especially when blue dyes such as RH 1838, 1691, and 1692 are used (Shoham et al. 1999). Still given this, the signal may also be subjected to equipment noise, fluctuations in the number of photons due to variations in the illumination source (Wu et al. 1999a).

Still if one assumes the noise sources are invariant and the pulsation artifact removed, it is not possible to measure depolarization and hyperpolarization in vivo. In the strict sense, depolarization is an increase in the membrane potential of a cell. This means that depolarization is defined and measured as an increase in mV from the resting potential, that is, the membrane potential of a neuron without any synaptic input. Consequently, the definition of depolarization and hyperpolarization will not work in vivo.

The $\Delta F/F_{xy}$ is a difference signal between the signal introduced by the background or no-stimulus condition and the stimulus + background condition that is made relative due to the division by the resting light intensity $F_{0x,y}$. If the F/F is >0 it means that the cortex from which the signal originates is relatively more depolarized during the stimulus condition, than during the condition when only the background is exposed to the animal. If the $\Delta F/F < 0$ the cortex during background condition is relatively more depolarized than it is during the stimulus condition. Furthermore, as the component from glia cells is moderate and has a much slower time course compared to the neuronal changes in membrane potentials (Konnerth and Orkand 1986; Lev-Ram and Grinvald 1986;

Konnerth et al. 1988; Bergles and Jahr 1997), fast changes of $\Delta F/F < 50$ ms lasting may be ascribed to the neurons. That the $\Delta F/F$ is a reliable measurement of the relative changes in population membrane potentials of supragranular neurons is also verified by simultaneous in vivo measurements of the $\Delta F/F$ and the membrane potentials of neurons in layers II and III (Petersen et al. 2003a, b; Ferezou et al. 2006). Moreover, as discussed by Destexhe et al. (2003), the cortical neurons are usually depolarized in vivo during spontaneous activity if one compares the membrane potential in mV with the resting membrane potential in a slice of the cortex. This is thoroughly dependent on the type and level of anesthesia, as slow waves give periodic and marked hyperpolarizations. Anesthesia with isoflurane in combination with N_2O gives 3–4 Hz regular delta–theta waves of the EEG, without frequent slow waves. Under such aesthetic conditions and in awake animals, we suggest that the term relative depolarization may be used for increases in ($\Delta F/F_{xy}$) (Eriksson et al. 2008).

Notably, although the action potentials are easily detected in vitro in single neurons or a few neurons in culture, it is impossible to detect action potentials in vivo with voltage sensitive dyes (Petersen et al. 2003a). One reason is probably that the contribution of action potentials to the $\Delta F/F_{xy}$ is so small compared with the changes in membrane potentials coming from dendrites and axon terminal branches. The axon terminals and especially the dendrites constitute a membrane surface that is several orders of magnitude larger than that constituting the axon hillock, the cell bodies, and unmyelinated axons in the cortex. Therefore, the signal from action potentials may be buried in the large signals from the neuropil even at high resolution (Ferezou et al. 2006). Another reason is that presumably many more neurons are depolarized, than are actually spiking (see below). A third reason is that neuronal firing under most physiological conditions may be sparse (Olshausen and Field 1996).

9.2 How Has Voltage Sensitive Dye Imaging Changed Our View on How the Brain Works?

9.2.1 Spontaneous Activity, Up-States, Down-States

It has been known for some time that single neurons tend to have their membrane potential fluctuations correlated with the fluctuations in the cortically evoked electroencephalogram, the EEG (Creutzfeldt et al. 1966). With the voltage sensitive dye techniques one samples from a cortical area varying between 50 and 150 μm in diameter depending on the equipment. This would correspond to sampling from cortex containing the cell bodies of 100–800 neurons, somewhat depending on species and cortical area. However, the small sampled tissue cone with a cortical surface diameter of 50 μm will contain not only all the dendrites and axon terminals of the neurons having their cell bodies within the tissue cone sampled, but also the dendrites and axons of neurons surrounding the tissue cone. If one assumes that the average dendritic span of a layer II or III neuron is 600 μm (Lübke et al. 2003), the small tissue cone will contain the dendrites of 40,000 neurons, plus a wealth of axon terminals from nearby and further away neurons. The measurement of the voltage dye signal is thus a weighted population average of the membrane potentials of local neurons (the minor part) and of neurons having their cell bodies in the surround (the major part). In vivo, this population membrane potential fluctuates spontaneously. In the awake state, the spontaneous membrane fluctuations of individual neurons have a root mean square variance of some 2–5 mV (Destexhe et al. 2003). The single neurons are depolarized to around 60 mV and the EEG typically shows activity of 10–40 Hz. These membrane fluctuations, however, are smaller than those seen under anesthesia when the EEG has a slower activity 2–4 Hz (Destexhe et al. 2003).

Amos Arieli and associates examined the spontaneous activity of the population membrane potentials in anesthetized cats with voltage sensitive dyes and found that the spontaneous population membrane signal was correlated over large distances, albeit not homogenously (Arieli et al. 1995). Compared to the single trial $\Delta F/F$ stimulus evoked response, the spontaneous activity had an amplitude which was 54% of the evoked response to a full screen drifting grating. Eighty percent of the variance in single trials of the evoked responses to the grating could be explained by the fluctuations of the spontaneous activity (Arieli et al. 1996). Averaging the evoked activity over trails removed this variability, but single trial spatial averaging over larger cortical space did not remove this variance. Arieli et al. (1995) then looked at the firing of neurons in the supragranular layers and found that the spiking activity was correlated with the $\Delta F/F$ cortical pattern as far away as 6 mm on the cortex. This was a surprise and paradigm shift, since neurons hitherto had been considered as independent units firing somewhat unreliable to sensory stimuli due to noise in the brain (Softky and Koch 1993; Shadlen and Newsome 1994). Again the correlations were not homogenously distributed over the cortex. Arieli et al. (1995, 1996) used a spike averaging technique. First a spiking neuron that responded to well to the drifting grating was selected. Then the $\Delta F/F$ was recorded in a single long trial without any stimulation during which also the action potentials of the neuron were recorded. Then a spike triggered averaging was done, off line. If the trial contained many spikes one got an average showing the $\Delta F/F$ pattern that was strongly correlated with the firing of that neuron. When Arieli et al. (1996) compared this pattern to the $\Delta F/F$ pattern evoked by the drifting grating in a single trial, it was possible to predict the evoked response with great certainty, provided that the initial $\Delta F/F$ pattern at time zero for the stimulus start was known. The evoked pattern could be predicted up to 50 ms after the start of the grating stimulus (Arieli et al. 1996). This behavior of the cerebral cortex is similar to what characterize complex dynamic systems, in which the value of a variable, in this case the relative population membrane potential, changes according to a function that only depends on the initial value.

Tsodyks et al. (1999) and Kenet et al. (2003) carried the examination of the spontaneous activity further by selecting a neuron with strong orientation tuning and finding the stimulus that maximally exited the neuron. Then they recorded the

$\Delta F/F$ pattern associated with the maximal firing of the neuron, which they called the preferred pattern of the neuron. Then they recorded the spontaneous activity with the voltage sensitive dye and noticed that whenever the spontaneous $\Delta F/F$ pattern was nearly identical to the preferred pattern, the neuron fired. Kenet et al. (2003) then made evoked $\Delta F/F$ patterns corresponding to different orientations of the drifting grating. Then they calculated the spatial correlation between the evoked orientation patterns, thresholded to show only the orientation modulation and thresholded the spontaneous $\Delta F/F$ patterns the same way. To their surprise they found that the spontaneous activity frequently showed a significant correlation with the evoked orientation maps (Kenet et al. 2003). This means that the population membrane activity in the visual cortex spontaneously goes through states strongly resembling those associated with different orientations of an external grating. They even found that the vertical and horizontal orientations were overrepresented among the states (Kenet et al. 2003).

These results fit well with results showing that the membrane potentials in distal dendrites often correlate over larger distances and also correlate with the cortically recorded EEG (Destexhe and Paré 1999; Paré et al. 1999; Svoboda et al. 1999). The single neurons often fire on top of an EPSP (Destexhe and Paré 1999; Paré et al. 1999; Svoboda et al. 1999). The spontaneous fluctuations of the single neuron membrane potentials and the spontaneous fluctuations of the population membrane potentials have frequencies corresponding to those in the EEG.

Up-states and down-states also occur spontaneously under anesthesia. The up-states are long lasting, typically several seconds long states, during which the neurons are more depolarized to around −50 mV and the local EEG shows fast activity as in the awake state even if the animal is anesthetized (Paré et al. 1999; Petersen et al. 2003b; Crochet and Petersen 2006; Ferezou et al. 2006). In down-states the neurons are hyperpolarized with membrane potentials of −70 to −90 mV (Paré et al. 1999; Petersen et al. 2003b; Crochet and Petersen 2006; Ferezou et al. 2006). The transition to an upstate is dependent on glutamate release and down-states are dependent on GABA release. The up-states or transition between states may be associated with spontaneously progressing relative depolarizations apparent as waves or wave fronts. These are typically slow moving over the cortex at 0.001–0.05 mm/ms. They have different origins and directions. Their shape varies from plane wave fronts to spiral waves (Prechtl et al. 1997; Petersen et al. 2003a, b; Crochet and Petersen 2006; Ferezou et al. 2006; Lippert et al. 2007; Xu et al. 2007). As up-states and down-states can even exist in slices, they are regarded as network properties of spontaneous nature (Cossart et al. 2003).

9.2.2 Propagating Waves in Cortex

Propagating waves ("traveling waves," Ermentrout and Kleinfeld 2001) in cortex have been extensively examined in brain slices in vitro (Chervin et al. 1988; Chagnac-Amitai and Connors 1989; Tanifuji et al. 1994; Albowitz and Kuhnt 1995;

Wadman and Gutnick 1993; Golomb and Amitai 1997; Demir et al. 1998; Tsau et al. 1998; Friedrich and Korsching 1997; Wu et al. 1999a, b, 2001; Bao and Wu 2003; Huang et al. 2004). Since their velocity and direction of propagation are highly dynamic (Fig. 9.7), the ability to record from single trials is essential for studying the propagation of spontaneous waves.

Evoked and spontaneous cortical activities are also organized as propagating waves or wave-fronts in vivo. After the advent of the blue dyes it is possible to do VSD imaging of sensory evoked waves in single trials in the mammalian brain (Derdikman et al. 2003; Petersen et al. 2003a, b; Ferezou et al. 2006; Roland et al. 2006; Xu et al. 2007). Sensory evoked waves and wave-fronts follow a consistent propagation pattern. In the primary sensory cortex the wave-fronts or waves are

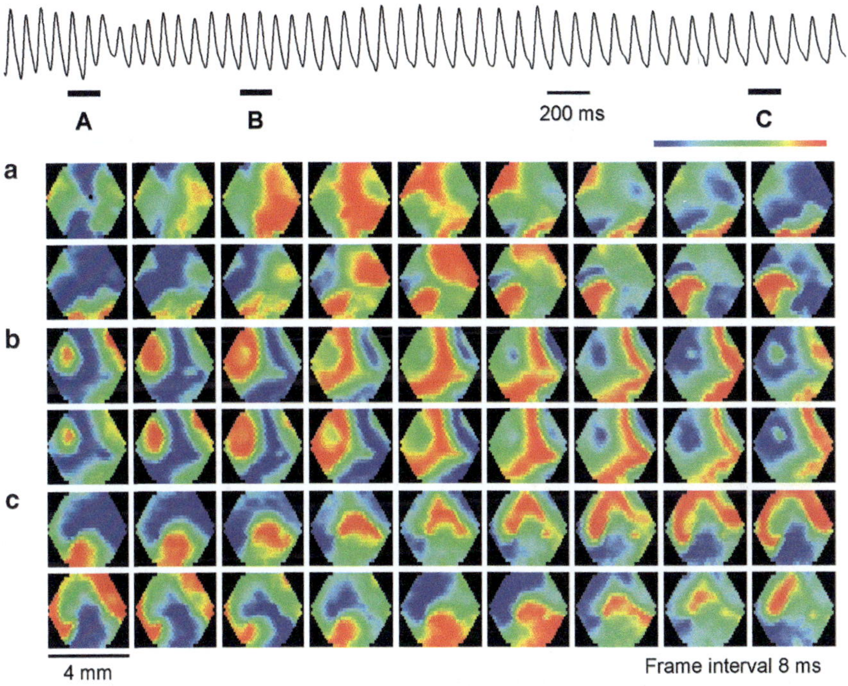

Fig. 9.7 Spatiotemporal wave dynamics in cortical slice. *Top trace:* Voltage sensitive dye signal of ~10 Hz spontaneous oscillations in a cortical slice. The dye signal is a population average of the neuronal depolarization, from a cortical area of 130 μm in diameter. **a**, **b**, **c** are three example sections of the oscillation with the spatiotemporal patterns shown in *bottom panels*. *Bottom panels:* Snapshots of the voltage sensitive dye imaging of the slice. The hexagon is the imaging field, about 4 mm from end to end. The signal amplitude is color coded (*peak-red*, *valley-blue*) according to the linear color scale on the *top right*. (**a**) Two cycles of oscillation with irregular patterns. The waves are initiated from different locations and propagating in the slice with anisotropic velocity and directions. (**b**) Two cycles of ring waves. The waves started from a locus at the left–upper Quadrant and propagated outward. (**c**). Two cycles of spiral patterns. A rotating wave around a center occurred at one turn per oscillation cycle. Patterns occurred in **a–c** are spontaneous with no manipulations (For details see Huang et al. 2004.)

initiated from the location of the thalamic afferents and then spread to cover large areas of the primary sensory cortex. In the barrel cortex, the whisker evoked wavefronts initiated from within the corresponding barrel and propagated to the entire barrel cortex (Derdikman et al. 2003; Petersen et al. 2003a, b, Fig. 9.6). In the visual cortex, evoked waves initiated from the retinal representation site in V1 are propagated to adjacent visual areas (Roland et al. 2006; Xu et al. 2007, Figs. 9.8 and 9.9). Feedback waves traveling from areas 21 and 19 towards area 18 and 17 were recently reported by voltage sensitive dye imaging in ferrets (Roland et al. 2006) (Fig. 9.9). Wave compression and reflection were observed in visual cortex during grating stimulation (Xu et al. 2007, Figs. 9.8 and 9.10). While marked differences in latency and propagating velocity were seen between these two papers

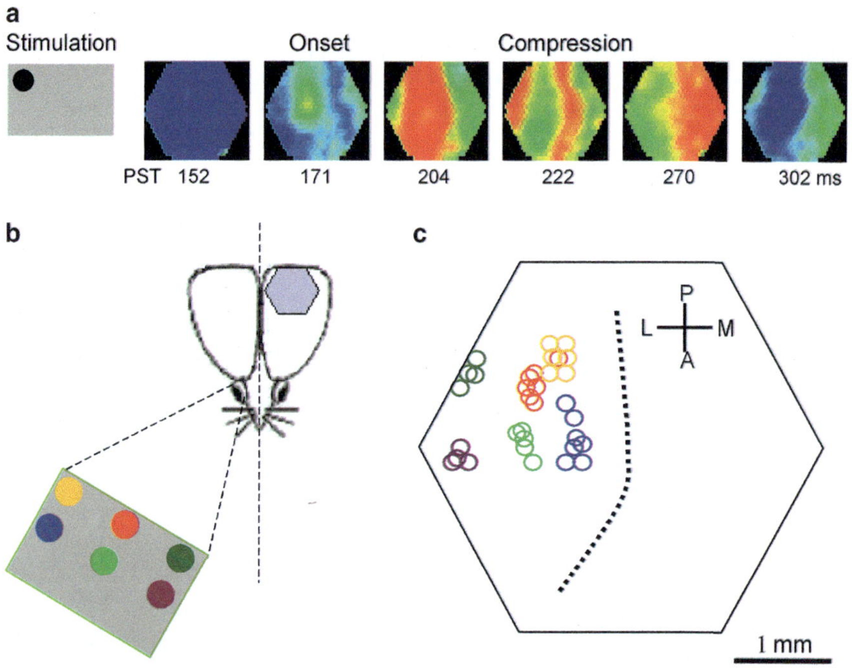

Fig. 9.8 Retinotopic map in V1. (**a**) An example trial showing a visual stimulus from a small field (6° in size) evoking a wave propagating in V1 and towards V2. Six snapshots are selected at the six stages: before onset, onset of primary wave, onset of compression, full compression, reflection and the end of the primary/reflection wave complex. The number under each image is the post-stimulus time (PST) in millisecond. (**b**) Visual stimulation. The stimuli were projected onto a screen of 10×7 in. placed at 20 cm in front of the animal's contralateral eye. The retinotopic map was made by presenting the drifting pattern (6° in size) at six locations on the screen (*colored dots*). (**c**) Retinotopic map in V1. Each *circle* represents the location of the response onset. Color of the *circles* represents the location of the visual stimulation in the field of view. The approximate coordinates of the stimulation sites are (in degree from the center): *yellow* (–19, 11); *red* (0, 11); *dark green* (19, 11); *blue* (–19, 0); *light green* (0, 0); and *purple* (19, 0). All the data are from the same animal. The *broken line* marks the approximate position of V1/V2 border (Modified, with permission, from Xu et al. 2007.)

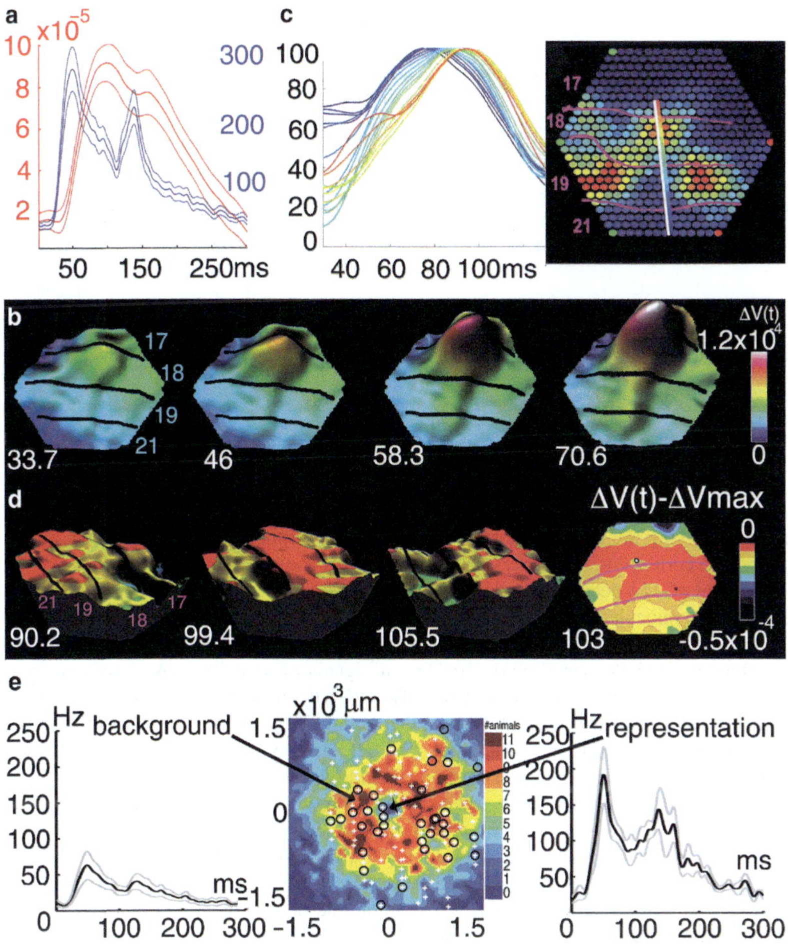

Fig. 9.9 Cortical dynamics evoked by a stationary object. The cortical dynamics evoked by a small stationary luminance defined square. The voltage sensitive dye measurements stem from cytoarchitectural areas 17, 18, 19, and 21 of the ferret. (**a**) The average firing of multiunits in areas 17 and 18 (*blue*) to a square of duration 83 ms. The average $\Delta F/F$ (*red*) from areas 17 and 18 evoked by the same square. Time after start of the stimulus in ms in all panels. *Red* ordinate $\Delta F/F$, *blue* ordinate firing frequency in Hz. (**b**) The relative depolarization of the upper layers of the visual cortex evoked by the square stimulus shown in a three-dimensional display. Note the local increase of $\Delta F/F$ at the retinotopic site of the square mapped at the areal border between areas 17 and 18, the feed-forward to the retinotopic site at the area 19/21 border and the lateral spreading depolarization. (**c**) The feedback as a traveling wave from area 21 to area 17, along the path indicated by color-coded stick to the *right*. (**d**) Three-dimensional display of top of the feedback wave. To the *right*, the segmentation of the object from its background dome by the feedback wave in interaction with the neurons at the retinotopic site of the square at the area 17/18 border. (**e**) The electrode penetration sites in areas 17 and 18 and their relation to the mean location of the retinotopic segmentation of the square from background. Below the firing rates of all neurons firing significantly at the representation and in the cortex representing the object background (Modified from Eriksson and Roland 2006 and Roland et al. 2006.)

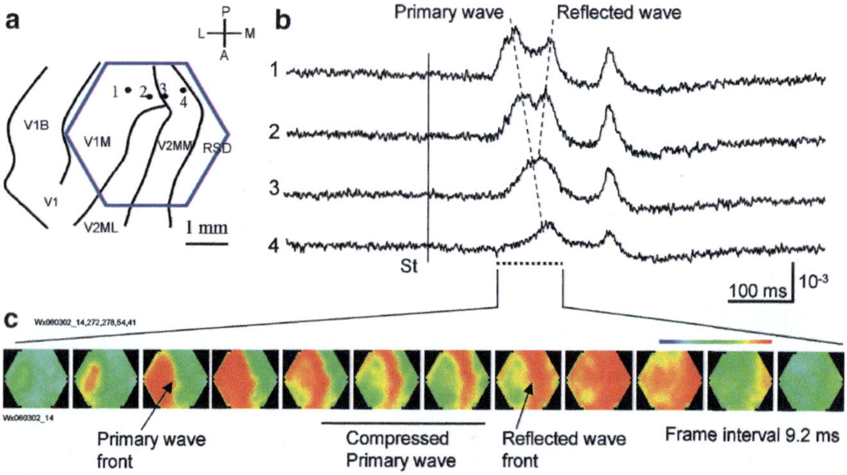

Fig. 9.10 Wave compression and reflection. (**a**) Schematic drawing of the imaging field (*blue* hexagon) overlying the map of the visual areas (left hemisphere, the map is shown as a mirror image of the cortex due to conversion in the macroscope). Four optical detectors, 1–4, were selected (total of 464 detectors) and their signal traces are shown on the *right* (**b**). V1B, V1M: binocular and monocular areas of V1; V2MM, V2ML: mediomedial and mediolateral areas of V2; RSD: retrosplenial dysgranular cortex. The map shown was made according to the stereotaxic map of Paxinos and Watson (2005). (**b**) Optical signals of visually evoked activity from four detectors (1–4). A grating (0.05 cycle/degree, 40w × 30 h degrees of viewing angle) was constantly presented to the contralateral eye. Drifting of the grating (3 cycles/s) was used as visual stimulus, with onset time marked by the vertical line (St). The peak of the activity occurred sequentially from detector 1 to 4, indicating a forward propagating wave (primary wave) from V1 to V2 (*left broken line*). A reflected wave can be seen starting from detector 3 and propagating backward to detector 1 (*right broken line*). The two waves can be clearly seen in the bottom images. (**c**) The pseudo-color images (0.6 ms snap shots) of the initial section of the evoked response. Twelve images (time marked by the *doted line* under the traces) are shown from a total of 8,192 frames in a 5 s recording trial. On each detector, the amplitude of the signal was converted to pseudo-color according to a linear color scale (peak *red*; baseline *blue*). The first image was taken when the evoked primary wave first appeared in the V1 monocular area, approximately 104 ms after the grating started to drift (Reprint, with permission, from Xu et al. 2007.)

(see Figs. 9.9 and 9.10), in general, both forward and backward waves were observed, thus suggesting that propagating waves are common phenomena during visual processing.

9.2.3 *Columns or No Columns, That Depends*

The somatosensory cortex and the visual cortex are the two domains of the cerebral cortex making the classical examples of cortex having a columnar organization. The idea of columnar organization is based mainly on microelectrode recordings, by which one can record action potentials of one or a few neurons very close to the electrode. The fundamental discovery was that neurons recorded in penetrations orthogonal to the surface responded in a striking similar way to external stimuli (Mountcastle 1957; Mountcastle and Powell 1959). Throughout the years this observation has been largely unchallenged. However, the columnar theory of cortical organization further states that the columns, each of some 250–600 μm in diameter spanning the space from the cortical surface to the underlying white matter, are the input–output devices of the cerebral cortex. In cortex having a topological organization, the large columns of 250–600 μm in diameter can overlap. These larger columns, called macro-columns or functional columns are surrounded by cortex in which synaptic inhibition predominates. The macro-columns thus are the functional elements of the cerebral cortex (Mountcastle 1971, 1997).

The experimental evidence for the existence of columns in the cerebral cortex is, not surprisingly, mainly from imaging experiments focused on the barrel cortex and the visual cortex showing columnar like preference for certain orientations and ocular input of the neurons in the granular and supragranular layers (deoxyglucose, intrinsic optical imaging; Ca^{2+} imaging see chapters 1,3,4). However by using voltage sensitive dyes one was able to test the columnar theory more rigorously by studying the changes in membrane potentials of the cortical neurons in real time. This enabled the scientists using this method to evaluate the stimulus attribute preferences, e.g., the orientation preference developed and whether it was restricted to columns or patches. In these experiments the authors typically record the VSD signal in response to a steady state stimulus, typically drifting gratings of different orientations. The VSD signals associated with each orientation then can either be compared to a neutral stimulus, a screen without any gratings but of the same average luminance, or to a recording of the orthogonal orientation. Alternatively, one can make a composite file of all other orientations except the one that needs to be examined and after division with the number of orientations subtract one this resulting "cocktail file" from the orientation one wants to examine. Subtraction of either the orthogonal or the cocktail file gives as a result a map of the orientation preferences just as it is shown in intrinsic imaging (see figs 1.2, 3.1, and 4.10 in Chapters 1, 3, and 4). In the original paper in which this was done, the $\Delta F/F$ signal appeared slowly increasing over a couple of seconds (Blasdel and Salama 1986), which indicates that the recordings could be a mixture of the VSD signal and the intrinsic signal which is also detected in the $\Delta F/F$ response. However, the authors also notice (p. 582) that when they subtracted the file obtained when there was only a gray background of average equal luminance, the cortex showed "diffuse activity-dependent" increase in the intrinsic signal (obtained with a tv-camera).

Grinvald and colleagues used equipment which allowed them to examine the development of the VSD signal in real time to drifting gratings of different sizes (Grinvald et al. 1994). To their surprise the found that, even to very small gratings, the $\Delta F/F$ signal spread out from a maximally relative depolarized retinotopical site of V1 to cover most of the exposed part of the primary visual cortex. The speed of this laterally spreading depolarization in the monkey V1 was 0.1–0.25 mm/ms (Grinvald et al. 1994). This means that the stimulus evoked $\Delta F/F$ activity is not confined to the retinotopical site of the primary visual cortex and hence not to any columns. This finding has since been replicated by all groups working on the visual cortex in cats, monkeys, ferrets, and rats with voltage sensitive dyes (Sharon and Grinvald 2002; Roland 2002; Slovin et al. 2002; Jancke et al. 2004; Chen et al. 2006; Roland et al. 2006; Benucci et al. 2007; Lippert et al. 2007; Sharon et al. 2007; Xu et al. 2007). Also it does not matter whether the stimulus is a stationary object (Chen et al. 2006; Roland et al. 2006) or a drifting grating limited in the visual field. One example is shown in Fig. 9.9. Chen et al. (2006) recently showed that this lateral spreading depolarization entailed that the variance of the VSD signal over an 8×8 mm region of monkey V1 was nearly constant. Further, this large region signaled the presence of their small 0.7° target stimulus. The spatial correlation in the VSD signal extends far within V1 and is independent of whether a target object is present or not (Chen et al. 2006). The correlation thus in the membrane potentials is high despite that the correlation between neighboring neurons is low. This is most likely because the VSD signal is a weighted average of all membrane potentials in layers I–III of the cortex and thus at each cortical point stems from ten thousands of neurons mixing their axon terminals and dendrites extensively. The lateral spreading depolarization in response to recently introduced objects in the visual field starts immediately after the neurons at retinotopic site in visual area 17 increase their membrane potentials and continues up to 70 ms after the start of the stimulus (Grinvald et al. 1994; Roland 2002; Slovin et al. 2002; Roland et al. 2006; Sharon et al. 2007) (Fig. 9.9). The velocity of the lateral spreading depolarization in area 17 varies between 0.1 and 0.2 mm/ms depending on the species (monkey, cat, ferret). There may be differences between primates and carnivores on one site and the rodents on the other (Xu et al. 2007).

It is not possible to find any evidence for columnar or patchy responses to localized stimuli in areas 17 and 18 in either single trial subtractions of the VSD signal obtained with a uniform surround or in averaged $\Delta F/F$ (Grinvald et al. 1994; Roland 2002; Slovin et al. 2002; Roland et al. 2006; Sharon et al. 2007). However, if one sets a threshold so as to include only the top 10–15% of the $\Delta F/F$ signal one may be able to see columnar like regions of orientation preference (Slovin et al. 2002; Kenet et al. 2003; Sharon et al. 2007). Neither is it possible by increasing the spatial resolution to find evidence for peri-columnar inhibition in the VSD signals from supragranular layers in either anesthetized animals or awake animals (Grinvald et al. 1994; Roland 2002; Slovin et al. 2002; Roland et al. 2006; Sharon et al. 2007). These columns only appear clearly in the $\Delta F/F$ signal when one subtracts files of stimuli with different orientation. Then what about the somatosensory barrel cortex, which is anatomically organized in columns?

Already London et al. (1989) noticed that the mechanical stimulation of one whisker evoked a $\Delta F/F$ signal that spread out from the corresponding cortical barrel. This observation was subsequently confirmed by other groups stimulating one whisker and recording a lateral spreading depolarization starting from 12 to 18 ms after the start of the brief stimulus. At some 40 ms after the stimulus start, the whole barrel cortex came depolarized (Kleinfeld and Delaney 1996; Petersen et al. 2003a, b; Derdikman et al. 2003; Civillico and Contreras 2005; Ferezou et al. 2006; Reidl et al. 2007). Also stimulating the inter-vibrissal fur results in a spreading depolarization outside the somatotopical site of the cortex corresponding to the fur stimulated. However, in this case the spreading depolarization avoids the barrel field (Takashima et al. 2005). Under conditions with a brief, 0.5 ms, strong mechanical stimuli to a whisker or during mechanical stimulation of the adjacent skin, there is first short latency strong (relative) depolarization that peaks after some 12 ms and then the depolarization rapidly decreases and give rise to a net (relative) hyperpolarization (Derdikman et al. 2003). The hyperpolarization engages the barrel of the stimulated whisker but then progresses asymmetrically into the surrounding cortex. After a while, at 120–130 ms the hyperpolarization turned into a longer lasting (relative) depolarization (Derdikman et al. 2003).

One may argue that the depolarizations spreading outside the retinotopic site of the object and the depolarizations spreading outside the single barrel probably are so modest that they cannot drive the neurons to fire. This is not so. Neurons outside the retinotopical site corresponding to the retinal part stimulated do fire statistically significantly more ($p<0.001$) when an object is presented in the visual field, as compared with stimulation of the visual field without objects (Fig. 9.9) (Roland et al. 2006; Eriksson and Roland 2006; Nauhaus et al. 2009). This firing and the lateral spreading depolarization seems to be driven by the neurons firing in the part of cortex representing the object, as the firing of the neurons in the supragranular layers outside the representation of the object is significantly delayed by a few ms (Nauhaus et al. submitted). As the infragranular neurons fire later in both the cortex representing the objects and the cortex representing the object background , it is likely that the neurons at the representations in layers II and III drive the neurons outside the representations through the many laterally horizontal axons in these layers (Nautaus et al. 2009).

The $\Delta F/F$ signal is thus not columnar or patchy in response to localized objects or gratings or physiological stimulation of whiskers. The $\Delta F/F$ signal shows a net relative depolarization spreading over larger parts of the sensory cortex when the control condition is a blank screen or the spontaneous prestimulus activity. This does not exclude that there might be – and probably is – also substantial inhibitory activity as part of the net signal (Monier et al. 2003). This inhibitory activity, however, is not strong enough to produce columns or patches of net excitation alternating with patches of net inhibition. Orientation preference columns or patches can be seen drifting to grating stimuli by thresholding the $\Delta F/F$ signal to show only its maximal amplitudes, by subtraction of signals obtained with stimuli of other orientations, or by integrating the signal over longer time (Ohki et al. 2005). However, none of these procedures reflect the real time computations in the cortex. And, after all, orientation and ocular input is only two aspects of a vast number of stimulus attributes in the visual world.

9.3 The Visual Cortex

9.3.1 Dynamics in V1/area 17 Evoked by Natural Visual Stimulation

For more than 50 years one has studied the activity of single neurons in the primary visual cortical area 17 in cats and monkeys with the purpose of understanding its contributions to visual perception (Jung et al. 1952; Hubel and

Wiesel 1962). As the neurons in area 17 respond readily to moving stimuli and because steady state stimuli give quasi-linear responses, drifting gratings or moving bars have been the most popular classes of stimuli. In particular, one has tried to establish relations between the particular properties of the surroundings such as orientation and contrast and the firing rates of area 17 neurons. As the neurons in area 17 receive most of their afferents from other neurons located within area 17 or from other cortical areas as well and no more than 1–2% coming from the lateral geniculate nucleus, one would expect that the contributions of area 17 to perception are mainly of cortical origin (Ahmed et al. 1994; Arieli et al. 1996; Binzegger et al. 2004; Bruno and Sakmann 2006). Moreover, the responses of the area 17 neurons change over time, such that one cannot in a spike train expect any fixed relations over time between the firing of neurons and features in the surround (Richmond and Optican 1990; Romo and Salinas 2003; Eriksson and Roland 2006). This implies that visual perception may be highly dynamic. Although only 1.5–2% of the synapses in area 17 are made by afferents from the lateral geniculate, and although the spontaneous activity in area 17 accounts for 85% of the variance (Arieli et al. 1995, 1996) and although the correlations between population neuron membrane potentials in vivo are relatively independent of the geniculate input (Chen et al. 2006), it is still possible for visual stimuli to drive the dynamics of the visual cortex. To detect the dynamics that is characteristic for different types of stimuli one must look at the stimulus evoked activity. For simplicity then one could divide this section according to the type of stimuli that have been used experimentally to drive the dynamics of the visual cortex. Implicit in this choice is the notion that if the membrane dynamics determine and reflect the computations done in V1/area17, then different stimuli should evoke different spatiotemporal dynamics of the neuronal populations in area 17 (Roland 2002).

9.3.1.1 Flash

A brief flash of light in front of a turtle evokes a local depolarization, a wavefront, and travelling waves in the turtle brain (Senseman 1996, 1999; Senseman and Robbins 2002). The same series of events can be elicited from the rat cortex (Lippert et al. (2007). In the rat area 17 cortex first a retinotopically located relative depolarization emerges, then a lateral spreading depolarization and subsequently a depolarization wave moves into adjacent areas (Lippert et al. 2007). Waves, wavefronts, and travelling waves are in the strict sense analogies from physics. Therefore, they have no direct neurophysiologic interpretation as the analogy presupposes that the wave/ wavefront propagates without the transport of matter (but not necessarily in a homogenous medium). Also waves may be signs of oscillations, which are phase-locked among neurons or propagating pulses in an excitable neuron network (Prechtl et al. 2000; Ermentrout and Kleinfeld 2001).

9.3.1.2 Stationary Objects, Bars, Gabor Patches

Small visual objects, for example a square exposed for a short duration, change the membrane potentials considerably in layers I–III of the visual cortex (Roland et al. 2006). Initially the firing of neurons in the supragranular layers increase simultaneously with an increased recruitment of more units in the supragranular layers at the retinotopic site of the square in area 17 (Eriksson et al. 2008). The relative depolarization of the neurons in the supragranular layers starts at the retinotopical site of the square some 30 ms after the start of the stimulus (Roland et al. 2006). This is slightly later than the start of the firing of the neurons which begin after some 22–30 ms (Fig. 9.9) (Roland et al. 2006). This delay of the depolarization in the supragranular layers may be explained by the action potentials from the lateral geniculate nucleus first affecting the layer IV spiny stellate neurons, and the spiny stellate neurons thereafter depolarizing the layer III and II neurons (Ahmed et al. 1994; Eriksson et al. 2008). This is what one may term a feed-forward input to the area 17 neurons. At the retinotopical site of the square in area 17, the neurons continue to further depolarize (Fig. 9.9). In the interval 20–60 ms, the time derivative of the $\Delta F/F$ signal follows and is correlated with the instantaneous firing rate (Eriksson et al. 2008). After 60 ms the $\Delta F/F$ signal continues to increase up to 100 ms, despite a simultaneous decrease in the instantaneous firing rate (Fig. 9.9).

Quite soon, after the start of the depolarization at the retinotopical site in area 17, the neurons outside the square representation increase their membrane potentials (Fig. 9.9). This is the lateral spreading depolarization (Roland et al. 2006). The lateral spreading depolarization is an automatic consequence of the feed-forward input to the supragranular layers in all (sensory) areas and is supposed to be mediated by axons running as lateral horizontal connections in layers II and III out from the retinotopical site (Ferezou et al. 2006; Grinvald et al. 1994; Petersen et al. 2003a; Roland et al. 2006; Slovin et al. 2002). As the spread of depolarization in supragranular layers in vitro is mediated by horizontally glutamatergic axons from the stimulation site (Tanifuji et al. 1994), one may assume that the neurons firing at the square retinotopic site may cause the lateral spreading depolarization through axons or axon collaterals extending into the surrounding cortex. The surrounding cortex represents the object background. In the surrounding cortex neurons start to fire significantly more action potentials at the time of the lateral depolarization spreads out from the square retinotopical site (Roland et al. 2006). This firing extends as far out as 10° from the object ($p<0.001$).

At 65–75 ms after the start of the stimulus, a broad peak of relative depolarization travels from temporal lobe visual areas via areas 21 and 19 towards areas 18 and 17 (Roland et al. 2006). It covers most, if not all, of the exposed surface of these visual areas (Roland et al. 2006). For this reason it was named a feedback wave (Fig. 9.9). The feedback wave contributes the maximal increase of the membrane potential at the retinotopical site(s) for the square (Fig. 9.9) and contributes to the firing of the supra- and infra-granular neurons here. Thereafter, the feedback wave in interaction with the depolarization and lateral spreading depolarization at the retinotopical site of the square immediately decrease the membrane potentials in the supragranular layers here (Roland et al. 2006). This creates an aperture, matching the expected position and size of the square retinotopical site (Fig. 9.9). This is presumably of importance for figure–ground segmentation (Roland et al. 2006).

The firing and the $\Delta F/F$ signal changes following the offset of the stimulus resembles the on-responses with the exception of the absence of a feedback from areas 19 and 21 (Eriksson et al. 2008). In an independent component analysis of the spike trains in area 17 and 18 of the ferret, the spike trains had three reproducible components having their maxima at 40, 55, and 105 ms after the start of the presentation of the stimulus. The time course of the third component was significantly correlated with the population membrane potential in the supragranular layers of areas 17 and 18. The first spike train component was interpreted as the feed-forward response, the second spike train component as driving the laterally spreading depolarization and the third spike train component as the firing caused by the lateral spreading and the feedback depolarization (Eriksson and Roland 2006).

Chen et al. (2006) measured the relative changes in membrane potentials of neurons in V1 of monkeys trained to detect a small 0.7° Gabor patch in the visual field. As mentioned above, the Gabor patch evoked a strong relative depolarization spreading laterally to cover major parts of the primary visual cortex. When the authors decreased the contrast of the patch the $\Delta F/F$ signal latency increased and the amplitude of the signal decreased. Trials in which the monkey successfully detected the patch and trials in which the monkey failed, were associated with $\Delta F/F$ signal of identical amplitude and localization (Chen et al. 2006). The authors also examined the optimal rule for distinguishing trials with a patch present from trials without patch. This is somewhat complicated because the noise is spatially correlated in much the same way in the two conditions (see above). Surprisingly a linear optimization rule that antagonistically weighted the information from the neurons at the object representation against the neurons in the cortex constituting the object background was by far superior (and outperformed the monkey). This may be comparable to the interaction the feedback wave makes with the neuronal population at the object representation by creating a segmentation of the object from the background based on the direction of change in the membrane potentials at the representation and the membrane potentials in the cortex representing the background (Fig. 9.9).

9.3.1.3 Moving Stimuli

Drifting Gratings

Drifting black and white gratings or colored gratings are popular stimuli because they evoke relatively strong firing in area 17. Further, they are steady state stimuli, the responses to which can often be modeled with linear models of good approximation. In a seminal study Grinvald et al. (1994) showed that even a very small

1° moving grating was able to increase the $\Delta F/F$ signal over area 17 in the monkey covering a cortex housing more than 10,000,000 neurons. It came as a big surprise that the borders of an, in the visual field, sharply defined grating did not have retinotopic sharp borders in the primary visual area. Grinvald et al. (1994) also demonstrated that surround stimuli reduced the $\Delta F/F$ increase to the central stimulus. For stimulation gratings with a hole, the hole was also mapped by the $\Delta F/F$ signal as a region of relative depolarization of less amplitude.

The dynamics of the $\Delta F/F$ signal to drifting gratings is different from the dynamics of the $\Delta F/F$ signal to stationary objects. The $\Delta F/F$ signal to drifting gratings peaks first between 220 and 350 ms in monkeys and cats (Grinvald et al. 1994; Shoham et al. 1999; Sharon and Grinvald 2002; Slovin et al. 2002; Sharon et al. 2007). In monkeys the $\Delta F/F$ signal in V2 is delayed by some 5–15 ms. In cats the LGN projects to both areas 17 and 18. In the rat the signal appears quite late, only after some 100 ms a relative depolarization appears in area 17 (Xu et al. 2007). Localized drifting gratings and stationary objects both evoke laterally spreading depolarizations (Fig. 9.8). In the experiment of Xu et al. (2007) a localized grating was presented to the contralateral eye of a rat. The drift of the grating (3 cycles/s) reliably evoked a propagating wave in the visual cortex (Figs. 9.8 and 9.10). The evoked wave started with a latency of ~100 ms (99.8 ± 18.2 ms, mean ± SD, $n=115$ trials) after the onset of the drifting, and the activity was seen in all optical detectors with a small time difference between each detector (Fig. 9.10b, traces 1–4). The pseudo-color images show that the evoked wave initiated in the monocular area of V1 (V1M) propagated as a circular spreading wave in both directions to the V1 binocular area (V1B) and to V2. This evoked a wave, referred to as the primary wave, was initiated by the visual stimulus and quickly expanded into the entire V1 area at a propagation velocity of 50–70 mm/s. It has not yet been examined whether drifting gratings are associated with feedback from higher order visual areas (but see further below).

The issue of when the orientation preference appears has been disputed for several decades. As the lateral geniculate neurons show no preferences for stimulus orientation, the preferences should either be made by the way the afferent axons connect to the spiny stellate and inhibitory neurons in area 17, or it could be a result of cortical computations. Sharon and Grinvald (2002) stimulated anesthetized cats with a full screen 100% contrast square gratings of different orientations. If the afferent connectivity from the lateral geniculate nucleus determined the preferences, the orientation preference should appear immediately when the spiny stellate neurons exited the pyramidal and inhibitory neurons in the supragranular layers. If the preferences developed as results of cortical computations in area 17 (and elsewhere) they should develop over time. The preference in the $\Delta F/F$ signal developed early, just 40 ms after the start of the drifting grating stimulus. The preference was measured as the difference in the VSD signal between two conditions, one with horizontal drifting gratings and one with vertical gratings. This difference reached maximum 100 ms after the start of the gratings. However, at this time-point, the orientation preferential modulation was only 15% of the total signal amplitude

(Sharon and Grinvald 2002; Slovin et al. 2002). After 100 ms, the orientation preferential part in the $\Delta F/F$ decreased to reach a steady level (Sharon and Grinvald 2002; Slovin et al. 2002). In a recent paper, the appearance of the relative depolarization in the supragranular layers to the drifting grating stimulus was described in 3D plots as a plateau in area 17 presumably representing the part of the visual field stimulated. Outside the plateau, the $\Delta F/F$ rapidly declined (Sharon et al. 2007). This is somewhat similar to the retinotopical mapping of luminance contrast defined objects in area 17 (Roland et al. 2006; Eriksson et al. 2008; Fig. 9.9b). However, unlike luminance contrast defined objects, the plateau contained some peaks related to the orientation of the drifting grating. These small peaks appeared maximally at 100 ms (Sharon et al. 2007). This is in accordance with another recent paper by Benucci et al. (2007) showing a somewhat slow latency of 82 ms of the bulk of the orientation response. The evolution of the orientation preference activity was localized by the authors described as a standing wave; whereas the lateral spreading depolarization did not contain any orientation information (Benucci et al. 2007). The plateau rim location and amplitude was independent of the orientations of the grating (Sharon et al. 2007). Benucci et al. (2007) used an alternating sinusoidal grating pattern. This would excite and inhibit simple cells at the stimulation frequency. Simple cells of opposite polarity located close together would cancel out, as their membrane changes would be of opposite polarity. In accordance with this prediction the authors did not find any contribution of simple cells, as their VSD signals oscillated at the double frequency, suggesting responses of complex cells (Benucci et al. 2007).

It is still not known which cortical areas in cats and monkeys may influence the computations of orientation preference in the primary visual areas of cats and monkeys. In both species, however, many cortical areas have neurons reacting to drifting gratings. So it is likely that these areas may also influence the computations of orientation in the V1/area 17. Sharon and Grinvald (2002) observed what they called a DA notch (deceleration–acceleration) 60 ms after the start of the drifting grating stimulus. The deceleration corresponds in time with the decrease in the initial firing at the representation site of the stimulus (Sharon and Grinvald 2002). This is also the case for luminance contrast defined stimuli (Roland et al. 2006; Eriksson et al. 2008) (Fig. 9.9a). This indicates that the later further increase in the $\Delta F/F$ signal to drifting gratings may come from elsewhere outside the representation.

Drifting Lines and Moving Objects

Prechtl et al. 1997 moved a white brick in front of an eye of a turtle connected to the live isolated turtle brain. This evoked a wavefront starting at the entrance of the fibers to the lateral visual area and propagating with the front orthogonal to the direction of the fibers.

There are two ways to perceive motion of objects in the field of view. Either one can follow the object with the eyes, or one can fixate the eyes on another, stationary object. In the last case the object moving in the surroundings will move over the retina. All retinal signals pass through the lateral geniculate nucleus (LGN) before reaching the primary visual cortex. The LGN, contains no directionally selective neurons and cannot compute object motion in carnivores and primates (Hochstein and Shapley 1976; Kaplan and Shapley 1982). So in general it is believed that the first computations of visual object motion takes place in the primary visual cortex, area 17. If an object moves over the retina the retinal ganglion cells will fire when the image of the edge of the object pass over that part of the retina. This means that the ganglion cells will fire sequentially in the direction of motion, as will the neurons carrying these signals from the LGN to area 17.

A drifting line or a slim grating was used to map the retinotopy of early visual areas 17 and 18 stained with voltage sensitive dyes in anesthetized cats and monkeys trained to fixate (Benucci et al. 2007; Yang et al. 2007). When the phase of the response was mapped it was a traveling wave moving over the cortex in retinotopical coordinates (Benucci et al. 2007; Yang et al. 2007). Or more precisely, a moving excitation in the supragranular layers in the 17 or 18 having its maximum where the cortex presumably represented the line/slim grating in retinal coordinates (see below).

Jancke et al. (2004) presented a small square moving in front of an anesthetized cat. This stimulus evoked a relative depolarization wavefront progressing over the cortex. In addition the most depolarized region persisted for some time behind the wavefront. This was interpreted as a motion streak, or tail due to the relatively fast movement of the square $32° s^{-1}$. Jancke et al. (2004) also observed that the latency to peak depolarization of moving object was shorter, compared with a stationary square flashed at the identical position. This was thus a clear sign of nonlinearity of the evoked response to a moving object.

Harvey et al. (2009) examined the motion of a small $2° \times 1°$ bar entering from the peripheral field of view of an anesthetized and paralyzed ferret. The bar moved either upwards or downwards along the vertical meridian at $24.5° s^{-1}$. This evoked two moving relative depolarizations, one following the cytoarchitectural border between areas 17 and 18, and another moving in parallel following the cytoarchitectural border between areas 19 and 21 (Fig. 9.11).

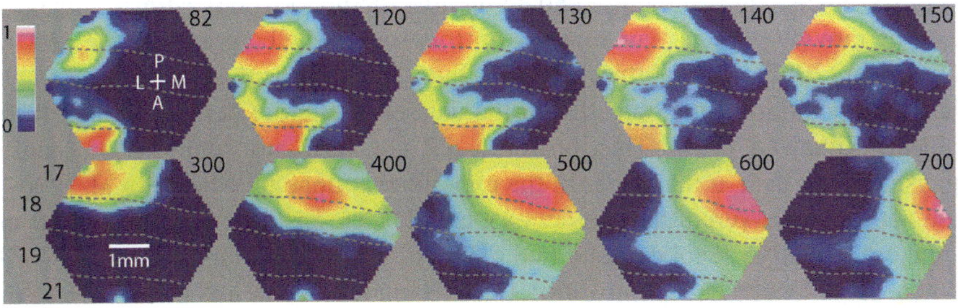

Fig. 9.11 Mapping of a moving bar. Motion prediction. The $\Delta F/F$ dynamics evoked by a small bar $2° \times 1°$ moving from the top of the visual field towards the bottom of the visual field along the vertical meridian at $24.5° s^{-1}$. Note the depolarization at 150 ms predicting the future path of the object (Modified from Harvey et al. 2009.)

In these experiments the retinae were always stationary, and therefore the line or object moved over the cortex in retinotopical coordinates. All authors assume that the maximum relative depolarization match the retinotopical position of the moving object on the cortex. This may be reasonable as Harvey et al. (2009) showed that the maximal firing and the maximal $\Delta F/F$ coincided in time and position in area 17 and 18. Another way is to map the receptive field of the neurons in the supragranular layers of cortex and compare the time when a neuron at a given position of the cortex (x, y) starts to fire significantly with the position of the object in the visual surroundings. However, irrespective of whether the maximum $\Delta F/F$ or the time the neurons start to fire in the cortex or fire maximally, the significant $\Delta F/F$ increases or firing will always be delayed in time with respect to the position of the object in visual space.

One way to get an estimate of this delay (and at the same time an estimate of the position of the object representation in the cortex) is to use the strategy of Kalatsky and Stryker (2003). This was done by Yang et al. (2007) and Harvey et al. (2009). Assume that the object moves over the display screen with a velocity v over the screen, starting at 0 ms at the bottom of the screen and ending at Tms at the top of the screen. After a certain time, t ,the square at position vt on the screen is mapped to a cortical position (x, y). The neurons at position (x, y) reach their maximal depolarization after a certain delay Δt after the object arrived to position vt on the screen. Thus the maximal depolarization will be reached after $tA = t + \Delta t$. When the object moves in the other direction, it will reach the same position (x, y) after $(T-t)$ and hence the maximal depolarization will be reached after $tB^{-} = (T-t) + \Delta t$ ms. Now if one reverses the time of one of the conditions, one will end up with two identical stimuli, i.e., $T-(T-t) = t$. However the maximal response will instead come Δt ms prior to the stimulus arrival at (x, y), i.e., $T-((T-t) + \Delta t) = t-\Delta t$. Now the time when the stimulus passes the position vt can be found by averaging tA and $T-tB^{-}$, i.e., $tA + (T-tB))/2 = (t + \Delta t + T-((T-t) + \Delta t))/2 = t$. Also the delay of the maximal response Δt is found by subtracting tA and $T-tB^{-}$, i.e., $(tA^{+}-(T-tB^{-}))/2 = (t + \Delta t-(T-((T-t) + \Delta t)))/2 = \Delta t$. Harvey et al. (2009) estimated the retinal and neuronal delay to 50 ± 4 ms in the ferret.

Obviously, the object will also move further during this delay of 50 ms. This factual delay in the visual system has been the basis of a wealth of psychophysical experiments and an intense debate on how the brain would estimate the position of moving objects. For reviews on this see Krekelberg and Lappe (2001), Nijhawan (2002), and Whitney (2002). One proposal was that the brain extrapolated the position of the object from the data on the initial perceived path of the object (Nijhawan 1994). This, however, did not, have any support from neurophysiological data at that time. In the experiments done by Harvey et al. (2009), the authors discovered that the area 19 and 21 neurons after a delay of 150 ms send out a relative depolarization in the direction of object motion. Almost simultaneously, a feedback depolarization went from area 19 towards area 17. Here the neurons repeated the behavior of area 19 and 21 neurons and send out a fast, directed depolarization in the average 8° ahead of the cortical position of the object. This directed depolarization in area 17 predicted the future trajectory of the object, such that the object representation after a delay of some 240 ms actually moved to exactly the cortical points exited previously by the predictive depolarization (Fig. 9.11). The neurons in the retinotopical part of the cortex excited by the predictive trajectory depolarization fired significantly in the supragranular layers 6° ahead of the object representation, but more moderately than the firing associated with the object representation (Harvey et al. 2009). The visual areas, are thus capable of computing a prediction of future events. As the animal was anesthetized, the predictive mechanism was an automatism – independent of any reasoning or previous experience.

9.3.1.4 Visual Illusions: The Line Illusion, Apparent Motion

Mechanisms of visual illusions may reside in the dynamics of the visual cortical neurons. The dynamics in turn are likely to be constrained by the connectivity of the neurons in the visual areas. Surprisingly, the dynamics driving the brain to compute visual illusions seems preserved even in anesthetized animals. Jancke et al. 2004 investigated the Hikosaka illusion or line motion illusion (Hikosaka et al. 1993a, b; Shimojo et al. 1997) by staining area 17 of an anesthetized cat with a voltage sensitive dye. Jancke et al. (2004) first presented a small 1.5° × 1.5° square for 50 ms, then after a delay of 60 ms a rectangle with one end overlapping the site of the visual field where the small square was presented. This gives, in humans, an illusion of the rectangle rapidly growing out from the square site. The illusion can be seen in the supplementary movies accompanying the published paper on the website of the journal Nature. In the cat, the cortex at the retinotopic site of the square rapidly depolarized in the direction set by the long sides of the rectangle. In other words, the depolarization grew out from the cortical retinotopical site of the square with a velocity corresponding to the motion of an object with 16–32° s^{-1} (Jancke et al. 2004). This growing depolarization, could, at least partly, be explained by the small square stimulus predepolarizing the cortical site corresponding to the proximal end of the rectangle. This depolarization would go as a laterally spreading depolarization in all directions out from the retinotopic sites of the square. When the depolarization induced by the rectangle is added to this predepolarization, the level of depolarization will grow in the direction of the long axis of the rectangle. There are, however, details such as a shorter latency for the rectangle response and steeper, linearly predicted depolarization, which pointed to additional nonlinear mechanisms (Jancke et al. 2004).

This illusion is a phenomenon quite different from apparent motion. The line motion is a perception of motion in a visual object, and not a perceived motion between objects (Hikosaka et al. 1993a, b; Shimojo et al. 1997). Both onset and offset of a visual stimulus produced the Hikosaka illusion (Hikosaka et al. 1993a, b; Shimojo et al. 1997). The line motion illusion is affected by attention (Hikosaka et al. 1993b), and can occur between different multi-sensory modalities, such as, somatosensory, auditory and visual (Shimojo et al. 1997). It is also present at iso-luminance. The line motion illusion, furthermore, has quite a different time course than an apparent motion as it can be elicited with more than 500 ms between cue and object (the line) (Hikosaka et al. 1993a). For these reasons the Hikosaka illusion or line motion illusion is separate from apparent motion and is likely to depend on another type of computational mechanism.

Humans and animals can perceive motion either because an object moves continuously or when static images are presented in rapid succession as in movies, television, and video-displays. This illusion is termed apparent motion and is computed in such a way that it is inseparable from perception of real motion. When the visual system is shown a briefly presented sequence of stationary images to induce apparent motion, the images can be presented over a short range (less than a degree of the visual field) or over a longer range (several degrees). Short range apparent motion depends on stimulus conditions different from those causing long range apparent motion (Braddick 1980; Anstis and Mather 1985; Chubb and Sperling 1988). Both of these illusions are very robust in their basic form: it is impossible by conscious effort or attention to inhibit or change the direction, speed and vividness with which the stationary stimuli appear to move (Palmer 1992). As the stimuli producing apparent motion are stationary on the retina, the brain must compute the apparent motion perception. Ahmed et al. (2008) showed three squares, briefly flashed, at three successive positions along the vertical meridian of the visual field to anesthetized ferrets (Fig. 9.12). This, humans perceive as apparent motion. The squares were initially mapped at the cytoarchitectural border between

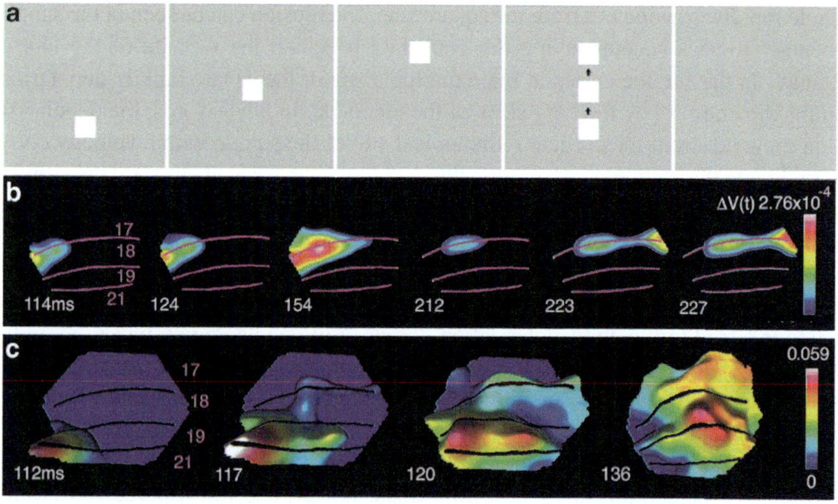

Fig. 9.12 Cortical dynamics subserving apparent motion. The cortical dynamics evoked by apparent motion. (**a**) Shows the experimental conditions. (**b**) Compared to the blank condition, the $\Delta F/F$ evoked to the rapid sequential presentation of the $2° \times 2°$ squares in **a** gives rise to a fast, 0.22 mm/ms moving depolarization between the retinotopic sites of the squares at 33 ms after the offset of the lower/central square. (**c**) The dynamics of the motion feedback depolarization from areas 19/21 to area 17. Time derivative of difference between AM and (control). Note motion feedback depolarization at 117–120 ms and the subsequent start of the moving depolarization wave-fronts at 17/18 from 126 ms, and the motion of the position of the square in area 19/21, 117–136 ms (Modified after Ahmed et al. 2008)

areas 17 and 18 as separate stationary representations. But time locked to the offset of the first stimulus, an increase in the membrane potential of the neurons traversed areas 19 and 21 in the direction of apparent motion. Simultaneously, a motion dependent feedback emanating from these areas increased the membrane potential of the neurons in the path between the first stimulus representation and the next along the area 17/18 border (Fig. 9.12c). Finally, a second increase of membrane potentials, traveling always from the representation of the first stimulus to the representation of the next or succeeding stimuli, relatively depolarized and fired the neurons situated between these stimulus representations along the 17/18 area border (Fig. 9.12b).

If apparent motion is always dependent on the visual cortex computing an activation wave-front from the retinotopic site of one stimulus to the next, a stimulation in which one square is followed by two oppositely positioned squares, should produce two wave-fronts, each travelling in opposite directions but towards the retinotopic sites of the two new squares. Indeed, that was the finding of Ahmed et al. (2008). In humans this is associated with the perception of one object moving while being split into two. These mechanisms are examples of motion computation that binds populations of neurons in the visual areas to interpret motion out of stationary stimuli.

9.3.2 *Communications Between Visual Areas*

So far, we have only discussed changes in $\Delta F/F$ within a cortical area and given examples from SI and V1 in rodents, carnivores, and primates. If one accepts that a $\Delta F/F$ increase is a relative increase in the population membrane potential in the supragranular layers, the question about how to interpret such population membrane potential increases moving as wave-fronts or waves across visual areas arises. It was mentioned that Senseman (1996, 1999), Prechtl et al. 1997, and Senseman and Robbins 2002, in response to a flash and a moving object, observed wavefronts and plane waves appearing after 150 ms and moving into the lateral visual area in the turtle brain. Thereafter, the wave moved first rostro-caudally in the lateral visual area with a velocity of 0.01–0.04 mm/ms and after some 200 ms moved over to the medial visual area and moving rostro-caudally there. In some cases even these waves were reflected, i.e., changed to the opposite direction at the borders within the lateral and medial visual cortex respectively (Senseman 1999; Senseman and Robbins 2002). Senseman and Robbins (2002) could by principal component analysis of their $\Delta F/F$ data, separate the component of communication between the lateral and the medial visual area. Prechtl et al. 1997, in addition to these slow waves, also found faster gradients more likely to emerge from underlying neuronal oscillators within the visual areas.

In general, such orderly moving membrane depolarizations over large populations of neurons could be due to (1) independent oscillations of the membrane potentials of neurons which are weakly phase coupled over shorter or longer distances. (2) Nearest neighbor coupling between neurons connected with short axons

(3) spontaneous up-states (4) thalamo-cortical input targeting cortical neurons in a specific order or changing the time-constant of the neurons in a particular order (5) excitatory input from axons of another part or area of cortex targeting neurons in a particular order (6) propagation through gap-junctions. This list is (probably) not exhaustive as little is yet known about cortical dynamics.

In mammals, each visual area has a full representation of the visual field. The neurons in all visual areas do not fire simultaneously to input from the lateral geniculate nucleus (Schroeder et al. 1998). Consequently, the visual areas must communicate to compute responses to sensory stimuli. They can do this in several ways, but it is most common to subdivide the communications into feed-forward connections, lateral connections and feed-back connections after Felleman and van Essen (1991). As the feed-forward axons mainly target the granular layer, and the dye signal originates from the supragranular layers, one can only indirectly measure the effect of feed-forward excitatory input to another area, as a relative depolarization of the supragranular layers. Feed-back axons, however, target supragranular and infragranular layers. The excitatory input to the supragranular layers can be measured directly as increases of the membrane potentials of the cell populations in the target region.

It is already mentioned that the presentation of a simple luminance contrast square evoked a lateral spreading depolarization that progressed as far as to the second representation of the square at the area 21 and 19 border (Fig. 9.9; Roland et al. 2006; Eriksson and Roland 2006).

Xu et al. (2007) observed a depolarization wave in the primary visual cortex of the rat initiated ~100 ms after the onset of a drifting grating (Fig. 9.10). This depolarization wave, referred to as primary wave, quickly expanded in the entire V1 area at a propagation velocity of 50–70 mm/s and towards the secondary visual area where it slowed down temporarily (getting compressed). When reaching the V1–V2 border, the leading edge of the wavefront slowed down abruptly (the velocity around the V1/V2 border was about 5 mm/s). Meanwhile the trailing edge of the wave was still in V1 and maintained a higher speed (50–70 mm/s). As a result, a thin band of compressed activity formed along the V1/V2 border (Fig. 9.10c). The compression and the resulting thin band sustained for a relatively long time compared with the time taken for the activity expanding within V1. A reflected wave was initiated after compression and propagated backward to V1 (Fig. 9.10c). The primary and reflected waves can be identified in the signal traces of individual detectors as double peaks (Fig. 9.10b). The spatiotemporal sequence of the wave compression/reflection can also be clearly seen when the data is presented in the movie (Supplemental movies of Xu et al. 2007).

This compression/reflection pattern was reliably observed in different recording trials. However, since there are large variations in the latency of the second wave, in averaged data the reflected wave is not obvious. In different animals, similar primary waves, compression bands, and reflected waves were observed with similar spatiotemporal patterns, but the locations of the compression and the shape of the compression band varied slightly, probably reflecting individual variability in the borders of the visual areas. Therefore, the ability of recording single trials without averaging between trials or cross animals is important in revealing dynamics of wave patterns.

The wave compression of evoked waves is likely to be related to the neuroanatomical structure of the cortex, especially the border between V1 and V2. The V1/V2 border can be estimated by visualizing corpus callosum fiber bundles because the latter are abundant near the V1/V2 border (Olavarria and Hiroi 2003). If electrical shocks were applied to the visual cortex contralateral to the imaging side, action potentials may reach the imaging side through the callosal fibers and be visualized with voltage sensitive dye imaging. Indeed, electrical stimulation in the contralateral V1M area with a moderate intensity evoked a localized activity on the imaging side. In the same animal, visually evoked waves compressed adjacent to the activity evoked by contralateral electrical shocks (Fig. 3 of Xu et al. 2007), indicating that the compression occurred near the V1/V2 border.

Xu et al. (2007) also found a second compression along the propagating path that was more medial to that of the first compression, probably at the border between V2MM and RSD areas. Between the two compressions there was a narrow gap, correlated well with the V2MM area on the map of the rodent, visual cortex (Fig. 4 of Xu et al. 2007). Again the onset time of the second compression was more variable from trial to trial so the ability to imaging single trial without averaging is important.

$GABA_A$ inhibition in the local cortical network is important to the wave compression. After testing several potential agents, Xu et al. found that $GABA_A$ inhibition in the local cortical network is important to the wave compression. Applying a $GABA_A$ receptor antagonist bicuculline epidurally with a low concentration of 3–5 μM (below the threshold of interictal-like spikes), can completely abolish the wave compression without significantly changing the speed of wave propagation within V1, suggesting that inhibition in the local circuit plays a major role in the wave compression. Compression bands reappeared after bicuculline was washed out, suggesting that elimination of the compression band does not require a permanent change in the cortical circuit. Under low dose of bicuculline perfusion, the propagating velocity across the V1/V2 border was same as that within V1 and V2, suggesting that changes in the excitatory connections at the border do not play a major role in the compression. Spontaneous waves do not have compressions at the V1/V2 border, which suggested that during evoked activity GABAergic inhibition is enhanced at the borders between visual areas.

The compression/reflection wave, would allow V1 and V2 to be depolarized together within 10 ms, following the compression. This distinct temporal pattern provides a mechanism for simultaneously depolarizing neurons in several visual areas. Neurons in two different visual areas may simultaneously increase their firing probability by the wave, during a particular period after receiving visual stimulus, thus facilitating the information exchange between these areas.

The feedback wave that was observed at 70–100 ms after stimulation with a luminance contrast defined object (Fig. 9.9) propagated with a constant velocity of 0.22 ± 0.06 mm/ms (Roland et al. 2006). During its course the feedback wave fired neurons in cortex representing the object background. The feedback wave entered the early visual areas from the visual part of the temporal lobe and propagated with constant velocity and without major changes in shape (Fig. 9.9) within as well as across visual areas 21, 19, 18, and 17 (Roland et al. 2006). This made it unlikely that the feedback wave could be explained as depolarizations of populations of cells with different time constants in each of the areas 21, 19, 18, and 17. The feedback wave is present in data averaged from the stimulus onset, making it unlikely that the feedback is just a systematic phase shift of oscillating membrane potentials over the areas 21, 19, 18, and 17 or due to spontaneous up-states.

Wehner and Roland (unpublished) presented two squares, a luminance contrast square, and a square defined by random motion of random dots. These two squares were shown in identical positions of the visual field. The dynamics of the $\Delta F/F$ evoked by the presentation of the square defined by motion of random dots, was different from the dynamics evoked by the luminance contrast square. The form from luminance square evoked immediate firing and $\Delta F/F$ increase in area 17 and 18, a lateral spread of the $\Delta F/F$ and a feed-forward signal to areas 19 and 21 and one feedback at 70–110 ms as in Fig. 9.9. The form from motion in contrast, evoked an initial $\Delta F/F$ increase in area 21, multiple feedbacks over the next 400 ms running over areas 19, 18, to area 17, where at the retinotopic site of the square, eventually a weak and delayed firing and ($V(t)$) increases appeared (Fig. 9.13).

These transient population membrane potential increases, appearing as feedbacks from area 21, represent possible communications between area 21 and lower order visual areas 19,18, and 17. As the depolarizations were all evoked, i.e., present in $\Delta F/F$ averaged 20–50 times and hence time-locked to the stimulus presentations, it is unlikely that they could reflect spontaneous up-states, noise, and phase shifts of oscillating membrane potentials. The feedbacks cannot be the effect of a gradual and general increase of $\Delta F/F$ spreading from area 21, i.e., an "iceberg effect." An iceberg increase of $\Delta F/F$ is associated with a time derivative of $\Delta F/F$ propagating in all directions from the steep part of the iceberg, while its center peak is stationary on the cortex. A wave-front is associated with a unidirectionally moving time derivative of $\Delta F/F$ followed by a zero time derivative. A wave is also uni-directionally moving with a positive time derivative of $\Delta F/F$ followed by a maximum and a negative time derivative. In this respect the feedbacks evoked by the form from random dot motion square in physical analogy must be wave-fronts.

The feedbacks observed are single, a-periodic, fast, and traveling relative depolarizations of the membrane potentials of a large population of neurons situated between areas 21 and 17. This excitatory activity must be propagated either by gap-junctions or by synaptic activity in the supragranular layers. The speed of the feedbacks 0.25 mm/ms, their propagation as net excitatory activity, their location in supragranular layers, and the large distance involved speak against gap-junctions (Hass et al. 2006). Also their speed, 0.25 mm/ms, was faster than possible for a multisynaptic propagation through large populations of neurons between area 21 and area 17 (Tanifuji et al. 1994; Pinto et al. 2005).The feedbacks, then most likely, represent a net increase of excitation propagating as a result of axonal communication from area 21 across the supragranular layers of areas 19 and 18 to area 17.

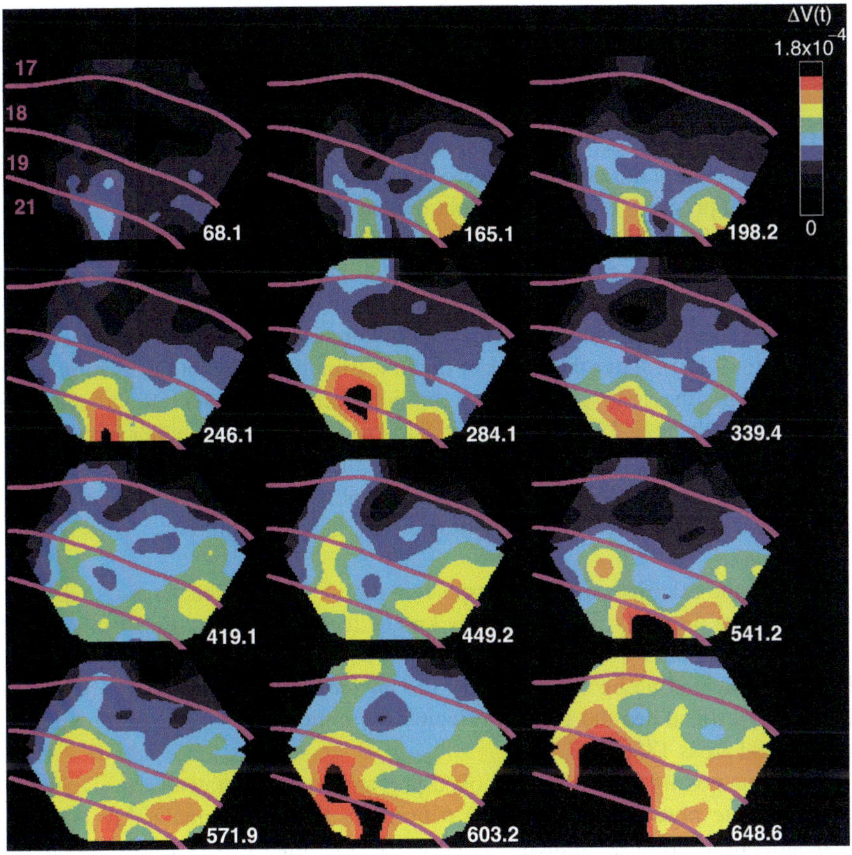

Fig. 9.13 Form from random dot motion evokes multiple feedbacks. Form from random dot motion and form from luminance contrast drive the visual areas with different dynamics. (**a**) Cartoon of the three conditions; Form from luminance, form from motion and stationary random dot screen. (**b**) Multiple feedback waves driven by the form from motion. Note that the representation at the area 17/18 border increases gradually in amplitude. Concomitant with this increase of the population membrane potential in area 17/18 the neurons at this site after being silent start to fire with increasing rates (Wehner and Roland unpublished)

The feedbacks propagated orthogonal to the direction of the feedback axons from area 21 and 19 towards areas 18 and 17 (Cantone et al. 2005, 2006).

There is also a feedback from area 21 towards area 17, in response to objects moving in the visual field. Invariably the predictive trajectory depolarization in the experiment by Harvey et al. (2009) was preceded by a feedback depolarization originating from the bar representation at the area 19/21 border and traversing with an average velocity of 0.12 mm/ms toward the bar representation at the 17/18 border (Fig. 9.11). This feedback depolarization began at 141 ms after stimulus onset and arrived at the 17/18 border at 165 ms, on average 25 ms prior to the maximal protrusion of the predictive trajectory depolarization (Fig. 9.11) (Harvey et al. 2009).

These events were quite similar to the physiological events associated with apparent motion conditions in the experiments by Ahmed et al. (2008). The travelling wave-fronts like increases in the population membrane potentials in the apparent motion conditions traversed across the cortex at the same velocity of 0.21–0.23 mm/ms (Fig. 9.12). Increasing the distance between the positions of the flashed squares did not change the velocity (Ahmed et al. 2008). This suggests that the maximal progress of relative depolarizations progressing continuously over the cortex supragranular layers is less 0.3 mm/ms in the ferret and cat. Invariably, each transverse increase in excitation of the population membrane potentials of the cell spanning the cortex between the retinotopic sites of the square stimuli were preceded by a feedback from the area 21 and 19 border towards area 17 (Ahmed et al. 2008) (Fig. 9.12c). After the offset of the first stimulus, a moving activation wave-front appears in areas 19/21 and a motion feedback is sent to areas 17/18. Immediately, the area 17/18 wave-front starts and progresses together with the area 19/21 wave-front in the direction of AM. In the time interval during which the 17/18 wave-front traverses from one retinotopic to the next retinotopic site, the neurons located between these retinotopic sites generate spike responses (Ahmed et al. 2008).

Also in the examples of real and apparent motion, the feedback relative depolarizations were probably propagating by known feedback connections from area 19 and 21 towards area 17 (Cantone et al. 2005, 2006). Specifically, the motion feedbacks propagated along the iso-elevations in the retinotopic map of the cortex. Figure 9.14 illustrates the general principle in motion feedback between areas 19 and 21 and areas 18 and 17.

It is thus possible to measure evoked propagations of (relative) depolarizations of membrane potentials of (mainly) neurons in the supragranular layers. These depolarizations are time limited lasting rarely more than 15–25 ms. In 3D plots of the amplitudes over cortex they present as waves or wave-fronts travelling across cytoarchitectural areas. The communications can be of different types, reflecting feed-forward or feedback communications.

The feed-forward communications from area 17 to area 19 and 21 was 0.1 mm/ms, starting as a lateral spreading depolarization within area 17 and 18 spreading to area 19 and 21 in the ferret (Fig. 9.9d) and a spread along supragranular layers in the rat from V1m to V2ml and V2mm (Fig. 9.10). In both cases the depolarization slowed down when passing a cytoarchitectural border indicating multisynaptic spread.

The feedbacks was fast 0.2–0.25 mm/ms and passing cytoarchitectural borders without slowing down indicating monosynaptic activation of the neurons along the feedback path. For stationary objects they start from the temporal cortex (Roland et al. 2006). For form from motion the feedbacks started from the activations at the suprasylvian and parietal lobe cortex (Fig. 9.13). In the conditions with moving objects or apparently moving objects the feedback(s) started at the moving retinotopical representation at the border between area 21 and area 19 (for objects moving along the vertical meridian) (Ahmed et al. 2008; Harvey et al. 2009). For the reasons mentioned above, these evoked feedbacks cannot be

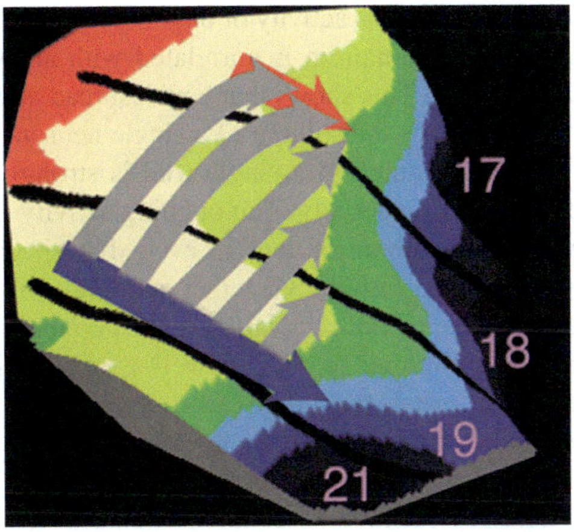

Fig. 9.14 Motion feedback. Time order of changes in the membrane potentials of the neurons in the supragranular layers of areas 21, 19, 18, and 17 at the time of the object motion or apparent motion. The object representation in cytoarchitectural areas 19/21 traverses to the next retinotopic site. Simultaneously, the neurons in 19/21 send motion feedbacks towards the corresponding retinotopic site at the 17/18 border. The motion feedbacks may also slightly depolarize the space between these borders as the iso-elevation domains of the retinotopical map in the ferret are in the direction of the motion feedbacks (Manger et al. 2002). Arriving at the retinotopic site and depolarizing the neurons at 17/18, these neurons start the 17/18 moving depolarization (apparent motion)/predictive trajectory depolarization (object motion). This cartoon explains the dynamics as seen in a phase plot of real apparent motion data (Modified from Ahmed et al. 2008.)

interpreted as oscillations in individual neurons which are phase coupled either as a population, or with nearest neighbor couplings. Neither were they spontaneous waves or depolarizations of different areas with different time constants. Therefore, it is possible that the feedback (relative) depolarizations are initiated by excitatory feedback axons making synapses on the neurons which depolarize.

9.4 Conclusions

In vivo voltage sensitive dye signals stem from thousands of cells in the upper layers of cortex. To remove the pulsation artifact, one must subtract the dye signals from different trials. The resulting $\Delta F/F$ then will show the changes in the membrane potentials over time for thousands of neurons in the upper layers. As the cortex is depolarized in the awake state and under appropriate anesthesia, a $\Delta F/F > 0$ indicates further net depolarization; whereas a $\Delta F/F < 0$ indicates either a reduction of excitation or a net inhibition. Such in vivo studies of changes in the membrane potentials of populations of neurons have fundamentally altered scientific concepts and views of cortical function.

During the so called spontaneous activity or ongoing activity unrelated to sensory stimuli, the spiking of a single neuron is correlated with an $\Delta F/F$ pattern that is spread widely over the cortex. This shows that a widespread and very large population of other neurons influences the spiking of a single neuron. Also the $\Delta F/F$ pattern evoked by a sensory stimulus in a single trial is strongly dependent on the immediately preceding ongoing activity. The ongoing activity goes through stages resembling the population membrane potentials evoked by specific sensory stimuli. These results are incompatible with the view that ongoing activity represents noise in the brain. Furthermore, the spatial correlation of the population membrane potentials is widespread during ongoing activity, as well as in stimulus evoked activity, and relatively independent of sensory input.

The dynamics of the stimulus evoked changes in population membrane potentials seriously challenge the view that cortical computations of sensory stimuli are confined to the cortical representations of the stimuli and to columns or patches processing particular stimulus properties. Rather, the cortical dynamics of the evoked changes in the membrane potentials indicate that the whole sensory scene is computed in each of the early sensory areas. In the early visual areas, the dynamics of the population membrane potentials driven by the sensory stimuli, and hence the computations depend on the physical nature of the stimuli. For example, luminance contrast defined stationary objects, drifting gratings, moving objects and stationary objects defined by random motion of dots, each evoke different population membrane dynamics. All focal visual stimuli, though, initially evoke a feed-forward depolarization and lateral spreading depolarization in the area(s) reacting to the stimuli. The further computations and the engagement of different visual areas depends on the initial computations and involve within area dynamics as well as dynamics of the population membrane potentials driven from other visual areas. This implies that the computations are not confined to particular cortical subregions, but highly dependent on intra-area as well as inter-area communications.

Several voltage sensitive dye studies indicate that it is possible, to some extent, to detect communication between areas as certain characteristics of the inter-area population membrane dynamics. This applies especially for intra-cortical spreading depolarizations crossing area borders and feedback from higher order visual areas.

References

Ahmed B, Anderson JC, Douglas RJ, Martin KA, Nelson JC (1994) Polyneuronal innervation of spiny stellate neurons in cat visual cortex. J Comp Neurol 341:39–49

Ahmed B, Hanazawa A, Undeman C, Eriksson D, Valentiniene S, Roland PE (2008) Cortical dynamics subserving visual apparent motion. Cerebral Cortex 18:2796–2810

Albowitz B, Kuhnt U (1995) Epileptiform activity in the guinea-pig neocortical slice spreads preferentially along supragranular layers–recordings with voltage-sensitive dyes. Eur J Neurosci 7:1273–1284

Anstis SM, Mather G (1985) Effects of luminance and contrast on direction of ambiguous apparent motion. Perception 14:167–179
Arieli A, Shoham D, Hildesheim R, Grinvald A (1995) Coherent spatiotemporal patterns of ongoing activity revealed by real-time optical imaging coupled with single-unit recording in the cat visual cortex. J Neurophysiol 73:2072–2093
Arieli A, Sterkin A, Grinvald A, Aertsen A (1996) Dynamics of ongoing activity: explanation of the large variability in evoked cortical responses. Science 273:1868–1871
Bao W, Wu JY (2003) Propagating wave and irregular dynamics: spatiotemporal patterns of cholinergic theta oscillations in neocortex in vitro. J Neurophysiol 90:333–341
Benucci A, Frazor RA, Carandini M (2007) Standing waves and traveling waves distinguish two circuits in visual cortex. Neuron 55:103–117
Bergles DE, Jahr CE (1997) Synaptic activation of glutamate transporters in hippocampal astrocytes. Neuron 19:1297–1308
Binzegger T, Douglas RJ, Martin KAC (2004) A quantitative map of the circuit of cat primary visual cortex. J Neurosci 24:8441–8453
Blasdel GG, Salama G (1986) Voltage-sensitive dyes reveal a modular organization in monkey striate cortex. Nature 321:579–585
Braddick O (1980) Low-level and high-level processes in apparent motion. Philos Trans R Soc Lond B Biol Sci 290:137–151
Bruno RM, Sakmann B (2006) Cortex is driven by weak but synchronously active thalamocortical synapses. Science 312:1622–1627
Cantone G, Xiao J, McFarlane N, Levitt JB (2005) Feedback connections to ferret striate cortex: direct evidence for visuotopic convergence of feedback inputs. J Comp Neurol 487: 312–331
Cantone G, Xiao J, Levitt JB (2006) Retinotopic organization of ferret suprasylvian cortex. Vis Neurosci 23:61–77
Chagnac-Amitai Y, Connors BW (1989) Horizontal spread of synchronized activity in neocortex and its control by GABA-mediated inhibition. J Neurophysiol 61:747–758
Chen Y, Geisler WS, Seidemann E (2006) Optimal decoding correlated neural population responses in the primate visual cortex. Nat Neurosci 9:1412–1420
Chervin RD, Pierce PA, Connors BW (1988) Periodicity and directionality in the propagation of epileptiform discharges across neocortex. J Neurophysiol 60:1695–1713
Chubb C, Sperling G (1988) Two motion perception mechanisms revealed through distance-driven reversal of apparent motion. Proc Natl Acad Sci USA 86:2985–2989
Cinelli AR, Neff SR, Kauer JS (1995) Salamander olfactory bulb neuronal activity observed by video rate, voltage-sensitive dye imaging. I. Characterization of the recording system. J Neurophysiol 73:2017–2032
Civillico EF, Contreras D (2005) Comparison of responses to electrical stimulation and whisker deflection using two different voltage-sensitive dyes in mouse barrel cortex in vivo. J Membr Biol 208:171–182
Cohen LB, Lesher S (1986) Optical monitoring of membrane potential: methods of multisite optical measurement. Soc Gen Physiol Ser 40:71–99
Cohen LB, Keynes RD, Hille B (1968) Light scattering and birefringence changes during nerve activity. Nature 218:438–441
Cohen LB, Salzberg BM, Davila HV, Ross WN, Landowne D (1974) Changes in axon fluorescence during activity: molecular probes of membrane potential. J Membr Biol 19:1–36
Cossart R, Aronow D, Yuste R (2003) Attractor dynamics of network UP states in the neocortex. Nature 423:283–288
Creutzfeldt O, Watanabe S, Lux HD (1966) Relations between EEG phenomena and potentials of single cortical cells: II. Spontaneous and convulsoid activity. Electroencephalogr Clin Neurophysiol 20:19–37
Crochet S, Petersen CCH (2006) Correlating whisker behavior with membrane potential in barrel cortex of awake mice. Nat Neurosci 9:608–610

Delaney KR, Gelperin A, Fee MS, Flores JA, Gervais R, Tank DW, Kleinfeld D (1994) Waves and stimulus-modulated dynamics in an oscillating olfactory network. Proc Natl Acad Sci USA 91:669–673

Demir R, Haberly LB, Jackson MB (1998) Voltage imaging of epileptiform activity in slices from rat piriform cortex: onset and propagation. J Neurophysiol 80:2727–2742

Derdikman D, Hildesheim R, Ahissar E, Arieli A, Grinvald A (2003) Imaging spatiotemporal dynamics of surround inhibition in the barrels somatosensory cortex. J Neurosci 23:3100–3105

Destexhe A, Paré D (1999) Impact of network activity on the integrative properties of neocortical pyramidal neurons in vivo. J Neurophysiol 81:1531–1547

Destexhe A, Rudolph M, Paré D (2003) The high-conductance state of neocortical neurons in vivo. Nat Rev Neurosci 4:739–751

Eriksson D, Roland P (2006) Feed-forward, feedback and lateral interactions in membrane potentials and spike trains from the visual cortex in vivo. J Physiol Paris 100:100–109

Ermentrout GB, Kleinfeld D (2001) Traveling electrical waves in cortex: insights from phase dynamics and speculation on a computational role. Neuron 29:33–44

Ferezou I, Bolea S, Petersen CC (2006) Visualizing the cortical representation of whisker touch: voltage-sensitive dye imaging in freely moving mice. Neuron 50:617–629

Felleman DJ, Van Essen DC (1991) Distributed hierarchical processing in the primate cerebral cortex. Cereb. Cortex 1:1–47

Friedrich RW, Korsching SI (1997) Combinatorial and chemotopic odorant coding in the zebrafish olfactory bulb visualized by optical imaging. Neuron 18:737–752

Golomb D, Amitai Y (1997) Propagating neuronal discharges in neocortical slices: computational and experimental study. J Neurophysiol 78:1199–1211

Grinvald A, Hildesheim R (2004) VSDI: A new era in functional imaging of cortical dynamics. Nat Rev Neurosci 5:5873–884

Grinvald A, Lieke EE, Frostig RD, Hildesheim R (1994) Cortical point-spread function and long-range lateral interactions revealed by real-time optical imaging of macaque monkey primary visual cortex. J Neurosci 14:2545–2568

Haas B, Schipke CG, Peters O, Söhl G, Willecke K, Kettenmann H (2006) Activity-dependent ATP-waves in the mouse neocortex are independent from astrocytic calcium waves. Cereb Cortex 16:237–246.

Harvey MA, Eriksson D, Valentiniene S, Roland PE (In preparation) Predictive object trajectories in the upper layers of the visual cortex

Harvey MA, Valentiniene S, Roland PE (2009) Cortical membrance potential dynamics and laminar firing during object motion. Frontiers in Systems Neuroscience 2. doi: 10.3389/neuro.06.007.2009.

Hikosaka O, Miyauchi S, Shimojo S (1993a) Focal visual attention produces illusory temporal order and motion sensation. Vision Res 33:1219

Hikosaka O, Miyauchi S, Shimojo S (1993b) Visual attention revealed by an illusion of motion. Neurosci Res 18:11–18

Hochstein S, Shapley RM (1976) Linear and nonlinear spatial subunits in Y cat retinal ganglion cells. J Physiol 262:265–284

Huang X, Troy WC, Yang Q, Ma H, Laing CR, Schiff SJ, Wu JY (2004) Spiral waves in disinhibited mammalian neocortex. J Neurosci 24:9897–9902

Hubel DH, Wiesel TN (1962) Receptive fields, binocular interaction and functional architecture in the cat's visual cortex. J Physiol 160:106–154

Jancke D, Chavane F, Naaman S, Grinvald A (2004) Imaging cortical correlates of illusion in early visual cortex. Nature 428:423–426

Jin W, Zhang RJ, Wu JY (2002) Voltage-sensitive dye imaging of population neuronal activity in cortical tissue. J Neurosci Methods 115:13–27

Jung R, Baumgarten RV, Baumgartner G (1952) Microleads of single nerve cells in visual cortex of cat: photo-activated neurons. Arch Psychiatr Nervenkr Z Gesamte Neurol Psychiatr 189:521–539

Kalatsky VA, Stryker MP (2003) New paradigm for optical imaging: temporally encoded maps of intrinsic signal. Neuron 38:529–545

Kaplan E, Shapley RM (1982) X and Y cells in the lateral geniculate nucleus of macaque monkeys. J Physiol 330:125–143

Kenet T, Bibitchkov D, Tsodyks M, Grinvald A, Arieli A (2003) Spontaneous emerging cortical representations of visual attributes. Nature 425:954–956

Kleinfeld D, Delaney KR (1996) Distributed representation of vibrissa movement in the upper layers of somatosensory cortex revealed with voltage-sensitive dyes. J Comp Neurol 375:89–108

Kleinfeld D, Delaney KR, Fee MS, Flores JA, Tank DW, Gelperin A (1994) Dynamics of propagating waves in the olfactory network of a terrestrial mollusk: an electrical and optical study. J Neurophysiol 72:1402–1419

Konnerth A, Orkand R (1986) Voltage-sensitive dyes measure potential changes in axons and glia of the frog optic nerve. Neurosci Lett 66:49–54

Konnerth A, Orkand PM, Orkand RK (1988) Optical recording of electrical activity from axons and glia of frog optic nerve: Potentiometric dye responses and morphometrics. Glia 1:225–232

Krekelberg B, Lappe M (2001) Neuronal latencies and the position of moving objects. Trends Neurosci 24:335–339

Lam YW, Cohen LB, Wachowiak M, Zochowski MR (2000) Odors elicit three different oscillations in the turtle olfactory bulb. J Neurosci 20:749–762

Lam YW, Cohen LB, Zochowski MR (2003) Odorant specificity of three oscillations and the DC signal in the turtle olfactory bulb. Eur J Neurosci 17:436–446

Lev-Ram V, Grinvald A (1986) Ca^{2+} and K^{+}-dependent communication between central nervous system myelinated axons and oligodendrocytes revealed by voltage-sensitive dyes. Proc Natl Acad Sci USA 83:6651–6655

Lippert MT, Takagaki K, Xu W, Huang X, Wu J-Y (2007) Methods for voltage-sensitive dye imaging of rat cortical activity with high signal-to-noise ratio. J Neurophysiol 98:502–512

London JA, Cohen LB, Wu JY (1989) Optical recordings of the cortical response to whisker stimulation before and after the addition of an epileptogenic agent. J Neurosci 9:2182–2190

Lübke J, Roth A, Feldmeyer D, Sakmann B (2003) Morphometric analysis of the columnar innervation domain of neurons connection layer 4 and layer 2/3 of juvenile rat barrel cortex. Cereb Cortex 13:1051–1063

Ma HT, Wu CH, Wu JY (2004) Initiation of spontaneous epileptiform events in the rat neocortex in vivo. J Neurophysiol 91:934–945

Manger PR, Kiper D, Masiello, Murillo L, Tettoni L, Hunyadi Z, Innocenti GM (2002) The representation of the visual field in three extrastriate areas of the ferret (Mustela putorius) and the relationship of retinotopy and field boundaries to callosal connectivity. Cereb Cortex 12:423–437

Momose-Sato Y, Sato K, Arai Y, Yazawa I, Mochida H, Kamino K (1999) Evaluation of voltage-sensitive dyes for long-term recording of neural activity in the hippocampus. J Membr Biol 172:145–157

Monier C, Chavane F, Baudot P, Graham LJ, Frégnac Y (2003) Orientation and Direction Selectivity of Synapic Inputs in Visual Cortical Neurons: A Diversity of Combinations Produces Spike Tuning. Neuron 37:663–680

Mountcastle VB (1957) Modality and topographic properties of single neurons of cat's somatic sensory cortex. J Neurophysiol 20:408–434

Mountcastle VB (1971) An organizing principle for cerebral function: the unit module and the distributed system. In: Edelman GM, Mountcastle VB (eds) The mindful brain. MIT, Cambridge, MA, pp 7–50

Mountcastle VB (1997) The columnar organization of the neocortex. Brain 120:701–722

Mountcastle VB, Powell TPS (1959) Central nervous mechanisms subserving position sense and kinesthesis. Bull Johns Hopkins Hosp 105:173–200

Nauhaus I, Busse L, Carandini M, Ringach DL (2009) Stimulus contrast modulates functional connectivity in visual cortex. Nat Neurosci 12:70–76

Nijhawan R (1994) Motion extrapolation in catching. Nature 370:256–257

Nijhawan R (2002) Neural delays, visual motion and the flash-lag effect. Trends Cogn Neurosci 6:387–393

Ohki K, Chung S, Ch'ng YH, Kara P, Reid RC (2005) Functional imaging with cellular resolution reveals precise micro-architecture in visual cortex. Nature 433:597–603

Olavarria JF, Hiroi R (2003) Retinal influences specify cortico-cortical maps by postnatal day six in rats and mice. J Comp Neurol 459:156–172

Olshausen BA, Field JD (2006) What is the other 85 percent of VI doing? In: van Hemmen, JL and Sejnovski TJ (eds) 23 problems in systems neuroscience. Oxford Univ. Press. Oxford. pp 182–211

Palmer SE (1992) Common region: A new principle of perceptual grouping. Cogn Psychol 24:436–447

Paré D et al (1999) Impact of spontaneous synaptic activity on the resting properties of cat neocortical pyramidal neurons in vivo. J Neurophysiol 81:1531–1547

Paxinos G, Watson C (2005) The Rat Brain in Stereotaxic Coordinates, Fifth Edition (New York: Elsevier Academic Press)

Pesaran B, Sornborger AT, Nishimura N, Kleinfeld D, Mitra P (2005) Analysis of dynamic optical imaging data. In: Yuste R, Konnerth A (eds) Imaging in neuroscience and development – a laboratory manual. Cold Spring Harbor Laboratory Press, Cold Spring Harbor, NY, pp 815–826

Petersen CC, Grinvald A, Sakmann B (2003a) Spatiotemporal dynamics of sensory responses in layer 2/3 of rat barrel cortex measured in vivo by voltage-sensitive dye imaging combined with whole-cell voltage recordings and neuron reconstructions. J Neurosci 23:1298–1309

Petersen CCH, Hahn TTG, Mehta M, Grinvald A, Sakmann B (2003b) Interaction of sensory responses with spontaneous depolarization in layer 2/3 barrel cortex. Proc Natl Acad Sci USA 100:13638–13643

Pinto DJ, Patrick SL, Huang WC, Connors BW (2005) Initiation, propagation, and termination of epileptiform activity in rodent neocortex in vitro involve distinct mechanisms. J Neurosci 25:8131–8140

Prechtl JC, Cohen LB, Pesaran B, Mitra PP, Kleinfeld D (1997) Visual stimuli induce waves of electrical activity in turtle cortex. Proc Natl Acad Sci USA 94:7621–7626

Prechtl JC, Bullock TH, Kleinfeld D (2000) Direct evidence for local oscillatory current sources and intracortical phase gradients in turtle visual cortex. Proc Natl Acad Sci USA 97:877–882

Reidl J, Starke J, Omer DB, Grinvald A, Spors H (2007) Independent component analysis of high-resolution imaging data identifies distinct functional domains. NeuroImage 34:94–108

Richmond BJ, Optican LM (1990) Temporal encoding of two-dimensional patterns by single units in primate primary visual cortex. II. Information transmission. J Neurophysiol 64:370–380

Roland PE (2002) Dynamic depolarization fields in the cerebral cortex. Trends Neurosci 25:183–190

Roland PE, Hanazawa A, Undeman C, Eriksson D, Tompa T, Nakamura H, Valentiniene S, Ahmed B (2006) Cortical feedback depolarization waves: A mechanism top-down influence on early visual areas. Proc Natl Acad Sci 103:12586–12591

Romo R, Salinas E (2003) Flutter discrimination: neural codes, perception, memory and decision making. Nat Rev Neurosci 4:203–218

Ross WN, Salzberg BM, Cohen LB, Grinvald A, Davila HV, Waggoner AS, Wang CH (1977) Changes in absorption, fluorescence, dichroism, and Birefringence in stained giant axons: optical measurement of membrane potential. J Membr Biol 33:141–183

Schroeder CE, Mehta AD, Givre SJ (1998) A spatiotemporal profile of visual system activation revealed by current source density analysis in the awake macaque. Cereb Cortex 8:575–592

Senseman DM (1996) Correspondence between visually evoked voltage-sensitive dye signals and synaptic activity recorded in cortical pyramidal cells with intracellular microelectrodes. Vis Neurosci 13:963–977

Senseman DM (1999) Spatiotemporal structure of depolarization spread in cortical pyramidal cell populations evoked by diffuse retinal light flashes. Vis Neurosci 16:65–79

Senseman DM, Robbins KA (1999) Modal behavior of cortical neural networks during visual processing. J Neurosci 19:RC3

Senseman DM, Robbins KA (2002) High-speed VSD imaging of visually evoked cortical waves: decomposition into intra-and intercortical wave motions. J Neurophysiol 87:1499–1514
Shadlen MN, Newsome WT (1994) Noise, neural codes and cortical organization. Curr Opin Neurobiol 4:569–579
Sharon D, Grinvald A (2002) Dynamics and constancy in cortical spatiotemporal patterns of orientation processing. Science 295:512–515
Sharon D, Jancke D, Chavane F, Na'aman S, Grinvald A (2007) Cortical response field dynamics in cat visual cortex. Cereb Cortex 17(12):2866–2877
Shimojo S, Miyauchi S, Hikosaka O (1997) Visual motion sensation yielded by non-visually driven attention. Vision Res 37:1575–1580
Shoham D, Glaser DE, Arieli A, Kenet T, Wijnbergen C, Toledo Y, Hildesheim R, Grinvald A (1999) Imaging cortical dynamics at high spatial and temporal resolution with novel blue voltage-sensitive dyes. Neuron 24:791–802
Slovin H, Arieli A, Hildesheim R, Grinvald A (2002) Long-term voltage-sensitive dye imaging reveals cortical dynamics in behaving monkeys J Neurophysiol 88:3421–3438
Softky WE, Koch C (1993) The highly irregular firing of cortical cells is inconsistent with temporal integration of random ESPSs. J Neurosci 13:334–350
Svoboda K et al (1999) Spread of dendritic excitation if layer 2/3 pyramidal neurons in rat barrel cortex in vivo. Nat Neurosci 2:65–73
Takashima I, Kajiwara R, Iijima T (2005) Voltage-sensitive dye imaging of intervibrissal fur-evoked activity in the rat somatosensory cortex. Neurosci Lett 281:258–263
Tanifuji M, Sugiyama T, Murase K (1994) Horizontal propagation of excitation in rat visual cortical slices revealed by optical imaging. Science 266:1057–1059
Tasaki I, Watanabe A, Sandlin R, Carnay L (1968) Changes in fluorescence, turbidity, and birefringence associated with nerve excitation. Proc Natl Acad Sci USA 61:883–888
Tsau Y, Guan L, Wu JY (1998) Initiation of spontaneous epileptiform activity in the neocortical slice. J Neurophysiol 80:978–982
Tsodyks M, Kenet T, Grinvald A, Arieli A (1999) Linking spontaneous activity of single cortical neurons and the underlying functional architecture. Science 286:1943–1946
Wadman WJ, Gutnick MJ (1993) Non-uniform propagation of epileptiform discharge in brain slices of rat neocortex. Neuroscience 52:255–262
Wehner S, Roland PE Different feed-forward feedback dynamics from shape from motion and shape from luminance. (unpublished)
Whitney D (2002) The influence of visual motion on perceived position. Trends Cogn Sci 6:211–216
Wu JY, Cohen LB (1993) Fast multisite optical measurement of membrane potential. In: Mason WT (ed) Biological techniques: fluorescent and luminescent probes for biological activity. Academic, New York, NY, pp 389–404
Wu J-Y, Cohen LB, Falk CX (1999a) Fast multisite optical measurement of membrane potential, with two examples. In: Mason WT (ed) Fluorescent and luminescent probes, 2nd edn. Academic, New York, NY
Wu JY, Guan L, Tsau Y (1999b) Propagating activation during oscillations and evoked responses in neocortical slices. J Neurosci 19:5005–5015
Wu JY, Guan L, Bai L, Yang Q (2001) Spatiotemporal properties of an evoked population activity in rat sensory cortical slices. J Neurophysiol 86:2461–2474
Xu W, Huang X, Takagaki K, Wu J-Y (2007) Compression and reflection of visually evoked cortical waves. Neuron 55:119–129
Yang Z, Heeger DJ, Seidemann E (2007) Rapid and precise retinotopic mapping of the visual cortex obtained by voltage-sensitive dye imaging in the behaving monkey. J Neurophysiol 98:1002–1014

Chapter 10
Fast Optical Neurophysiology

Jennifer L. Schei and David M. Rector

Abstract Simultaneous high spatial and temporal resolution brain imaging is the ultimate goal for many neurophysiological studies; however, no existing technique can provide all the information currently needed. Intrinsic optical properties of neural tissue have the potential to someday provide images of fast responses corresponding directly to neuronal activation with high spatial resolution. Most intrinsic optical procedures use slow metabolic changes in absorption or scattering. However, birefringence imaging techniques rely on changes in light polarization to record neural activity related to membrane potential. The largest responses result when the incident light is linearly polarized at 45 degrees with respect to the long axis of an isolated nerve. Further system optimizations can use multiple angles and incoherent light sources with narrow bandwidths in order to reduce background noise and increase signal contrast. We have explored methods for obtaining fast optical signals in vivo occurring tens to hundreds of milliseconds after stimulation based on several physiological components that may contribute to the optical changes. Our aim is to employ birefringence techniques in vivo in order to reduce background noise, increase signal contrast, and record optical responses concomitant with the electrical correlates of neural activation.

10.1 Where Science Meets Science Fiction

Imagine a scenario where patients lie on an exam table and all of their bodily functions and structures are imaged with no discomfort or awareness. Often, advancements in science and technology are inspired by science fiction. Yesterday's science fiction is meeting today's forefront of science and technology. This scenario may be possible using electromagnetic radiation of varying wavelengths, including light, which gives rise to information about all the tissue structures in the body.

D.M. Rector (✉)
Department VCAPP, Washington State University, 205 Wegner Hall, Pullman, WA, 99164-6520, USA
e-mail: drector@vetmed.wsu.edu

A.W. Roe (ed.), *Imaging the Brain with Optical Methods*,
DOI 10.1007/978-1-4419-0452-2_10, © Springer Science+Business Media, LLC 2010

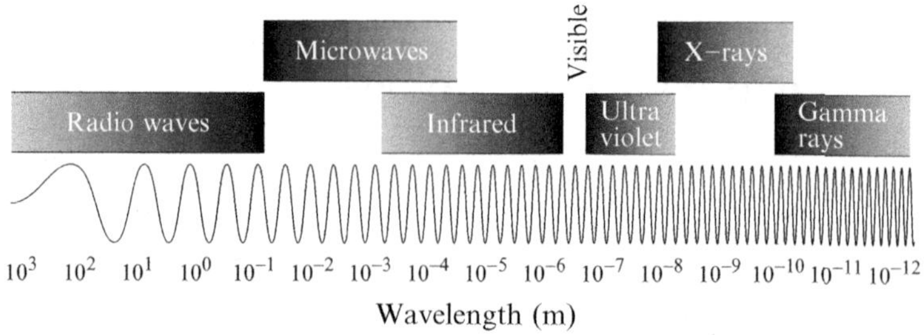

Fig. 10.1 Spectrum of electromagnetic radiation as a function of wavelength. Several current imaging technologies use a variety of electromagnetic radiation sources in order to create structural and functional images of the body

Electromagnetic radiation consists of perpendicular electric and magnetic fields propagating through space. The electromagnetic wave spectrum is mapped out as a function of wavelength in Fig. 10.1. Wavelengths in the infrared, visible, and ultraviolet range are commonly referred to as light. Ever since the description of light as electromagnetic wave propagation by Maxwell in the 1800s (Maxwell 1865), the scientific community has utilized disturbances of these waves as an avenue to probe, measure, and record a wide variety of physical processes and properties within the human body (Sochurek and Miller 1987). For example, x-rays penetrate soft tissues and scatter off denser objects such as bone. X-ray scattering is used to produce visible images of the skeletal structure on radiographic film. Magnetic resonance imaging (MRI) uses pulsed radio waves at a specific region in the body to compile structural and functional images of various soft tissues. Positron emission tomography (PET) imaging relies on an injected short-lived tracer isotope that emits a positron upon decay. Eventually the positron collides with an electron and the two annihilate, producing a pair of gamma-rays that propagate in opposite directions. A detector collects gamma-rays and reconstruction techniques trace the location of radioactivity and maps the concentrations within the tissue. Electroencephalography (EEG) and magnetoencephalography (MEG) rely on synchronous neural activity in order to produce ionic currents large enough to generate detectable electrical and magnetic signals. Ultrasound waves, sound at frequencies above human detection, directed into tissues generate echoes at regions where the tissue density changes. The time delay, amplitude, and frequency of the echoes are used to construct low-resolution images of tissue structures.

Most types of electromagnetic radiation used in noninvasive measurements use high energy levels and often cause tissue damage with prolonged exposure. Light, on the other hand, provides information using lower energy photons and causing less tissue damage. Optical imaging of the human brain provides a noninvasive means for mapping functional activity. Light in the near-infrared (NIR) range has been used in vivo since it penetrates the skin and skull (Eggert and Blazek 1987; Villringer and Chance 1997) and interacts with blood displaying signals correlated with hemodynamic changes as a consequence of neural activity and metabolism (Grinvald et al. 1986; Frostig et al. 1990; Mayhew et al. 2000; Sheth et al. 2003;

Devor et al. 2007). Since these signals are consequences of the hemodynamic responses, they occur several seconds after the stimulus.

Fast intrinsic optical imaging uses scattered light changes concomitant with neural activation. Optical responses occurring tens of milliseconds after the stimulus may be more closely related to electrical events of neural activation (Rector et al. 1997, 2005; Maclin et al. 2004). Birefringence imaging techniques utilize polarization optics to measure changes in light polarization consequential of neural activation. The information extracted from these signals provides temporal resolution on the order of milliseconds, concurrent with electrical activity. While birefringence imaging technologies have farther to go before being applicable in vivo, our purpose is to compile and review the current state of these fast optical signals as well as discuss the limitations and future approaches of these techniques. By utilizing fast optical imaging techniques, perhaps we will be able to lay a patient on a table, bathe them with electromagnetic radiation, and functionally image every structure in the body with high temporal and spatial resolution.

10.2 History

Before the electrical properties of neurons were fully described, scientists were using light to study the properties of nerves. Over a century ago, Gustav Mann observed that active cells appear to be larger and swollen (Mann 1894). This simple yet profound discovery that cells undergo physiological change upon activation led to an explosion of experiments studying neural activity. Subsequently, the squid giant axon was discovered to be the largest and longest nerve in the animal kingdom making it an ideal model for studying neural properties. Some experimenters stretched and twisted squid giant axons to study the forces exerted by nerves during activation (Kornakova et al. 1947). Other experimenters emptied the contents of the axon and analyzed the chemical components (Flaig 1947; Cragg 1951; Hodgkin and Keynes 1955). In 1952, Hodgkin and Huxley characterized the electrochemical properties of the action potential (Hodgkin and Huxley 1952). Amidst the studies of axon properties, the use of scattered light to image neural activation was demonstrated by Hill and Keynes (1949). These ground-breaking experiments showed a correlation between the activation and opacity changes in nerve bundles. While further studies aimed at connecting the physical components responsible for the changes in scattered light, some investigators discovered that using cross polarizers to image changes in light polarization produced signals that were an order of magnitude larger than scattering signals (Cohen et al. 1968; Tasaki et al. 1968). Additional studies using polarization techniques have characterized the optical signal to determine the underlying source of light polarization changes and improve imaging technologies (Carter et al. 2004; Yao et al. 2005; Foust et al. 2005, 2008; Foust and Rector 2007). Several investigators have exploited optical imaging techniques in vivo in order to investigate brain structure and functionality (Hoshi and Tamura 1993; Gratton et al. 1997; Rector et al. 1997; Mayhew et al. 2000;

Steinbrink et al. 2000; Maclin et al. 2004; Culver et al. 2005; Chen-Bee et al. 2007; Devor et al. 2007). Many of these imaging modalities rely on light scattered from the cortex that provides information related to the hemodynamic consequences of neural activity. Our aim, however, is to use birefringence techniques to record information related to the electrical consequences of neural activity more closely. Since cross polarizers reject nonspecific background light, signal contrast is improved, allowing for more robust responses.

10.3 Electromagnetic Phenomena

As light waves interact with different physical media, they have the potential to undergo alterations or perturbations. In order to use light to probe neural activity, we must first understand the various ways that light interacts with different materials. Some processes that contribute to changes in electromagnetic radiation are shown in Fig. 10.2.

10.3.1 Elastic Scattering

Scattering, in the simplistic case, is the redirection of light rays. During elastic scattering events, the energy of the photon is conserved while undergoing some change in direction. While conserving energy, light interacting with a medium can either

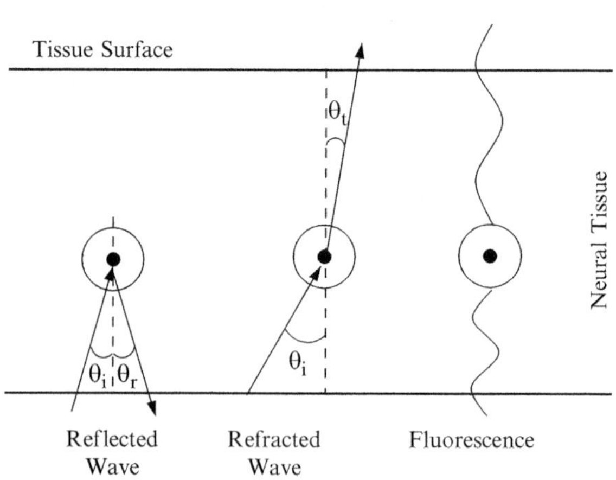

Fig. 10.2 Types of electromagnetic disturbances. (**a**) Rays reflect from the surface of a medium at an angle equal to the incident angle: $\theta_r = \theta_i$. (**b**) Rays passing between media will change propagation direction due to a change in velocity. The angle of refraction can be calculated using Snell's law (10.3). (**c**) Florescent materials absorb a particular wavelength of radiation and emit an wave of lower energy or longer wavelength

reflect or refract (Fig. 10.2a, b). The law of reflection states that the angle of incidence is equal to the angle of reflection.

$$\theta_i = \theta_r \tag{10.1}$$

In other words, we can direct light onto some medium at an angle θ_i and the light will reflect at the same angle $\theta_r = \theta_i$.

Refraction occurs when a light ray changes direction of propagation upon interaction with a nonconductive medium. We observe this when sticking a pencil into a glass of water. The pencil appears to bend between the air/water interface. This phenomenon occurs because the light wave changes propagation velocity as it enters a different medium. Velocity changes are proportional to the indices of refraction of the two media, where an index of refraction is defined as the speed of light propagating in a vacuum (c) divided by the speed of light propagating through the medium (v).

$$n = \frac{c}{v} \tag{10.2}$$

The angle at which the light wave will refract is given by Snell's law:

$$n_1 \sin \theta_1 = n_2 \sin \theta_2 \tag{10.3}$$

So, when we shine light through the air and onto a tissue, we expect that it will refract at some angle that can be calculated from Snell's law (10.3). Dispersive media display wavelength-dependent refraction. For example, shining white light onto a prism creates a rainbow since the prism causes different wavelengths refract at different angles. Activated tissue will change its index of refraction causing a small, but detectable, change in refracted light.

10.3.2 Inelastic Scattering

In the case of inelastic scattering, the energy of the photon is not conserved upon interaction with a material. These events cause the photon to lose some energy. Absorption occurs when a material takes in the energy of a photon. This phenomenon is wavelength-dependent, meaning that a material will best absorb waves near particular wavelengths. Molecules that absorb light at specific wavelengths, causing an apparent decrease in light intensity, are called chromaphores. Hemoglobin, albumin, water, amino acids, lipids, and many proteins are chromaphores available in tissue and serve as indicators for physiological processes.

Fluorescence occurs when a material absorbs light of a particular wavelength, exciting the electrons in the material to a higher energy state. The excitation is followed by a rapid jump of the electrons to a lower energy state. This electron transition results in the emission of a photon of lower energy (Fig. 10.2c).

The process of photon absorption and an appreciable delay of photon emission is called phosphorescence. Currently, many molecules that fluoresce exist as dyes but are toxic to cells. Endogenous fluorescent molecules such as nucleic acids, amino acids, NAD, and NADH, absorb ultraviolet light. These can be used as indicators of neural activity; however, extensive ultraviolet light exposure is damaging to the tissue.

10.3.3 Polarization

Electromagnetic waves consist of perpendicular electric and magnetic fields that oscillate perpendicular to the direction of propagation. They are typically represented by drawing the electric field component. Randomly polarized light is composed of several randomly oriented waves where, on average, there is no preferred direction. This is also commonly referred to as unpolarized light; although this term is somewhat misleading since each wave has a particular polarization angle. In the case of linear polarization, the net wave is oriented in a plane (Fig. 10.3). Elliptically polarized light contains two perpendicular wave components shifted out of phase as shown in Fig. 10.3.

10.3.4 Dichroism

A material that holds the same optical properties in all directions is termed optically isotropic. Conversely, a material that exhibits different optical properties depending on the direction of measurement is termed optically anisotropic. A dichroic material is

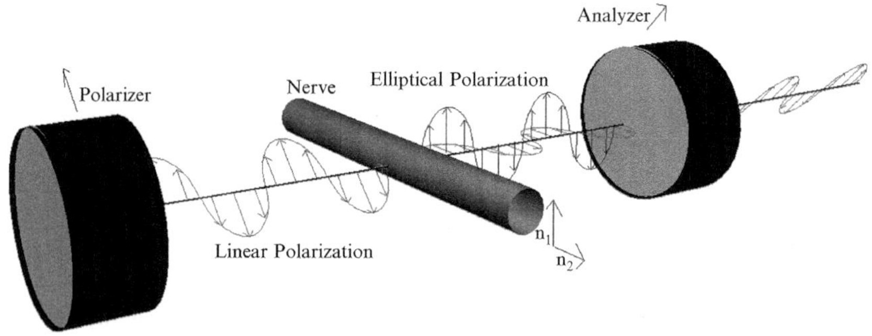

Fig. 10.3 Schematic of birefringence measurements. Incident light was linearly polarized 45 degrees with respect to the long axis of the nerve. The nerve bundle contains perpendicular, differing indices of refraction (n_1 and n_2). The components of the linearly polarized light propagate along the indices of refraction at different velocities, which make up elliptically polarized light. An analyzer, oriented 90 degrees out of phase with the first polarizer, transmits the components of the elliptically polarized light along the angle of polarization

an example of optical anisotropy where light is selectively absorbed in one direction and transmitted in the perpendicular direction. Although selective absorption properties may change during neural activity, they have not been differentiated from birefringence. One possible way to determine the dichroic properties of a material is to measure the response across the Mueller matrix parameters. The Mueller matrix is a mathematical representation of light polarization. Experimentally, we can measure these parameters by changing the incident polarization and measuring the transmitted and reflected polarizations. Ultimately, this approach may enable us to decompose the anisotropic optical properties during neural activation.

10.3.5 Birefringence

A birefringent material is an optically anisotropic material containing perpendicular indices of refraction that differ (n_1 and n_2 as shown in Fig. 10.3). Light incident upon a nonequivalent axis will refract into two rays with different velocities. Because of this phenomenon, birefringence literally means double refraction.

Incident linearly polarized light passed through a birefringent material will give rise to a wave component polarized along one index of refraction (n_1) and another wave component polarized along the other index of refraction (n_2). The phase shift between these two perpendicular waves, due to different propagation velocities, results in elliptically polarized light. A second polarizer, also called an analyzer, oriented 90 degrees with respect to the first polarizer transmits the components of the elliptically polarized light parallel to the analyzer. These two wave components combine, resulting in a superposition of linearly polarized light. The use of polarizers to reject waves not altered by neural activity allows for more robust signals.

The largest changes in polarization emerge when light is linearly polarized 45 degrees with respect to the indices of refraction. In the case of the lobster nerve, the largest signals emerge when light is polarized 45 degrees with respect to the long axis of the nerve (Cohen et al. 1968; Yao et al. 2005). Birefringence techniques provide more robust signals than scattered light measurements because the cross polarizers reject photons that are unaltered by neural activity, thus improving the size of the signal relative to background.

10.4 Temporal Signal Components

Stimulation of isolated lobster nerves with incident linearly polarized light elicits a decrease in light polarization changes and a decrease in 90 degree scattered light simultaneous to the action potential (Fig. 10.4a). The fractional changes in scattered light are small, on the order of 10^{-5}. The fractional changes in polarization, however, are an order of magnitude larger (10^{-4}) than those observed during scattering events (Cohen et al. 1968; Tasaki et al. 1968; Carter et al. 2004).

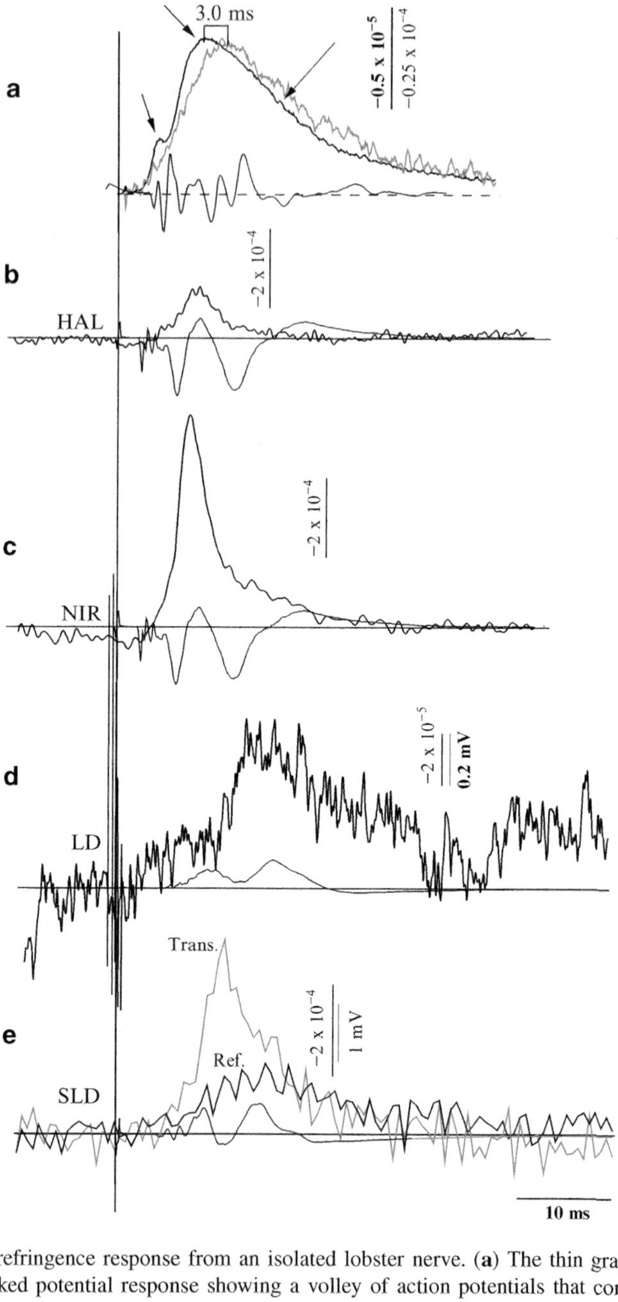

Fig. 10.4 Birefringence response from an isolated lobster nerve. (**a**) The thin gray trace is the electrical evoked potential response showing a volley of action potentials that corresponded to axons of different diameters. The thick black line is the changes in light polarization and the thick gray line is the 90 degree scattering response. Polarization changes occurred approximately 3 ms before scattering changes. Additionally, three temporal components (*arrows*) revealed signals from large, medium, and small diameter axons. (**b**) Polarization changes from a halogen (HAL) light source. (**c**) Polarization changes from a near-infrared (NIR) light source (880 nm).

Since the 90 degree scattering measurement experienced a delayed onset of 2–3 ms after the polarization signal, the mechanisms underlying the changes might be different. After averaging the optical signals across stimuli, the polarization signal exhibited a significantly higher signal-to-noise ratio (SNR) than the 90 degree scattering signal. The larger amplitude, earlier onset, and higher SNR characteristics of the polarization signal may be due to the polarization optics rejecting light unaltered by neural activity, and therefore, increasing the signal contrast. Scattered light and birefringence imaging techniques are useful because the signals are tightly coupled with the membrane potential, providing measurements with a temporal resolution on the order of milliseconds and spatial resolution of micrometers (Foust and Rector 2007). These techniques have the potential to provide information about neural activation in very specific tissue regions.

The electrical potential measured in Fig. 10.4 shows a volley of action potentials corresponding to axons with different diameters. The evoked polarization signal not only occured 2–3 ms before the scattering signal, but also exhibited temporal structure related to axon diameter (Carter et al. 2004). The initial peak in evoked response was associated with faster, larger diameter axons whereas the last peak was associated with slower, smaller diameter axons. The temporal structure of the polarization signals allows us to discriminate between different axon diameters. The scattering signal lacked the temporal structure observed in the polarization signal perhaps because the physiological mechanisms that produce the two optical signals differ.

10.5 Physiological Mechanisms for Polarization and Scattering Changes

Several mechanisms may be responsible for the decrease in polarized and scattered light associated with neural activation. Microtubules are birefringent (Oldenbourg et al. 1998), an influx of water causes axonal swelling and membrane unfolding which may contribute to the change in the evoked optical signal (Yao et al. 2003), and some investigators argue that the source of decreased polarized light is the reorientation of membrane dipoles (Cohen et al. 1968; Landowne 1985). Studies using pharmacological agents have further dissected the source of the elusive polarization signal. Toxins such as Tetrodotoxin (TTX), Tetraethylammonium chloride (TEA), and Dimethyl sulfoxide (DMSO) that disrupt in the membrane potential and concomitantly disrupt

Fig. 10.4 (continued) Wavelengths in the NIR range elicited the largest responses because the polarizers achieved better extinction coefficients at narrower wavelengths. (**d**) Polarization changes from a laser diode (LD) light source. Small shifts in laser speckle, due to the coherence of the light, contributed to the noisy signal. (**e**) Reflected (*thick black line*) and transmitted (*thick gray line*) polarization changes using a superluminescent diode (SLD) emitting NIR light was less noisy because there was much less speckle as with LDs. The amplitude of the transmitted SLD signal was three times larger than the amplitude of the reflected signal perhaps because fewer photons both change polarization and scatter a full 180 degrees in the reflection mode measurement signals from voltage sensitive dyes and the polarization signal, suggest that polarization changes are tightly coupled to the membrane potential. Large angle scattering has typically been thought to be resultant of cellular swelling. Pharmacological agents that altered the cellular swelling and refractive index had profound effects on the large angle scatting signal. These results affirm that optical scattering signals are closely dependent on changes in the interstitial spaces (Foust and Rector 2007).

10.6 Technological Issues

10.6.1 Light Sources

In order to optimize the fractional changes in the evoked optical signals, we studied the birefringence responses using different illumination sources. Preliminary experiments utilized halogen lights, emitting white light, to record changes in polarization (Fig. 10.4b). While these sources exhibit high intensity and stability, they also include a large range of wavelengths which are difficult to exclude using polarizers. This excess amount of light contributes to the background brightness during imaging. Light emitting diodes (LEDs) are also high intensity but specific to a narrow wavelength. Polarizers achieved better extinction coefficients (the relation between the incident and transmitted light) when using narrow bandwidth light sources. In other words, in the absence of tissue, cross polarizers exclude more light when the source is a specific wavelength. These better extinction coefficients reduced the background noise and contributed to larger fractional changes of the evoked responses as shown in Fig. 10.4a (Foust et al. 2005). Additionally, LED light sources exhibited lower low-frequency noise than halogen lamps (Salzberg et al. 2005). Wavelengths in the NIR spectrum elicited larger signals and better SNRs than wavelengths in the visible spectrum (Fig. 10.4c). Reducing noise sources and increasing SNRs is crucial for extracting and optimizing these small intrinsic optical signals. By using light sources that emit specific wavelengths with narrow bandwidths, we can achieve better extinction coefficients with the polarization optics and thereby optimize the signal.

10.6.2 Noise Sources In Vitro

While LEDs are superior to white light sources, laser diodes (LD) also have narrow band emission with high intensity. Many current optical imaging technologies rely on lasers to illuminate tissue because of their high intensity and small size. We characterized the instrumental noise components between LEDs and LDs in order to gain a better understanding of noise sources within the optical system. While a LD light source showed a definite change in polarized light correlated to the action potential, the signal was significantly noisier (Fig. 10.4d). Laser speckle, a noise source deriving from the coherence of the light, significantly increased the root mean square (RMS) noise, which detrimentally contributed to a lower SNR. Increasing the LD intensity resulted in speckle noise dominating the signal and therefore decreasing SNR. The LED source exhibited low RMS noise since it is a noncoherent light source and exhibited an order of magnitude larger SNRs with fewer averages. In fact, signals prevail on single pass trials. This may be due to a combination of brighter illumination and lower noise (Foust et al. 2008).

10.6.3 Noise Sources In Vivo

In order to further characterize the noise sources, we recorded and compared the scattered light signals from a stimulated rat cortex using the LD and LED (Fig. 10.5). The triphasic signals correlated to the hemodynamic response resulting from neural activation (Rector et al. 2001; Chen-Bee et al. 2007). The signals from both light sources experienced severe cardiac artifacts. When filtered out, the evoked optical signal from the LED source showed a SNR that is two to three times larger than the SNR from the LD.

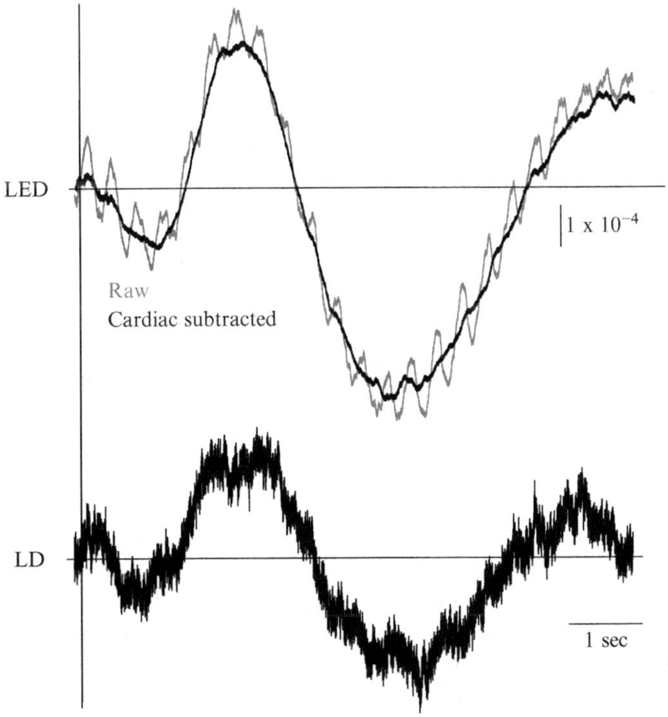

Fig. 10.5 Evoked optical response from the somatosensory cortex of a rat using two different light sources. In the *top panel*, the signal from an light emitting diode (LED) suffered sever cardiac artifacts (*gray line*) and was filtered (*black line*). In the *bottom panel*, a LD illuminated the cortex and exhibited much higher noise that even masked the cardiac artifact. The filtered LED signal exhibited a SNR that was two to three times larger than the LD signal

While the shape of the evoked optical signal from the LD light source was distinct, the noise overlying the signal was substantial. This speckle noise has the potential to swallow any smaller amplitude event-related responses. Since the fast optical signals are much smaller than the hemodynamic-based signals, significant improvement in SNR is necessary. Thus, LEDs are necessary as illumination methods because the noncoherent light allows for significantly larger SNRs both in vitro and in vivo.

10.7 Reflection Mode Imaging

While most birefringence studies focus on the transmitted signal, reflection mode imaging is required for in vivo imaging. Before employing birefringence techniques in vivo, we first optimized a reflection imaging modality. In both transmission and reflection imaging modalities there was a clear decrease in polarized light corresponding to the action potential (Fig. 10.4e). The amplitude of the evoked signal in reflection mode, however, was about one-third the amplitude of the signal observed in transmission mode. This smaller reflected signal may be due to fewer photons both changing polarization and scattering a full 180 degrees. In transmission mode imaging, many photons may be changing polarization while undergoing fewer scattering events, contributing to a larger response (Schei et al. 2008). While there is a distinct decrease in reflected polarized light corresponding to electrical activation, it is crucial to optimize the imaging techniques by utilizing procedures with good extinction coefficients and using noncoherent light sources to extract the small reflected changes in polarization.

10.8 Spatial Signal Components

The spatial components contributing to the polarization signal can resolve the neural constituents responsible for generating the signal. Spatial studies of the evoked polarization signals from lobster nerves revealed that the transmitted signals arose predominantly from the inner nerve bundle rather than the outer sheath. Cohen et al. (1968) hypothesized that the birefringence signal from squid giant axons arises from the edge of the axon. Lobster nerves, comprising hundreds of axons, elicited a transmitted signal that was evenly distributed throughout the inner nerve bundle (Schei et al. 2008). Within the bundle, the strongest absolute signals may be emanating from the outer sheath of individual axons, but the observed optical signal is an integration of each axon. Therefore, the largest change arose from the inner nerve bundle where there were more axons. Reflected polarization signals revealed that the largest spatial component of the signal emanates from the outer sheath of the nerve bundle because the outer sheath is more reflective (Schei et al. 2008). Birefringence techniques applied in vivo require the use of reflected light topologies and thus will rely on short range polarization changes.

10.9 Action Potential Movie

Using a camera to both spatially and temporally image the change in polarized light, we created a movie of a propagating action potential. Figure 10.6 shows successive frames of the change in polarized light. The initial response began on the left side, where the stimulus wires were placed, and corresponded to the activation of large diameter axons. The largest portion of the signal arose from medium diameter axons, although it involved components from axons of all sizes. The slow decay began on the left side and correspond to small diameter axons. Being able to resolve, both spatially and temporally, movies of action potential propagation demonstrates the advantages of fast intrinsic optical imaging. These imaging technologies have spatial resolution on the order of tens of micrometers and temporal resolution on the order of milliseconds, a combination that exceeds current imaging techniques.

10.10 Irregular Cell Orientation

Imaging the lobster nerve provides relatively simple experiments and data interpretation because the physiological noise is low and the cells are oriented in the same direction. Imaging in vivo, however, introduces complexities from random cell orientation and dendritic, somatic, and glial activation. In order to optimize our birefringence imaging techniques before employing them in vivo, we confronted random cell orientation by tying the nerve in an overhand knot and imaging the change in polarization using transmitted light. Figure 10.7a shows the image of the knotted nerve. Since the orientation is difficult to decipher, Fig. 10.7c shows a spatial map of the knot. The polarization signal at four regions of the knot are plotted along with an overlying fitting function in Fig. 10.7b. Recall that the strongest signals arise when the incident light is polarized at 45 degrees with respect to the nerve bundle. With that, the signal in region (a) arose from the underlying straight portion of the nerve where the light is incident at 45 degrees. The following sections b, c, and d elicited a temporal latency in the polarization changes according to the spatial path. Each region revealed a response had incident light oriented 45 degrees with respect to the nerve bundle (Schei et al. 2008). In theory, if the orientation of polarized light was rotated across different angles, different areas of the nerve would elicit signals.

Retinal imaging experiments using bright field, dark field, and polarization illumination techniques showed changes in scattered and transmitted light in response to optical stimulation. While optical responses using bright field illumination yielded single cell resolution images, dark field illumination techniques improved the contrast of the signal by reducing the background light. Polarization measurements showed similar signal contrast to dark field illumination images. The response was confined to the location of illumination whereas the dark field response was subject to responses from spreading neural activation beyond the

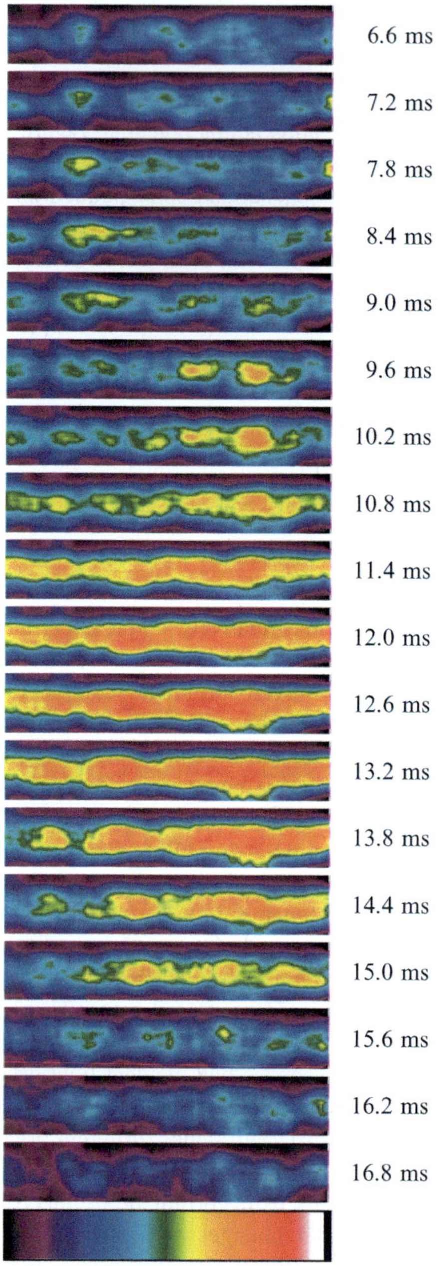

Fig. 10.6 Movie frames of an action potential propagating along an isolated lobster nerve. The time noted to the right of each frame is referenced to the time after the stimulus. Initial activation began on the left side, where the stimulus wires were placed, and corresponded to large diameter axons. Peak activation occurs at 12.6 ms where several components of the response, emanating from axons with different diameters, are merged together. Signal decay began on the left side and the residual response corresponded to the slow recovery of the smaller diameter axons (reproduced from Schei et al. (2008) with permission from Elsevier Limited)

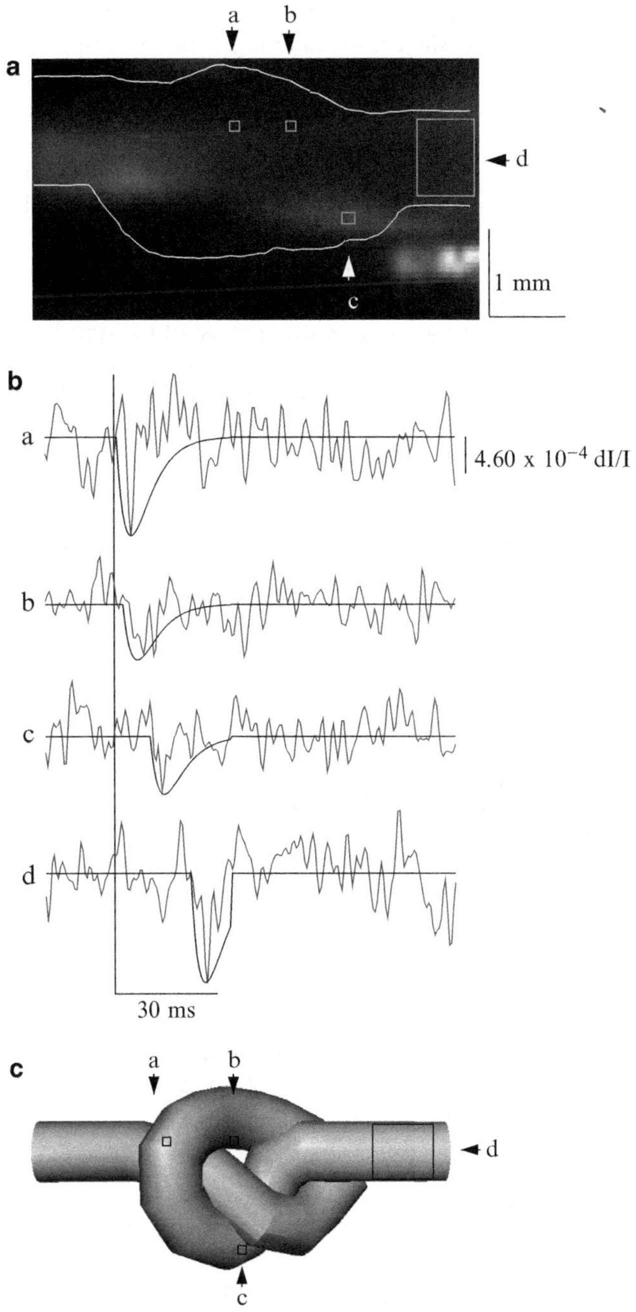

Fig. 10.7 Birefringence imaging from a knotted nerve. (**a**) A CCD image of the knotted nerve. The response from regions *a*, *b*, *c*, and *d* are shown in the traces (**b**). (**c**) An orientation diagram of the knot shows the regions *a*, *b*, *c*, and *d*, which are regions of the nerve that lie parallel to the long axis of the nerve and polarization signals appeared with delays corresponding to the action potential propagation along the nerve (reproduced from Schei et al. (2008) with permission from Elsevier Limited)

stimulated area (Yao and George 2006b). The temporal optical response using dark field illumination was subject to scattering events corresponding to cellular activation delays and metabolic responses. As a result, multiple temporally overlapping components convolve into the optical response (Yao and George 2006a).

Although fast optical scattering responses can be used to map highly localized neural activity, the signals remain small. Overcoming irregular cell orientations continues to be difficult since the responses emerge from very specific orientations of the nerve while the rest of the signals appear noisy. Further imaging developments are hindered by the ability to reject light that is unaltered by neural activity. Birefringence techniques could increase the contrast of the signal since the polarization optics reject light unaltered by neural activation. Additionally, birefringence techniques may be able to better discriminate between the multiple biophysical components embedded within the optical response.

10.11 In Vivo Applications and Fast Optical Responses

10.11.1 Hippocampus

Scattered light techniques can be used in vivo to image scattering changes of the multiple cascading events that arise from stimulation. In order to separate the different components of the optical response, we recorded scattered light changes from the hippocampus of a cat during Schaeffer collateral stimulation. A decrease in scattered light occurred concomitantly with population action potentials around 5 ms after the stimulus (Fig. 10.8). These initial scattering changes may be governed by cellular swelling events. While multiple peaks were recorded electrically, corresponding to population action potentials, only a single response was recorded optically. The EEG electrodes may be recording multiple electrical events from tissue volume conduction whereas the optical recording is from a spatially confined set of activated cells. A second dip in the optical response occurred around 100–500 ms after the stimulus, corresponding to excitatory postsynaptic potentials. These longer-latency scattering changes may be an aggregate of neural, glial, and fast hemodynamic events (Rector et al. 1997).

10.11.2 Barrel Cortex

We also recorded spatiotemporal fast scattering events from the rat whisker barrel cortex using a 660 nm LED and a CCD camera (Rector et al. 2001, 2005). Figure 10.9 shows the electrical evoked response potential (ERP) and the corresponding scattering changes (OPTICAL) to single whisker stimulation.

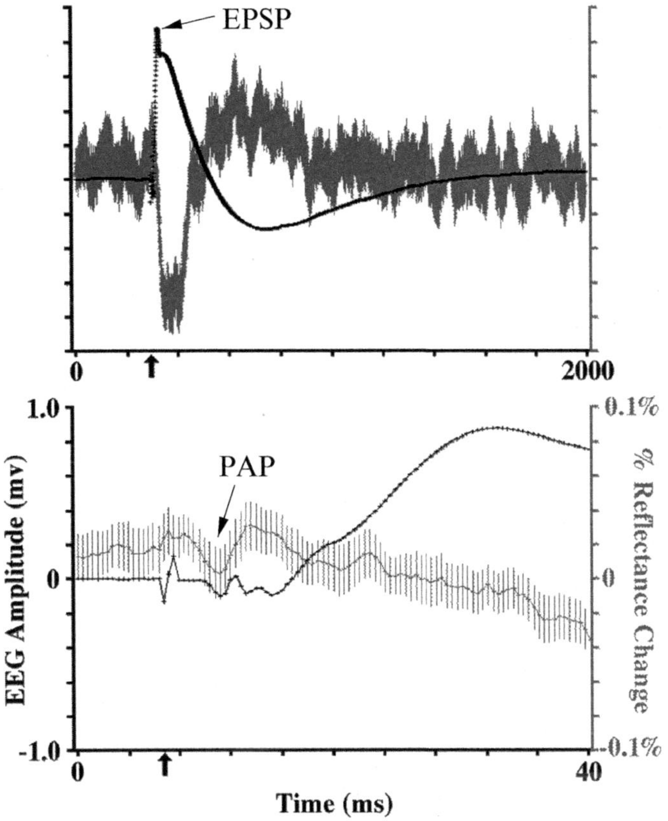

Fig. 10.8 Electrical (*thin black line*) and fast optical responses (*thick gray line*) from the hippocampus of a cat during Schaeffer collateral stimulation (*upward arrow*). The *top panel* shows a decrease in scattered light at 100–150 ms after the stimulus, concurrent with an excitatory postsynaptic potential (EPSP). The *bottom panel* shows a decrease in scattered light occurring around 5 ms after the stimulus and coinciding with population action potentials (PAP) (reproduced from Rector et al. (1997) with permission from American physiological society)

Spatial maps of the optical response show that the P30 component, corresponding to the presynaptic population spikes, is confined to the region of electrical stimulation whereas the N80 component, corresponding to postsynaptic potentials, is more diffuse across the cortex but with the peak response remaining over the barrel. Additional spatial maps of electrical and initial optical responses for single whisker stimulation are shown in Fig. 10.10. The P30 component of the optical response is mapped across multiple locations over the barrel cortex. High density electrode arrays map the ERP response for single whisker stimulation. These electrical maps, however, have lower spatial resolution than the early components of the optical maps due to electrical volume conduction (Rector et al. 2005).

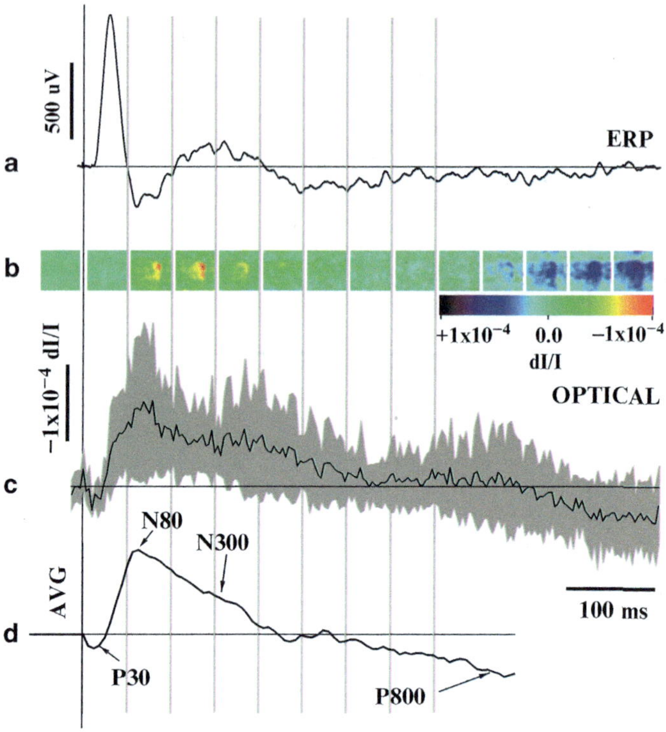

Fig. 10.9 Evoked responses from the whisker barrel cortex of a rat. (**a**) Evoked response potentials (ERP) to single whisker stimulation. (**b**) Spatially resolved fast images of the scattering response to a single whisker twitch using a CCD camera. An average of 25 frames before stimulation were subtracted from each frame in an image sequence compiled of 1,000 trials. An early response is confined to a specific location of activation corresponding to a single barrel and a later response is more diffuse across the cortex with peak activity remaining over the barrel. Warmer colors correspond to decreases in scattered light and cool colors correspond to increases in scattered light. (**c**) The inverted optical response (OPTICAL) averaged across all pixels of the CCD array. (**d**) A fitting paradigm identified four temporal components of the fast scattering signal. P30 is a positive-going response occurring about 30 ms after stimulation and N80 is a negative-going response occurring around 80 ms after the stimulus. The N300 and P800 peaks corresponded to fast and slow hemodynamic components of the response

10.11.3 Noninvasive Measurements

Some efforts have been made to uncover fast optical signals in humans using noninvasive imaging techniques. Fast backscattering light intensity changes have been shown during motor stimulation (Steinbrink et al. 2000). Other optical imaging techniques using modulated light sources show signal phase delays after stimulation. Photon migration methods measure relative photon delays during stimulation paradigms (Gratton et al. 1997; Maclin et al. 2004). Optical detections of fast physiological changes in vivo are easily swallowed by background noise corresponding to glial, synaptic, and hemodynamic events. Better signal optimization is required

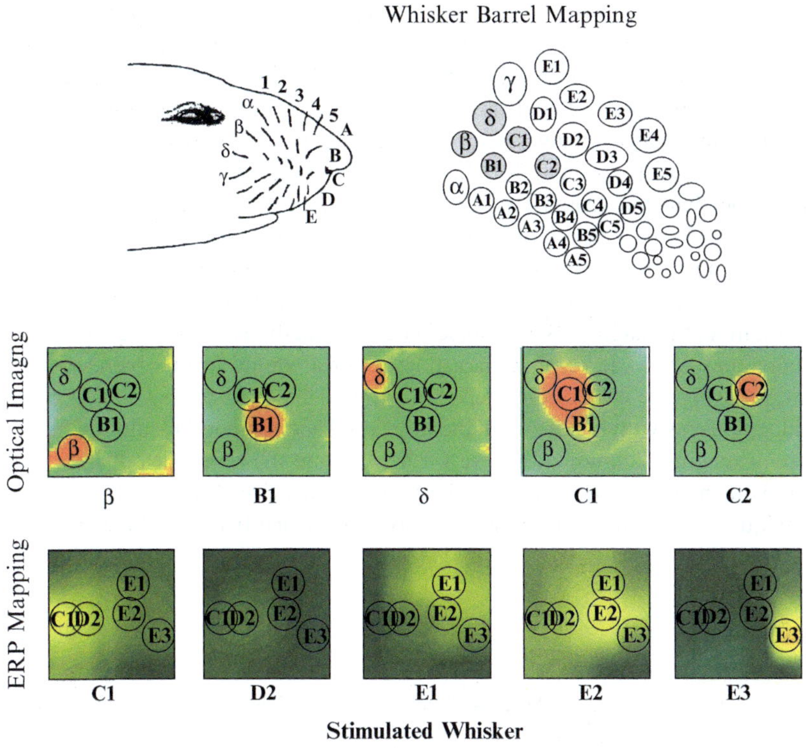

Fig. 10.10 Electrical and optical maps of single whisker stimulation. The *top panel* of images maps the early N80 component of the fast scattered signal for specific whisker stimulation. Each image is subtracted from an averaged baseline of 25 frames. *Circles* were drawn for spatial comparison. The *bottom panel* of images is the evoked electrical response potential (ERP) for single whisker stimulation mapped with an 8×8 high density electrode array

in order for fast optical responses to be readily applied in noninvasive imaging. Additionally, special care must be taken to exclude artifacts from vascular pulsations and vasomotion.

10.12 Future Approaches

Intrinsic optical imaging techniques have evolved significantly in the past half century and while in vivo optical imaging techniques are rapidly gaining popularity, these techniques typically use scattered light to record hemodynamic-related changes. Our research has focused on integrating birefringence techniques for in vivo imaging in order to measure electrical consequences of neural activity. Thus far, we have shown that birefringence imaging techniques can be utilized with irregular cell orientations. A more robust signal may be possible by decomposing the parameters of the Mueller matrix and then reconstructing the responses associated with the polarization changes. Although improvements were made by overcoming irregular cell orientation obstacles during optical imaging, other scattering events hinder the application of optical techniques in vivo. Somatic, dendritic, and glial activation may cause changes in optical activity, thus complicating the interpretations of the detected signal. Polarization optics increase the signal by an order of magnitude than the scattering signal; however, further increases in contrast are needed in order to measure optical responses consequential to electrical activity in fewer averages, especially when recording the signals noninvasively. Using modulated light sources or time-gated detectors would allow for photon path retracing and multidepth imaging. Better noise reduction methods such as background subtraction and EKG and respiration artifact removal would further increase the contrast of the signal. With birefringence imaging techniques, we have the potential to produce images with better simultaneous spatial and temporal resolution than other current imaging techniques. As we further explore the properties of light interactions with neural tissue, we step closer to noninvasive, high resolution imaging techniques and begin to close the gap between science fiction and technology.

References

Carter KM, George JS, Rector DM (2004) Simultaneous birefringence and scattered light measurements reveal anatomical features in isolated crustacean nerve. J Neurosci Methods 135:9–16

Chen-Bee CH, Agoncillo T, Xiong Y, Frostig RD (2007) The triphasic intrinsic signal: implications for functional imaging. J Neurosci 27(17):4572–4586

Cohen LB, Keynes RD, Hille B (1968) Light scattering and birefringence changes during nerve activity. Nature 218:438–441

Cragg BG (1951) The turgidity of giant axons. J Physiol 114:234–239

Culver JP, Siegel AM, Franceschini MA, Mandeville JB, Boas DA (2005) Evidence that cerebral blood volume can provide brain activation maps with better spatial resolution than deoxygenated hemoglobin. Neuroimage 27:947–959

Devor A, Tian P, Nishimura N, Teng IC, Hillman EMC, Narayanan SN, Ulbert I, Boas DA, Klienfeld D, Dale AM (2007) Suppressed neuronal activity and concurrent anteriolar vasoconstriction may explain negative blood oxygenation level-dependent signal. J Neurosci 27(16):4452–4459

Eggert HR, Blazek V (1987) Optical properties of human brain tissue, meninges, and brain tumors in the spectral range of 200 to 900 nm. Neurosugery 21(4):459–464

Flaig JV (1947) Viscosity changes in axoplasm under stimulation. J Neurophysiol 10(3):211–221

Foust AJ, Beiu RM, Rector DM (2005) Optimized birefringence changes during isolated nerve activation. Appl Opt 44(11):2008–2012

Foust AJ, Rector DM (2007) Optically teasing apart neural swelling and depolarization. Neuroscience 145:887–899

Foust AJ, Schei JL, Rojas MJ, Rector DM (2008) In-vitro and in-vivo noise analysis for optical neural recording. J Biomed Opt 13:044038

Frostig RD, Lieke EE, Ts'o DY, Grinvald A (1990) Cortical functional architecture and local coupling between neuronal activity and the microcirculation revealed by *in vivo* high-resolution optical imaging of intrinsic signals. Proc Natl Acad Sci U S A 87:6082–6086

Gratton G, Fabiani M, Corballis PM, Gratton E (1997) Noninvasive detection of fast signals from the cortex using frequency-domain optical methods. Ann N Y Acad Sci 820:286–298

Grinvald A, Lieke E, Frostig RD, Gilbert CD, Wiesel TN (1986) Functional architecture of cortex revealed by optical imaging of intrinsic signals. Nature 324(6095):361–364

Hill DK, Keynes RD (1949) Opacity changes in stimulated nerve. J Physiol 108:278–281
Hodgkin AL, Huxley AF (1952) Propagation of electrical signals along giant nerve fibres. Proc R Soc Lond B Biol Sci 140(899):177–183
Hodgkin AL, Keynes RD (1955) Active transport of cations in giant axons from Sepia and Loligo. J Physiol 128:28–60
Hoshi Y, Tamura M (1993) Dynamic mulitichannel near-infrared optical imaging of human brain activity. J Appl Physiol 75:1842–1846
Kornakova YE, Frank GM, Shteyngauz LN (1947) Structural processes in the nerve. Fiziol. J. USSR. 33:483–494
Landowne D (1985) Molecular motion underlying activation and inactivation of sodium channels in squid giant axons. J Membr Biol 88:173–185
Maclin EL, Low KA, Sable JJ, Fabiani M, Gratton G (2004) The event-related optical signal to electrical stimulation of the median nerve. Neuroimage 21:1798–1804
Mann G (1894) Histological changes induced in sympathetic, motor, and sensory nerve cells by functional activity. J Anat Physiol 29:100–108
Maxwell JC (1865) A dynamical theory of the electromagnetic field. Philos Trans R Soc London 155:459–512
Mayhew J, Johnston D, Berwick J, Jones M, Coffey P, Zheng Y (2000) Spectroscopic analysis of neural activity in brain: Increased oxygen consumption following activation of barrel cortex. Neuroimage 12:664–675
Oldenbourg R, Salmon ED, Tran PT (1998) Birefringence of single and bundled microtubules. Biophys J 74:645–654
Rector DM, Carter KM, Volegov PL, George JS (2005) Spatio-temporal mapping of rat whisker barrels with fast scattered light signals. Neuroimage 26:619–627
Rector DM, Poe GR, Kristensen MP, Harper RM (1997) Light scattering changes follow evoked potentials from hippocampal schaeffer collateral stimulation. Neurophysiology 78:1707–1713
Rector DM, Rogers RF, Schwaber JS, Harper RM, George JS (2001) Scattered-light imaging *in vivo* tracks fast and slow processes of neurophysiological activation. Neuroimage 14:977–994
Salzberg BM, Kosterin PV, Muschol M, Obaid AL, Rumyantsev SL, Bilenko Yu, Shur MS (2005) An ultra-stable non-coherent light source for optical measurements in neuroscience and cell physiology. J Neurosci Methods 141:165–169
Schei JL, McCluskey MD, Foust AJ, Yao XC, Rector DM (2008) Action potential propagation imaged with high temporal resolution near-infrared video microscopy and polarized light. Neuroimage 40:1034–1043
Sheth S, Nemoto M, Guiou M, Walker M, Paouratian N, Toga AW (2003) Evaluation of coupling between optical intrinsic signals and neuronal activity in rat somatosensory cortex. Neuroimage 19:884–894
Sochurek H, Miller P (1987) Medicine's new vision. National Geographic 171(1):2–40
Steinbrink J, Kohl M, Obrig H, Curio G, Syre F, Thomas F, Wabnitz H, Rinneberg H, Villringer A (2000) Somatosensory evoked fast optical intensity changes detected non-invasively in the adult human head. Neurosci Lett 291:105–108
Tasaki I, Watanabe A, Sandlin R, Carnay L (1968) Changes in fluorescence, turbidity, and birefringence associated with nerve excitation. Proc Natl Acad Sci U S A 61:883–888
Villringer A, Chance B (1997) Non-invasive optical spectroscopy and imaging of human brain function. Trends Neurosci 20(10):435–442
Yao XC, Foust A, Rector DM, Barrowes B, George JS (2005) Cross-polarized reflected light measurements of fast optical responses associated with neural activation. Biophys J 88:4170–4177
Yao XC, George JS (2006a) Dynamic neuroimaging of retinal light responses using fast intrinsic optical signals. Neuroimage 33:898–906
Yao XC, George JS (2006b) Near-infrared imaging of fast intrinsic optical responses in visible light-activated amphibian retina. J Biomed Opt 11(6):064030
Yao XC, Rector DM, George JS (2003) Optical lever recording of displacements from activated lobster nerve bundles and Nitella internodes. Appl Opt 42(16):2972–2978

Chapter 11
Two-Photon Laser Scanning Microscopy as a Tool to Study Cortical Vasodynamics Under Normal and Ischemic Conditions

Anna Devor, Andy Y. Shih, Philbert S. Tsai, P. Blinder, Peifang Tian, Ivan C. Teng, and David Kleinfeld

Abstract This review focuses on the application of two-photon laser scanning microscopy (TPLSM) for in vivo imaging of vascular reactivity, neurovascular communication and as interventional tool for altering vascular and neuronal networks.

A mechanistic understanding of hemodynamics requires a systematic analysis of microvascular responses to changes in underlying neuronal activity. This analysis needs to be carried at the single vessel level in the context of the angioarchitecture of vascular networks (Scremin 1995). Moreover, interventional strategies are needed to manipulate specific neuronal and vascular structures to test the role in neurovascular ensemble responses. In this context, recent advances in TPLSM (Denk et al. 1990) allows one to measure the activity of single, visually-identified blood vessels with the potential for simultaneous measurements of neuronal and glial activity using calcium indicators (Stosiek et al. 2003; Ohki et al. 2005; Garaschuk et al. 2006), and targeted occlusion of single blood vessels (Nishimura et al. 2006; Schaffer et al. 2006; Nishimura et al. 2007). While technical aspects of TPLSM have been recently extensively reviewed (Denk et al. 1990; Denk and Svoboda 1997; Svoboda and Yasuda 2006), here we focus on the application of TPLSM to study cortical vascular dynamics under normal conditions and following experimentally-induced ischemic stroke.

11.1 Macroscopic Imaging of Blood Oxygenation and Flow

Great progress in understanding the relationship between different aspects of cerebral hemodynamics, such as blood flow, volume, and oxygenation, has been made through noninvasive macroscopic functional imaging methods: functional magnetic resonance imaging (Kwong et al. 1992; Ogawa et al. 1992), positron emission

---

A. Devor (✉)
Department of Neurosciences, UCSD, La Jolla, CA, USA
e-mail: adevor@ucsd.edu

A.W. Roe (ed.), *Imaging the Brain with Optical Methods*,
DOI 10.1007/978-1-4419-0452-2_11, © Springer Science+Business Media, LLC 2010

tomography (Fox and Raichle 1986; Fox et al. 1988; Raichle and Mintun 2006) and near infrared imaging spectroscopy (Villringer et al. 1993; Obrig and Villringer 2003; Boas et al. 2004a, b). These macroscopic methods can measure blood volume, oxygenation, and flow, but cannot resolve individual vascular compartments: arterial, venous, and capillary. In addition, all these methods have a limited spatial resolution, i.e., usually >1 mm. Thus, physiological interpretation of these noninvasive signals is based on numerous assumptions (Buxton et al. 1998; Davis et al. 1998; Mandeville et al. 1999).

Invasive optical imaging of "intrinsic signals" using a charge-coupled device (CCD) camera as a detector has been successfully used to map hemodynamic activity on a finer scale (Grinvald et al. 1986; Malonek and Grinvald 1996; Vanzetta and Grinvald 1999). Since oxy- and deoxyhemoglobin have different absorption spectra at visible wavelengths, intrinsic imaging provides a quantitative measure of hemoglobin oxygenation if two or more illumination wavelengths are used; the same principle is used by near infrared imaging spectroscopy (Malonek and Grinvald 1996; Mayhew et al. 1998; Devor et al. 2003; Dunn et al. 2003; Devor et al. 2005; Dunn et al. 2005; Devor et al. 2007). Intrinsic imaging revealed a wealth of information regarding the relationship between blood oxygenation and volume (Berwick et al. 2002; Martindale et al. 2003; Sheth et al. 2003; Nemoto et al. 2004; Sheth et al. 2004a, b; Sheth et al. 2005) with inferences to blood flow and tissue metabolism (Vanzetta and Grinvald 1999, 2001). Attempts have been made to isolate signals from major surface arteries and veins, and the capillary bed (Vanzetta et al. 2005). However, light scattering and lack of depth resolution, inherent to CCD-based methods, result in the contamination of signals by the surrounding tissue and introduce an ambiguity in interpretation.

A new technique, laminar optical tomography (LOT) (Hillman et al. 2004; Hillman et al. 2007), has been developed based on a combination of confocal illumination and an array of detectors collecting light at different distances relative to the illumination spot. Through modeling of photon migration, the general practice in diffuse optical methods (Boas et al. 2004a, b), these distances are translated into cortical depths. Thus, LOT overcomes the lack of depth resolution by using calculated profiles of light penetration. LOT can also be used to estimate timecourses of oxygenation change across vascular compartments based on the signal timecourses isolated from major vessels (Hillman et al. 2007). Nevertheless, similar to the CCD-based optical imaging of intrinsic signals, LOT suffers the effects of light scattering.

Laser Doppler flowmetry (LDF), which determines the speed of red blood cells (RBCs) from a shift in the frequency of reflected light, provides a point measurement of blood flow that can be serially repeated at multiple locations. LDF is accurate only for detecting relative changes since absolute measures of flux cannot be obtained (Dirnagl et al. 1989; Barfod et al. 1997). A scanning version of the LDF probe has also been implemented to obtain 2D images of flow (Lauritzen and Fabricius 1995; Ances et al. 1999), as well as a recent interferometric method to simultaneously map flow across large area of cortex (Atlan et al. 2006). Laser speckle contrast imaging is an alternative method that allows 2D imaging of the flow. This method is based on the observation that the movement of RBCs through vessels will smear the speckle contrast when averaged over a short (~5 ms) time window (Dunn et al. 2001; Zhang and Murphy 2007). Thus, both laser Doppler and laser speckle contrast imaging reflect the speed of RBCs, and a good correspondence between the two methods has been reported (Dunn et al. 2001). As other optical methods mentioned above, LDF and speckle contrast imaging do not resolve vascular compartments. In addition, a quantitative translation of these signals into blood flow requires a number of assumptions (Dirnagl et al. 1989; Dunn et al. 2001).

11.2 Microscopic In Vivo Imaging of Vascular Reactivity on the Level of Single Blood Vessels

The imaging techniques mentioned above average blood oxygenation or flow over large voxels (or pixels) of tissue including a multitude of small and large blood vessels. However, blood vessels that belong to different vascular compartments: arterial, venous and capillary, have different properties. Arterial vessels can actively change their diameter and thus regulate blood blow since their walls contain smooth muscle cells, a substrate for active diameter control by neurovascular mediators; for a review see (Iadecola 2004; Lauritzen 2005; Hamel 2006). The regulation of capillary diameter by contractile pericytes has also been demonstrated (Peppiatt et al. 2006). Veins, in contrast, do not possess active constrictive elements and passively comply with changes in pressure (Buxton et al. 1998).

The first serious attempt to study dynamics of single cerebral blood vessels was undertaken long before the invention of two-photon laser scanning microscopy (TPLSM) (Wayland and Johnson 1967). The investigators used a CCD detector to measure flow in surface capillaries based on the movement of "void" segments with no RBCs. Later, capillary flow was also observed using conventional fluorescence microscopy (Hudetz 1997), and confocal laser scanning microscopy (Dirnagl et al. 1991; Villringer et al. 1994). Investigation of motion of RBCs in capillaries in superficial layers of rat cerebral cortex revealed a large variability in perfusion. However, a coherent response was observed following a hypercapnic challenge, measured as an overall increase in RBC flux (Villringer et al. 1994).

In contrast to confocal microscopy, TPLSM can image brain cortex up to 600 μm below the pia with micron resolution by using commercial low energy ultrafast laser oscillators (Kleinfeld et al. 1998), and up to 1,000 μm with high-energy laser amplifiers (Theer et al. 2003). The depth penetration can be further extended *ex-vivo* using "all-optical histology" (Tsai et al. 2003), which complements TPLSM functional approach by providing complete reconstruction of underlying vascular architecture (Fig. 11.1). All-optical histology takes advantage of the fact that ablation with ultrashort, near-infrared pulses is highly localized with minimal collateral damage (Tsai et al. 2003). Hence, it can iteratively image and ablate cortex to arbitrary depths while maintaining micrometer resolution and preserving the structural integrity of both the neurons and vessels.

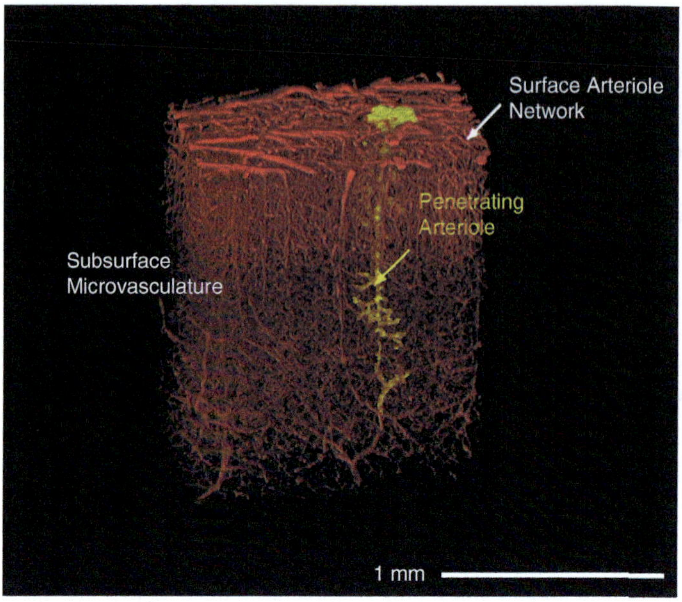

Fig. 11.1 Reconstruction of mouse cortical vasculature. The mouse cortical vasculature was labeled using a fluorescent perfusion cast. The surface vascular network includes pial arteries, arterioles, and veins, while the subsurface vasculature consists of dense capillary beds. Penetrating arterioles, one of which is shown in *yellow*, bridge the flow of blood from surface to subsurface vessels. The entire cortical thickness was imaged using iterations of TPLSM followed by laser ablation of the imaged tissue depth, i.e., all-optical histology (Tsai et al. 2003)

The use of TPLSM to measure blood flow allows 3D mapping of evoked vascular response in single vessels across different vascular compartments (Chaigneau et al. 2007; Devor et al. 2007). TPLSM has been employed to image vascular changes in single vessels in vitro (Simard et al. 2003; Zonta et al. 2003; Cauli et al. 2004; Mulligan and MacVicar 2004; Filosa et al. 2006; Rancillac et al. 2006) and in vivo (Faraci and Breese 1993; Kleinfeld et al. 1998; Chaigneau et al. 2003; Takano et al. 2006; Chaigneau et al. 2007; Devor et al. 2007; Zhang and Murphy 2007). The design of a two-photon laser scanning microscope and supporting software has been described previously (Tsai and Kleinfeld 2009; Nguyen et al. 2009).

11.3 Measurements of Diameter and RBC Speed Describe Blood Flow in Individual Vessels

Two complementary measures, vascular lumen diameter and RBC speed, are used to characterize the vascular dynamics of an individual blood vessel. The lumen diameter of capillaries is small, i.e., <10 μm in rat, and allows only single-file

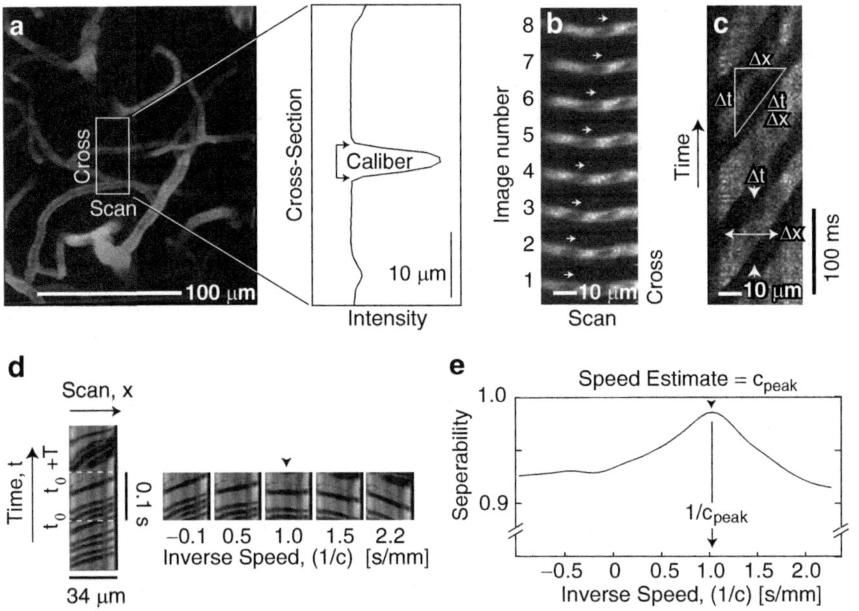

Fig. 11.2 Measurement of lumen diameter and RBC speed of individual capillaries. (**a**) Projection of TPLSM image stack showing subsurface capillaries labeled with fluorescein-dextran dye in vivo. The capillaries were 300–400 μm below the cortical surface. The *inset* shows the intensity profile along the cross section for the capillary in question. The lumen diameter is estimated from the number of pixels with intensity above the background level. (**b**) Successive planar images through a capillary, acquired every 16 ms. The change in position of an unstained RBC against a fluorescent background is indicated by the series of *arrows*; the speed of the RBC is +0.11 mm/s. (**c**) RBC speed was calculated from line-scans collected repeatedly along the central axis of the capillary, which were then stacked sequentially to form a space–time image. The streaks within the line-scan are caused by unstained RBCs moving through the capillary over time. The x-axis represents the distance traveled by the RBCs (x) and the y-axis represents time (t). RBC speed is then calculated from the inverse of the slope of the RBC streaks ($\Delta t/\Delta x$); the direction of flow is discerned from the sign of the slope. (**d**) Illustration of automated algorithm for finding slope of streaks formed by moving RBCs. Line-scan data from an epoch in time is transformed to a square matrix with normalized axes. In the *left image*, an abrupt change in the slope of the streaks can be due to a heartbeat or irregularity in capillary flow. The central region of the square matrix is rotated to find the angle that yields horizontal streaks, as in the middle panel (*arrowhead*). (**e**) Separability of line-scan data as a function of rotation angle; separability is maximal for vertical or horizontal streaks. Figure adapted from Kleinfeld et al. (1998)

passage of RBCs. Capillary flow is typically only described in terms of RBC speed or flux, i.e., number of RBCs per second. In order to measure RBC movement in vivo, dextran-conjugated fluorescent dyes are injected intravenously to label the plasma (Fig. 11.2a) leaving RBCs visible as dark shadows on bright fluorescent background (Fig. 11.2b). The speed of the RBCs is captured by repeated line-scans along the axis of the vessel lumen that form a space-time image when stacked sequentially and leads to the generation of streaks caused by the motion of RBCs (Fig. 11.2c). The speed of RBCs is given by the inverse of the slope of these streaks and the direction of flow is discerned from the sign of the slope. An algorithm based on singular value decomposition is used to automate the calculation of speed from the line-scan data (Fig. 11.2d, e) (Kleinfeld et al. 1998).

In larger vessels, such as surface pial arterioles (Fig. 11.3a), flow assumes a laminar profile with fastest flow in the center and slowest flow near the endothelial wall. Scanning is done along the central axis, where the speed is maximal (Fig. 11.3b), and RBC speed is also calculated from the inverse slope of the line-scan streaks (Fig. 11.3c), as with capillaries. The average speed of RBCs in the arteriole is theoretically equivalent to onehalf of the center-line speed, although this may be an underestimate for nonNewtonian fluids. The measurements from arterioles and capillaries typically show a periodic modulation due to the heartbeat (Fig. 11.3d). The parabolic curve in Fig. 11.3e represents the laminar flow profile (the fastest flow is in the middle of the vessel) that most closely matches the data.

Absolute vessel diameter is measured by obtaining a planar image stack, or by using continuous line-scans perpendicular to the vessel axis to gauge rapid changes over time (Figs. 11.4 and 11.5). By scanning longer distances across multiple vessels, one can obtain the diameter changes of multiple vessels in a single space-time image (Devor et al. 2007). In this case, the scanning direction may not always be perpendicular to vessel axis for each of the measured vessels. However, the fractional diameter change is unaffected and the absolute diameter D can be obtained by $D = DMsin(\theta)$, where DM is the measured diameter and the θ is the angle between the scan line and the axis of the vessel. In practice, the number of simultaneously measured vessels is limited to vessels in focus. One can overcome this limitation through development of more effective arbitrary line-scan algorithms (Gobel and Helmchen 2007; Gobel et al. 2007).

By measuring both diameter and speed changes in the same arteriole, the vascular responses can be represented by flux, F, defined by $F = A*V$ where A and V is the cross-section area and average RBC speed, respectively. As mentioned above, the measurement of flux is unambiguous in capillaries since cells can only move single-file. For arterioles, however, one can calculate flux as $F = (\pi/8)vd^2$, where v is the center-line speed and d is a diameter (Schaffer et al. 2006). While an increase in diameter or speed alone suggests but does not prove an increase in blood flow, e.g., an increase in diameter can be accompanied by a decrease in speed, F gives a complete description of the vascular response at the time of measurement. Fig. 11.4 shows a comparison of diameter, RBC speed, and flux in surface pial arterioles, penetrating arterioles, and capillaries. Despite the overlapping peaks in the diameter (Fig. 11.4a) and speed (Fig. 11.4b) profiles for the surface and penetrating arterioles, they differ significantly in flux (Fig. 11.4c). The capillary profiles are also better separated from those of the arterioles with regard to flux.

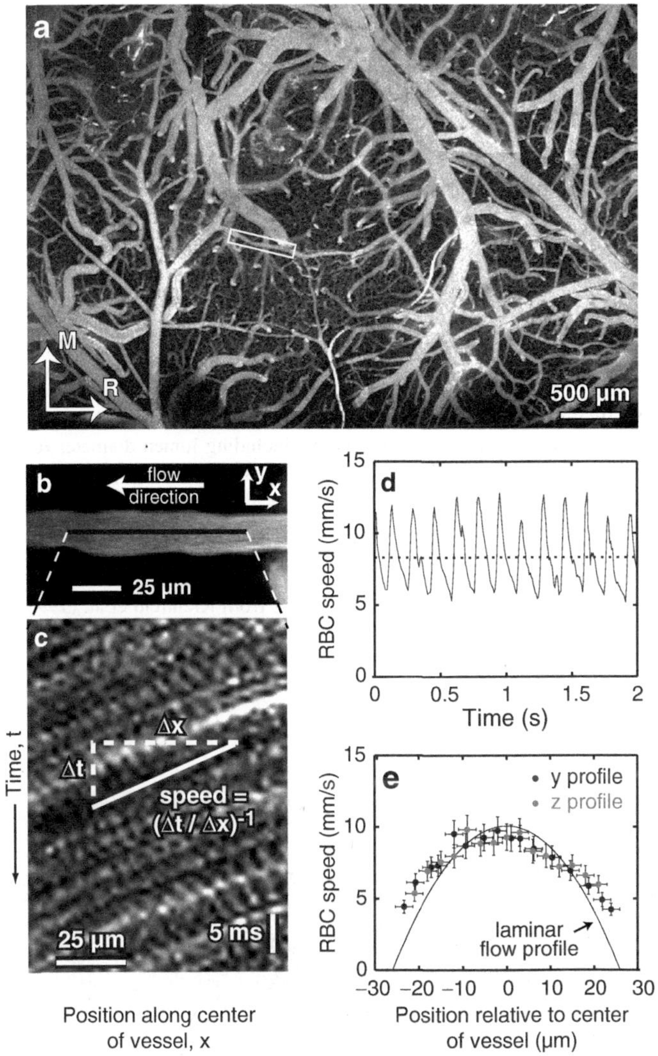

Fig. 11.3 Measurement of RBC speed of individual pial arterioles. (**a**) Low-magnification TPLSM image of pial vessels in rat cortex labeled with fluorescein-dextran dye in vivo. The axes indicate the rostral (R) and medial (M) directions. (**b**) Maximal projection of a TPLSM image stack through a surface arteriole identified in the *inset* in (**a**). The *dark line* indicates the location where the line-scan data were taken, and the *arrow* represents the direction of flow obtained from these scans. (**c**) Line-scan data from the vessel in (**b**) to quantify RBC speed. As with capillaries, RBC speed is measured from the inverse slope of the line-scan streaks; see Fig. 11.2c–e. (**d**) RBC speed along the central axis of the arteriole shown in (**b**) and (**c**) as a function of time. The periodic modulation of the RBC speed occurs at the approximately 6-Hz heart rate. The *dotted line* represents the temporal average of the speed. (**e**) The speed of RBCs in a different arteriole, averaged over 40 s, as a function of the transverse position in the vessel along horizontal (y) and vertical (z) directions. The parabolic curve represents the laminar flow profile that most closely matches the data. Figure adapted from Schaffer et al. (2006)

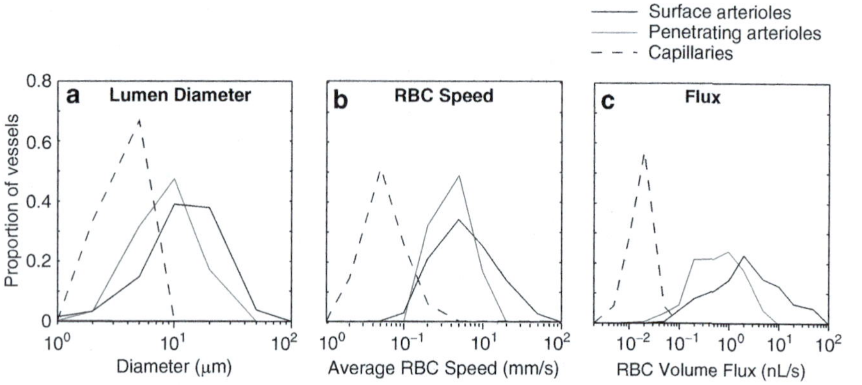

Fig. 11.4 Quantitative blood flow measurements from individual cerebral pial arterioles and capillaries. Compendium of in vivo measurements including lumen diameter (**a**), average RBC speed (**b**), and volume flux (**c**), collected from three distinct vessel types of the rat cortex. For penetrating arterioles, RBC speed was measured at locations where the vessels were oriented in *xy* plane. Arteriole flux was calculated using the equation $F=(\pi/8)vd^2$, where v is the center-line RBC speed and d is the lumen diameter. For capillaries, flux was calculated using the same equation minus a laminar flow constant, $F=(\pi/4)vd^2$. Arteriole data was taken from Schaffer et al. (2006) and Shih et al. (unpublished). Capillary data is from Kleinfeld et al. (1998). Surface arterioles ($n=454$), penetrating arterioles ($n=360$), and capillaries ($n=35$)

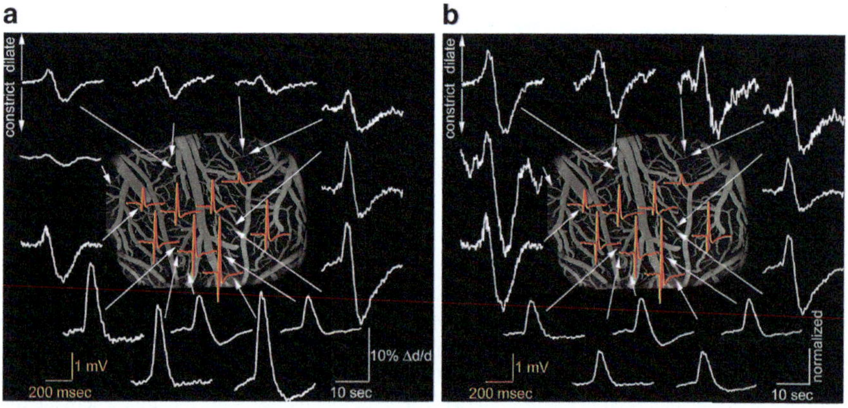

Fig. 11.5 TPLSM imaging of vascular response to a somatosensory stimulus. (**a**) Stimulus-evoked diameter changes in single arterioles. Increase in the diameter is plotted upward, and decrease is plotted downward. Electrophysiological response to the stimulus was mapped using a ball electrode. Recorded surface potentials (*red traces*) are overlaid on the image of the brain surface vasculature. Both vasodilatation and vasoconstriction are present at all locations. The degree of vasodilation decreases with an increase of distance from the center of neuronal activity, denoted by the largest surface potential amplitude. (**b**) Initial vasodilation in vascular diameter time-courses is normalized. This area corresponds to the center of the evoked neuronal activity. Note that vasoconstriction increases with an increase in the distance from the center

11.4 Simultaneous Measurements of Neuronal and Vascular Activity

TPLSM is strategically positioned to provide simultaneous measurements of the vascular and neuronal activity, crucial for understanding the basis of neurovascular coupling. In this respect, a study by our group (Devor et al. 2007) showed the spatial pattern of evoked diameter changes in surface arterioles as a function of distance from the focus of neuronal activity caused by somatosensory stimulation (Fig. 11.5). Neuronal activity was not imaged simultaneously with vascular measurements, but was mapped prior to TPLSM. The results show that at every location within the exposure, arteriolar response was composed of an initial dilatory and later constrictive phase (Fig. 11.5a). The relative strength of dilatation and constriction varied as a function of distance from the center of neuronal activity, where the strongest dilation was observed in the center and the strongest relative constriction was observed in the surround (Fig. 11.5b). Moreover, these TPLSM findings were consistent with imaging of intrinsic signals and also with voltagesensitive dyes (VSD) imaging. Comparison of TPLSM, intrinsic and VSD results demonstrated a correlation of the arteriolar constriction with enhanced neuronal inhibition. This study supports the hypothesis of neurogenic regulation of the arteriolar diameter, a key control parameter in the hemodynamic response (Cauli et al. 2004; Hamel 2004, 2006; Rancillac et al. 2006).

While our study was focused on surface arterioles to facilitate the comparison with intrinsic optical imaging that has high sensitivity to the cortical surface (Polimeni et al. 2005), we expect that multiple parameters, such as vessel-type, distance from the center of neuronal activity and connectivity contribute to the vascular response. Ideally, mapping should be achieved by measuring individual penetrating arterioles. This is because surface arterioles are highly redundant and, as a result, have less specificity to cortical columns (Iadecola 2004; Schaffer et al. 2006) than penetrating arterioles that feed well-defined columns of cortical tissue (Nishimura et al. 2007).

Neuronal activity can be imaged simultaneously with measurements of vascular dynamics and RBC speed by using calcium indicators (Chaigneau et al. 2007) (Fig. 11.6). A study by Chaigneau et al. investigated the relationship between the capillary blood flow and neuronal activity in the olfactory bulb in vivo using extrinsic calcium dyes in rats and genetically-encoded indicators in mice (Chaigneau et al. 2007). In rats, calcium indicator Oregon Green BAPTA-1 was loaded into mitral cells in the olfactory glomeruli intracellularly through a recording pipette, while in mice transgenic expression of G-CaMP2 served as the indicator (Diez-Garcia et al. 2005). In all experiments, blood plasma was labeled by intravenous injection of Texas Red dextran. The authors observed an increase in capillary RBC speed concurrent with an increase in calcium in dendritic branches of mitral cells following an odor stimulus (Fig. 11.6). This study demonstrated the feasibility of simultaneous imaging of neuronal calcium signals and capillary RBC speed using TPLSM. Furthermore, it supports the prior hypothesis that neuronal

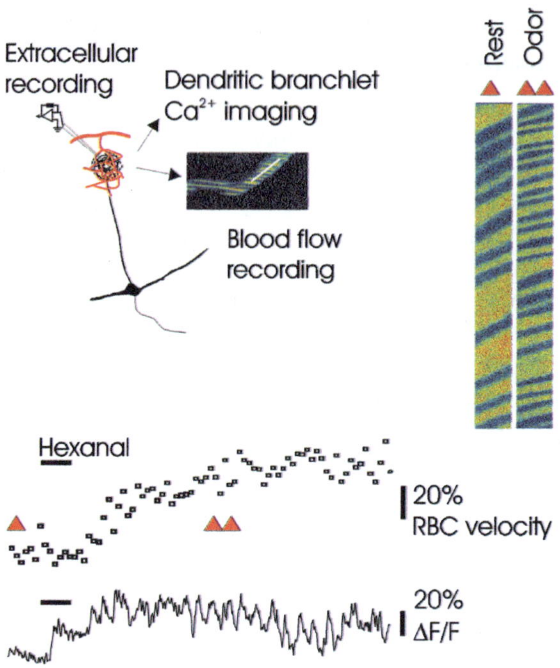

Fig. 11.6 Simultaneous TPLSM imaging of the capillary blood flow and neuronal activity in the olfactory bulb. A single mitral cell was labeled with calcium indicator Oregon Green BAPTA-1 through an intracellular pipette and imaged with TPLSM. Odor stimulus (hexanal 1.2%) generated calcium increases in the dendritic tuft and RBC speed increase in a nearby capillary. Calcium measurements were performed using line-scan acquisition, allowing simultaneous measurement of RBC speed. *Red triangles* under RBC speed time-course indicate times corresponding to blood flow recordings shown on the *right*. Figure adapted from Chaigneau et al. (2007)

intracellular calcium concentration plays a central role in neurovascular coupling (Lauritzen 2005).

An ability to image both neuronal and glial calcium transients is important for neurovascular research since both neurons and glia have been implicated in communication to the vasculature (Cauli et al. 2004; Mulligan and MacVicar 2004; Rancillac et al. 2006; Takano et al. 2006; Chuquet et al. 2007). Further, functional changes in calcium levels can also be detected in arteriolar smooth muscle, which appear to coincide with active changes in vascular diameter (Filosa et al. 2004). Therefore, through a combination of calcium imaging and direct blood flow measurements, TPLSM can be used to observe neurons, glia, and vessels that control blood flow. This approach is expected to aid the identification of cellular elements that control vascular response under normal conditions and possibly the mechanisms that lead to its failure during pathology.

11.5 TPLSM Guided Plasma Mediated Ablation as an Intervention Tool

TPLSM can be used not only to image neuronal and vascular activity but also to guide microscopic ablation of tissue as a means to manipulate neuronal circuits and vascular architectonics in vivo. Nishimura et al. have developed a method for targeted occlusion of single capillaries deep to the cortical surface by using two-photon irradiation (Nishimura et al. 2006). This method makes use of plasma-mediated ablation through the nonlinear absorption of 100-fs, ultrashort laser pulses by blood plasma or the vascular wall, generating a variety of injury models including intravascular clots, serum extravasation, and hemorrhage (Tsai et al. 2009). Another method for targeted occlusion of single vessels is based on focal photothrombosis induced by optical activation of the photosensitizer Rose Bengal (Schaffer et al. 2006; Nishimura et al. 2007; Sigler et al. 2008). Unlike the use of nonlinear absorption, photothrombotic clots are confined to the cortical surface since irradiation of Rose Bengal-dye is performed with a green laser. A recent study demonstrated that photothrombic occlusion of localized middle cerebral artery branches is sufficiently accurate to remove targeted regions of the somatosensory representation while leaving neighboring regions intact (Zhang and Murphy 2007). Similarly, the authors used this method to demonstrate that surface arterioles form complicated redundant interconnected networks (Schaffer et al. 2006), and therefore can tolerate a local damage, while penetrating arterioles define only partially overlapping vascular territories (Fig. 11.7a) (Nishimura et al. 2007). Figure 11.7 gives an example of single penetrating arteriole occlusion and illustrates the loss of flow in downstream microvessels and the resultant generation of a column of hypoxic tissue (Fig. 11.7b).

In future studies, TPLSM will also be used to guide plasma mediated ablation for targeted removal of individual neurons. A body of recent data increasingly suggests that differential vascular control originates from activation of specific neuronal subpopulations through release of specific vasoactive agents (Cauli et al. 2004; Hamel 2004, 2006; Rancillac et al. 2006). However, direct evidence showing that selective activation of known cell types correlates with vasodilation or vasoconstriction in vivo is missing. Ablation of single or multiple neurons with known vasoactive properties can be used to test the spatial precision of neurovascular control. In principle, TPLSM alone can also be used for activation of single cells. Selective in vivo photoactivation has been achieved by use of the light-gated ion channel Channelrhodopsin-2 (Arenkiel et al. 2007), that exhibits a nonselective cation flux when exposed to blue-green light (Sineshchekov et al. 2002; Nagel et al. 2003; Boyden et al. 2005). This and other recently developed methods for targeted photoactivation (Han and Boyden 2007; Szobota et al. 2007; Volgraf et al. 2007; Zhang et al. 2007) in current practice require one-photon excitation. Thus, the excitation cannot be limited to a single cell. In future, similar agents will be developed amendable to TPLSM.

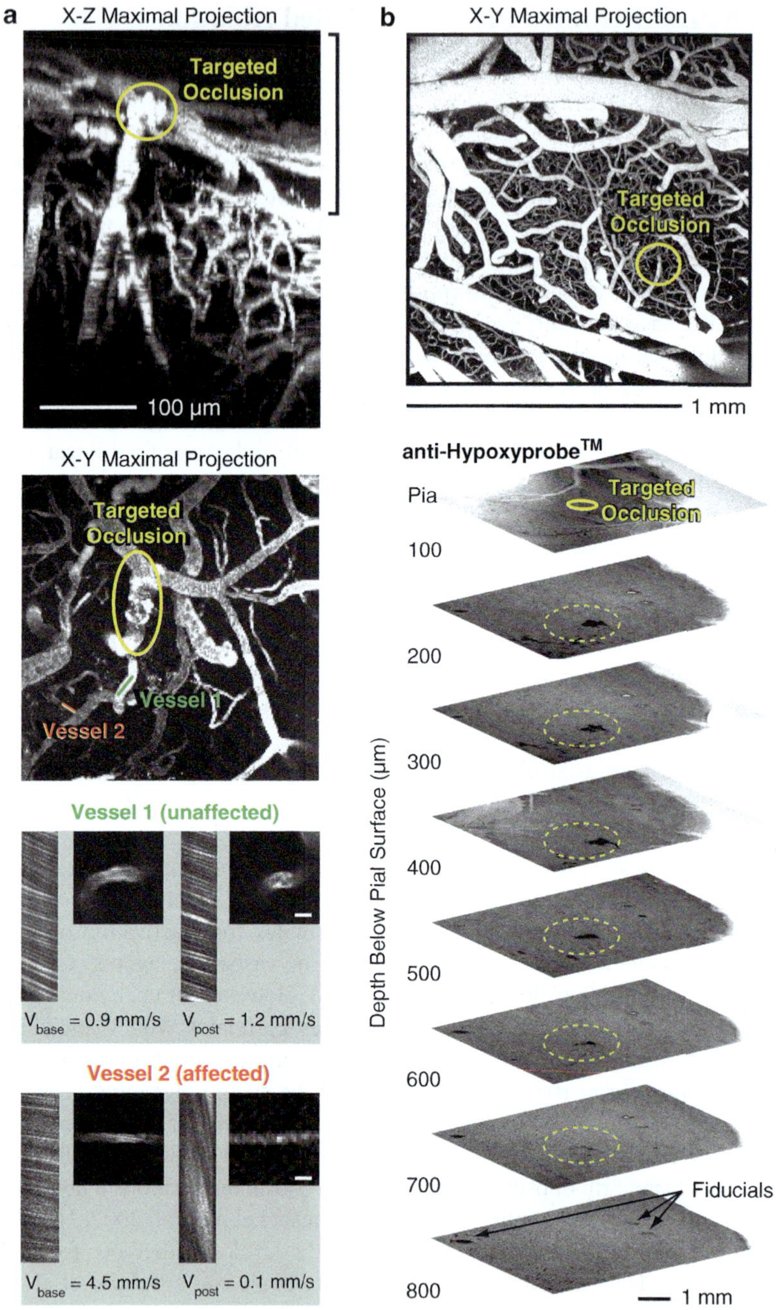

Fig. 11.7 Photo-thrombotic occlusion of penetrating arterioles using Rose Bengal. (**a**) Projected TPLSM image of a single penetrating arteriole occluded by photo-thrombosis (*yellow circle*) shown in the coronal plane (X-Z maximal projection), with surface arterioles near the top of the image. For the same field of interest, the surface vascular network is visible in the plane tangential to the

11.6 Conclusion

Multiple parameters, such as vessel-type, distance from the center of neuronal activity and connectivity, contribute to the vascular response. Although putative regulatory structures has been identified in ex vivo corrosion casts (Harrison et al. 1998), measurement of both diameter and RBC speed across vascular compartments and cortical depths are needed to address the question of spatial precision of the vascular functional control in vivo. Recent developments in TPLSM and optical probes have greatly advanced our understanding of relationship between the activity of neuroglial control elements and resultant vascular changes. Further technological developments are needed to estimate currently unknown variables such as intravascular blood pressure, shear stress, blood and tissue gas (O_2, CO_2, and NO) and pH, which are required for mechanistic modeling of neurovascular dynamics.

References

Ances BM, Greenberg JH, Detre JA (1999) Laser doppler imaging of activation-flow coupling in the rat somatosensory cortex. Neuroimage 10:716–723

Arenkiel BR, Peca J, Davison IG, Feliciano C, Deisseroth K, Augustine GJ, Ehlers MD, Feng G (2007) In vivo light-induced activation of neural circuitry in transgenic mice expressing channelrhodopsin-2. Neuron 54:205–218

Atlan M, Gross M, Forget BC, Vitalis T, Rancillac A, Dunn AK (2006) Frequency-domain wide-field laser Doppler *in vivo* imaging. Opt Lett 31:2762–2764

Barfod C, Akgoren N, Fabricius M, Dirnagl U, Lauritzen M (1997) Laser-Doppler measurements of concentration and velocity of moving blood cells in rat cerebral circulation. Acta Physiol Scand 160:123–132

Berwick J, Martin C, Martindale J, Jones M, Johnston D, Zheng Y, Redgrave P, Mayhew J (2002) Hemodynamic response in the unanesthetized rat: intrinsic optical imaging and spectroscopy of the barrel cortex. J Cereb Blood Flow Metab 22:670–679

Boas DA, Dale AM, Franceschini MA (2004a) Diffuse optical imaging of brain activation: approaches to optimizing image sensitivity, resolution, and accuracy. Neuroimage 23(Suppl 1): S275–288

Boas DA, Chen K, Grebert D, Franceschini MA (2004b) Improving the diffuse optical imaging spatial resolution of the cerebral hemodynamic response to brain activation in humans. Opt Lett 29:1506–1508

Fig. 11.7 (continued) cortical surface (X–Y maximal projection). The RBC speed in two neighboring surface vessels (Vessel 1 and 2) was measured using repeated line-scans. Vessel 1 shows no change in postocclusion RBC speed (v_{post}) compared to baseline (v_{base}), whereas Vessel 2 shows essentially no flow after the occlusion. (**b**) Examination of the hypoxic volume generated after a single penetrating arteriole was occluded (*yellow circle*). Pimonidazole hydrochloride (Hypoxyprobe™) was injected intravenously after clot formation and 1 h before transcardial perfusion. To locate the occluded vessel after brain extraction, fiducials markers were electrolytically placed near the target vessel before the brain was extracted from the skull. The ipsilateral cortex was then removed and flattened prior to immunohistology. Pimonidazole staining revealed a column of hypoxia, within the *dashed yellow circles*, that reached 700 μm below the pial surface. Panel (**a**) and (**b**) were adapted from Nishimura et al. (2007) and Kleinfeld et al. (2008), respectively

Boyden ES, Zhang F, Bamberg E, Nagel G, Deisseroth K (2005) Millisecond-timescale, genetically targeted optical control of neural activity. Nat Neurosci 8:1263–1268

Buxton RB, Wong EC, Frank LR (1998) Dynamics of blood flow and oxygenation changes during brain activation: the balloon model. Magn Reson Med 39:855–864

Cauli B, Tong XK, Rancillac A, Serluca N, Lambolez B, Rossier J, Hamel E (2004) Cortical GABA interneurons in neurovascular coupling: relays for subcortical vasoactive pathways. J Neurosci 24:8940–8949

Chaigneau E, Oheim M, Audinat E, Charpak S (2003) Two-photon imaging of capillary blood flow in olfactory bulb glomeruli. Proc Natl Acad Sci U S A 100:13081–13086

Chaigneau E, Tiret P, Lecoq J, Ducros M, Knopfel T, Charpak S (2007) The relationship between blood flow and neuronal activity in the rodent olfactory bulb. J Neurosci 27:6452–6460

Chuquet J, Hollender L, Nimchinsky EA (2007) High-resolution in vivo imaging of the neurovascular unit during spreading depression. J Neurosci 27:4036–4044

Davis TL, Kwong KK, Weisskoff RM, Rosen BR (1998) Calibrated functional MRI: mapping the dynamics of oxidative metabolism. Proc Natl Acad Sci U S A 95:1834–1839

Denk W, Svoboda K (1997) Photon upmanship: why multiphoton imaging is more than a gimmick. Neuron 18:351–357

Denk W, Strickler JH, Webb WW (1990) Two-photon laser scanning fluorescence microscopy. Science 248:73–76

Devor A, Dunn AK, Andermann ML, Ulbert I, Boas DA, Dale AM (2003) Coupling of total hemoglobin concentration, oxygenation, and neural activity in rat somatosensory cortex. Neuron 39:353–359

Devor A, Ulbert I, Dunn AK, Narayanan SN, Jones SR, Andermann ML, Boas DA, Dale AM (2005) Coupling of the cortical hemodynamic response to cortical and thalamic neuronal activity. Proc Natl Acad Sci U S A 102:3822–3827

Devor A, Tian P, Nishimura N, Teng IC, Hillman EM, Narayanan SN, Ulbert I, Boas DA, Kleinfeld D, Dale AM (2007) Suppressed neuronal activity and concurrent arteriolar vasoconstriction may explain negative blood oxygenation level-dependent signal. J Neurosci 27:4452–4459

Diez-Garcia J, Matsushita S, Mutoh H, Nakai J, Ohkura M, Yokoyama J, Dimitrov D, Knopfel T (2005) Activation of cerebellar parallel fibers monitored in transgenic mice expressing a fluorescent Ca2+ indicator protein. Eur J Neurosci 22:627–635

Dirnagl U, Kaplan B, Jacewicz M, Pulsinelli W (1989) Continuous measurement of cerebral cortical blood flow by laser-Doppler flowmetry in a rat stroke model. J Cereb Blood Flow Metab 9:589–596

Dirnagl U, Villringer A, Gebhardt R, Haberl RL, Schmiedek P, Einhaupl KM (1991) Three-dimensional reconstruction of the rat brain cortical microcirculation in vivo. J Cereb Blood Flow Metab 11:353–360

Dunn AK, Bolay H, Moskowitz MA, Boas DA (2001) Dynamic imaging of cerebral blood flow using laser speckle. J Cereb Blood Flow Metab 21:195–201

Dunn AK, Devor A, Dale AM, Boas DA (2005) Spatial extent of oxygen metabolism and hemodynamic changes during functional activation of the rat somatosensory cortex. Neuroimage 27:279–290

Dunn AK, Devor A, Bolay H, Andermann ML, Moskowitz MA, Dale AM, Boas DA (2003) Simultaneous imaging of total cerebral hemoglobin concentration, oxygenation, and blood flow during functional activation. Opt Lett 28:28–30

Faraci FM, Breese KR (1993) Nitric oxide mediates vasodilatation in response to activation of N-methyl-D-aspartate receptors in brain. Circ Res 72:476–480

Filosa JA, Bonev AD, Nelson MT (2004) Calcium dynamics in cortical astrocytes and arterioles during neurovascular coupling. Circ Res 95:e73–81

Filosa JA, Bonev AD, Straub SV, Meredith AL, Wilkerson MK, Aldrich RW, Nelson MT (2006) Local potassium signaling couples neuronal activity to vasodilation in the brain. Nat Neurosci 11:1397–1403

Fox PT, Raichle ME (1986) Focal physiological uncoupling of cerebral blood flow and oxidative metabolism during somatosensory stimulation in human subjects. Proc Natl Acad Sci U S A 83:1140–1144

Fox PT, Raichle ME, Mintun MA, Dence C (1988) Nonoxidative glucose consumption during focal physiologic neural activity. Science 241:462–464

Garaschuk O, Milos RI, Konnerth A (2006) Targeted bulk-loading of fluorescent indicators for two-photon brain imaging in vivo. Nat Protoc 1:380–386

Gobel W, Helmchen F (2007) New angles on neuronal dendrites in vivo. J Neurophysiol 98:3770–3779

Gobel W, Kampa BM, Helmchen F (2007) Imaging cellular network dynamics in three dimensions using fast 3D laser scanning. Nat Methods 4:73–79

Grinvald A, Lieke E, Frostig RD, Gilbert CD, Wiesel TN (1986) Functional architecture of cortex revealed by optical imaging of intrinsic signals. Nature 324:361–364

Hamel E (2004) Cholinergic modulation of the cortical microvascular bed. Prog Brain Res 145:171–178

Hamel E (2006) Perivascular nerves and the regulation of cerebrovascular tone. J Appl Physiol 100:1059–1064

Han X, Boyden ES (2007) Multiple-color optical activation, silencing, and desynchronization of neural activity, with single-spike temporal resolution. PLoS ONE 2:e299

Harrison RV, Harel N, Kakigi A, Raveh E, Mount RJ (1998) Optical imaging of intrinsic signals in chinchilla auditory cortex. Audiol Neurootol 3:214–223

Hillman EM, Boas DA, Dale AM, Dunn AK (2004) Laminar optical tomography: demonstration of millimeter-scale depth-resolved imaging in turbid media. Opt Lett 29:1650–1652

Hillman EM, Devor A, Bouchard MB, Dunn AK, Krauss GW, Skoch J, Bacskai BJ, Dale AM, Boas DA (2007) Depth-resolved optical imaging and microscopy of vascular compartment dynamics during somatosensory stimulation. Neuroimage 35:89–104

Hudetz AG (1997) Blood flow in the cerebral capillary network: a review emphasizing observations with intravital microscopy. Microcirculation 4:233–252

Iadecola C (2004) Neurovascular regulation in the normal brain and in Alzheimer's disease. Nat Rev Neurosci 5:347–360

Kleinfeld D, Friedman B, Lyden PD, Shih AY (2008) Targeted occlusion to surface and deep vessels in neocortex via linear and nonlinear optical absorption. In: Chen J, Xu Z, Xu X-M, Zhang J (eds) Animal models of acute neurological injuries. The Humana Press, Totowa, NJ

Kleinfeld D, Mitra PP, Helmchen F, Denk W (1998) Fluctuations and stimulus-induced changes in blood flow observed in individual capillaries in layers 2 through 4 of rat neocortex. Proc Natl Acad Sci U S A 95:15741–15746

Kwong KK, Belliveau JW, Chesler DA, Goldberg IE, Weisskoff RM, Poncelet BP, Kennedy DN, Hoppel BE, Cohen MS, Turner R et al (1992) Dynamic magnetic resonance imaging of human brain activity during primary sensory stimulation. Proc Natl Acad Sci U S A 89:5675–5679

Lauritzen M (2005) Opinion: Reading vascular changes in brain imaging: is dendritic calcium the key? Nat Rev Neurosci 6:77–85

Lauritzen M, Fabricius M (1995) Real time laser-Doppler perfusion imaging of cortical spreading depression in rat neocortex. Neuroreport 6:1271–1273

Malonek D, Grinvald A (1996) Interactions between electrical activity and cortical microcirculation revealed by imaging spectroscopy: implications for functional brain mapping. Science 272:551–554

Mandeville JB, Marota JJ, Ayata C, Zaharchuk G, Moskowitz MA, Rosen BR, Weisskoff RM (1999) Evidence of a cerebrovascular postarteriole windkessel with delayed compliance. J Cereb Blood Flow Metab 19:679–689

Martindale J, Mayhew J, Berwick J, Jones M, Martin C, Johnston D, Redgrave P, Zheng Y (2003) The hemodynamic impulse response to a single neural event. J Cereb Blood Flow Metab 23:546–555

Mayhew J, Zhao L, Hou Y, Berwick J, Askew S, Zheng Y, Coffey P (1998) Spectroscopic investigation of reflectance changes in the barrel cortex following whisker stimulation. Adv Exp Med Biol 454:139–148

Mulligan SJ, MacVicar BA (2004) Calcium transients in astrocyte endfeet cause cerebrovascular constrictions. Nature 431:195–199

Nagel G, Szellas T, Huhn W, Kateriya S, Adeishvili N, Berthold P, Ollig D, Hegemann P, Bamberg E (2003) Channelrhodopsin-2, a directly light-gated cation-selective membrane channel. Proc Natl Acad Sci U S A 100:13940–13945

Nemoto M, Sheth S, Guiou M, Pouratian N, Chen JW, Toga AW (2004) Functional signal- and paradigm-dependent linear relationships between synaptic activity and hemodynamic responses in rat somatosensory cortex. J Neurosci 24:3850–3861

Nguyen Q-T, Driscoll J, Dolnick EM, Kleinfeld D (2009) MPScope 2.0: A computer system for two-photon laser scanning microscopy with concurrent plasma-mediated ablation and electrophysiology. In: Frosting RD (ed) Methods for in Vivo Optical Imaging, Second Edition, CRC Press, Boca Raton, FL, pp 117–142

Nishimura N, Schaffer CB, Friedman B, Lyden PD, Kleinfeld D (2007) Penetrating arterioles are a bottleneck in the perfusion of neocortex. Proc Natl Acad Sci U S A 104:365–370

Nishimura N, Schaffer CB, Friedman B, Tsai PS, Lyden PD, Kleinfeld D (2006) Targeted insult to subsurface cortical blood vessels using ultrashort laser pulses: three models of stroke. Nat Methods 3:99–108

Obrig H, Villringer A (2003) Beyond the visible–imaging the human brain with light. J Cereb Blood Flow Metab 23:1–18

Ogawa S, Tank DW, Menon R, Ellermann JM, Kim SG, Merkle H, Ugurbil K (1992) Intrinsic signal changes accompanying sensory stimulation: functional brain mapping with magnetic resonance imaging. Proc Natl Acad Sci U S A 89:5951–5955

Ohki K, Chung S, Ch'ng YH, Kara P, Reid RC (2005) Functional imaging with cellular resolution reveals precise micro-architecture in visual cortex. Nature 433:597–603

Peppiatt CM, Howarth C, Mobbs P, Attwell D (2006) Bidirectional control of CNS capillary diameter by pericytes. Nature 443:700–704

Polimeni JR, Granquist-Fraser D, Wood RJ, Schwartz EL (2005) Physical limits to spatial resolution of optical recording: clarifying the spatial structure of cortical hypercolumns. Proc Natl Acad Sci U S A 102:4158–4163

Raichle ME, Mintun MA (2006) Brain work and brain imaging. Annu Rev Neurosci 29:449–476

Rancillac A, Rossier J, Guille M, Tong XK, Geoffroy H, Amatore C, Arbault S, Hamel E, Cauli B (2006) Glutamatergic control of microvascular tone by distinct GABA neurons in the cerebellum. J Neurosci 26:6997–7006

Schaffer CB, Friedman B, Nishimura N, Schroeder LF, Tsai PS, Ebner FF, Lyden PD, Kleinfeld D (2006) Two-photon imaging of cortical surface microvessels reveals a robust redistribution in blood flow after vascular occlusion. PLoS Biol 4:e22

Scremin OU (1995) The rat nervous system. In: Paxinos G (ed) Cerebral vascular system. Academic, New York, pp 3–55

Sheth S, Nemoto M, Guiou M, Walker M, Pouratian N, Toga AW (2003) Evaluation of coupling between optical intrinsic signals and neuronal activity in rat somatosensory cortex. Neuroimage 19:884–894

Sheth SA, Nemoto M, Guiou MW, Walker MA, Toga AW (2005) Spatiotemporal evolution of functional hemodynamic changes and their relationship to neuronal activity. J Cereb Blood Flow Metab 25:830–841

Sheth SA, Nemoto M, Guiou M, Walker M, Pouratian N, Toga AW (2004a) Linear and nonlinear relationships between neuronal activity, oxygen metabolism, and hemodynamic responses. Neuron 42:347–355

Sheth SA, Nemoto M, Guiou M, Walker M, Pouratian N, Hageman N, Toga AW (2004b) Columnar specificity of microvascular oxygenation and volume responses: implications for functional brain mapping. J Neurosci 24:634–641

Sigler A, Goroshkov A, Murphy TH (2008) Hardware and methodology for targeting single brain arterioles for photothrombotic stroke on an upright microscope. J Neurosci Methods 170:35–44

Simard M, Arcuino G, Takano T, Liu QS, Nedergaard M (2003) Signaling at the gliovascular interface. J Neurosci 23:9254–9262

Sineshchekov OA, Jung KH, Spudich JL (2002) Two rhodopsins mediate phototaxis to low- and high-intensity light in Chlamydomonas reinhardtii. Proc Natl Acad Sci U S A 99:8689–8694

Stosiek C, Garaschuk O, Holthoff K, Konnerth A (2003) In vivo two-photon calcium imaging of neuronal networks. Proc Natl Acad Sci U S A 100:7319–7324

Svoboda K, Yasuda R (2006) Principles of two-photon excitation microscopy and its applications to neuroscience. Neuron 50:823–839

Szobota S, Gorostiza P, Del Bene F, Wyart C, Fortin DL, Kolstad KD, Tulyathan O, Volgraf M, Numano R, Aaron HL, Scott EK, Kramer RH, Flannery J, Baier H, Trauner D, Isacoff EY (2007) Remote control of neuronal activity with a light-gated glutamate receptor. Neuron 54:535–545

Takano T, Tian GF, Peng W, Lou N, Libionka W, Han X, Nedergaard M (2006) Astrocyte-mediated control of cerebral blood flow. Nat Neurosci 9:260–267

Theer P, Hasan MT, Denk W (2003) Two-photon imaging to a depth of 1000 microm in living brains by use of a Ti:Al2O3 regenerative amplifier. Opt Lett 28:1022–1024

Tsai PS, Kleinfeld D (2009) In vivo two-photon laser scanning microscopy with concurrent plasma-mediated ablation: Principles and hardware realization. In: Frosting RD (ed) Methods for In Vivo Optical Imaging, Second Edition, CRC Press, Boca Raton, FL, pp 59–115

Tsai PS, Friedman B, Ifarraguerri AI, Thompson BD, Lev-Ram V, Schaffer CB, Xiong Q, Tsien RY, Squier JA, Kleinfeld D (2003) All-optical histology using ultrashort laser pulses. Neuron 39:27–41

Tsai PS, Blinder P, Migliori BJ, Neev J, Jin Y, Squier JA, Kleinfeld D (2009) Plasma-mediated ablation: An optical tool for submicrometer surgery on neuronal and vascular systems. Curr Opin Biotechnol 20:90–99

Vanzetta I, Grinvald A (1999) Increased cortical oxidative metabolism due to sensory stimulation: implications for functional brain imaging. Science 286:1555–1558

Vanzetta I, Grinvald A (2001) Evidence and lack of evidence for the initial dip in the anesthetized rat: implications for human functional brain imaging. Neuroimage 13:959–967

Vanzetta I, Hildesheim R, Grinvald A (2005) Compartment-resolved imaging of activity-dependent dynamics of cortical blood volume and oximetry. J Neurosci 25:2233–2244

Villringer A, Planck J, Hock C, Schleinkofer L, Dirnagl U (1993) Near infrared spectroscopy (NIRS): a new tool to study hemodynamic changes during activation of brain function in human adults. Neurosci Lett 154:101–104

Villringer A, Them A, Lindauer U, Einhaupl K, Dirnagl U (1994) Capillary perfusion of the rat brain cortex. An in vivo confocal microscopy study. Circ Res 75:55–62

Volgraf M, Gorostiza P, Szobota S, Helix MR, Isacoff EY, Trauner D (2007) Reversibly caged glutamate: a photochromic agonist of ionotropic glutamate receptors. J Am Chem Soc 129:260–261

Wayland H, Johnson PC (1967) Erythrocyte velocity measurement in microvessels by a two-slit photometric method. J Appl Physiol 22:333–337

Zhang F, Wang LP, Brauner M, Liewald JF, Kay K, Watzke N, Wood PG, Bamberg E, Nagel G, Gottschalk A, Deisseroth K (2007) Multimodal fast optical interrogation of neural circuitry. Nature 446:633–639

Zhang S, Murphy TH (2007) Imaging the impact of cortical microcirculation on synaptic structure and sensory-evoked hemodynamic responses in vivo. PLoS Biol 5:e119

Zonta M, Angulo MC, Gobbo S, Rosengarten B, Hossmann KA, Pozzan T, Carmignoto G (2003) Neuron-to-astrocyte signaling is central to the dynamic control of brain microcirculation. Nat Neurosci 6:43–50

Index

9781441904515